EPITHELIAL ANION TRANSPORT IN HEALTH AND DISEASE: THE ROLE OF THE SLC26 TRANSPORTERS FAMILY

The Novartis Foundation is an international scientific and educational charity (UK Registered Charity No. 313574). Known until September 1997 as the Ciba Foundation, it was established in 1947 by the CIBA company of Basle, which merged with Sandoz in 1996, to form Novartis. The Foundation operates independently in London under English trust law. It was formally opened on 22 June 1949.

The Foundation promotes the study and general knowledge of science and in particular encourages international co-operation in scientific research. To this end, it organizes internationally acclaimed meetings (typically eight symposia and allied open meetings and 15–20 discussion meetings each year) and publishes eight books per year featuring the presented papers and discussions from the symposia. Although primarily an operational rather than a grant-making foundation, it awards bursaries to young scientists to attend the symposia and afterwards work with one of the other participants.

The Foundation's headquarters at 41 Portland Place, London W1B 1BN, provide library facilities, open to graduates in science and allied disciplines. Media relations are fostered by regular press conferences and by articles prepared by the Foundation's Science Writer in Residence. The Foundation offers accommodation and meeting facilities to visiting scientists and their societies.

Information on all Foundation activities can be found at http://www.novartisfound.org.uk

Novartis Foundation Symposium 273

EPITHELIAL ANION TRANSPORT IN HEALTH AND DISEASE: THE ROLE OF THE SLC26 TRANSPORTERS FAMILY

2006

John Wiley & Sons, Ltd

Other Wiley Editorial Offices

John Wiley & Sons Inc., 111 River Street, Hoboken, NJ 07030, USA

Jossey-Bass, 989 Market Street, San Francisco, CA 94103-1741, USA

Wiley-VCH Verlag GmbH, Boschstr. 12, D-69469 Weinheim, Germany

John Wiley & Sons Australia Ltd, 33 Park Road, Milton, Queensland 4064, Australia

John Wiley & Sons (Asia) Pte Ltd, 2 Clementi Loop #02-01, Jin Xing Distripark,
Singapore 129809

John Wiley & Sons Canada Ltd, 6045 Freemont Boulevard Mississauga, Ontario, Canada L5R
4J3

Wiley also publishes its books in a variety of electronic formats. Some content that appears in
print may not be available in electronic books.

Novartis Foundation Symposium 273
x + 273 pages, 49 figures, 3 tables

British Library Cataloguing in Publication Data

A catalogue record for this book is available from the British Library

ISBN-13 978-0-470-01624-4 (HB)
ISBN-10 0-470-01624-8 (HB)

Typeset in 10½ on 12½ pt Garamond by SNP Best-set Typesetter Ltd., Hong Kong
Printed and bound in Great Britain by T. J. International Ltd, Padstow, Cornwall.
This book is printed on acid-free paper responsibly manufactured from sustainable forestry, in
which at least two trees are planted for each one used for paper production.

Contents

Participants

Seth L. Alper Molecular and Vascular Medicine and Renal Units, Beth Israel Deaconess Medical Center, Department of Medicine, Harvard Medical School, Boston, MA 02215, USA

Barry E. Argent Institute for Cell and Molecular Biosciences, University of Newcastle upon Tyne, Catherine Cookson Building, Framlington Place, Newcastle upon Tyne NE2 4HH, UK

Peter S. Aronson Section of Nephrology, Department of Medicine, Yale University School of Medicine, 1 Gilbert Street, TAC S-255, PO Box 208029, New Haven, CT 06520-8029, USA

Jonathan Ashmore Department of Physiology, University College London, Gower Street, London WC1E 6BT, UK

Maynard Case Faculty of Life Sciences, University of Manchester, Floor 2, Core Technology Facility, 46 Grafton St, Manchester M13 9NT, UK

Hsiao Chang Chan Epithelial Cell Biology Research Centre, Department of Physiology, The Chinese University of Hong Kong, Shatin, Hong Kong

Lorraine A. Everett Audiovestibular Genomics Group, The Wellcome Trust Sanger Institute, Wellcome Trust Genome Campus, Hinxton, Cambridge CB10 1SA, UK

Michael A. Gray Institute for Cell and Molecular Biosciences, University of Newcastle upon Tyne, Catherine Cookson Building, Framlington Place, Newcastle upon Tyne NE2 4HH, UK

Pia Höglund Hospital for Children and Adolescents, PO Box 281 (Stenbäckinkatu 11), 00029 Hus, Finland

Hiroshi Ishiguro Internal Medicine and Human Nutrition, Nagoya University School of Medicine, Showa-ku, Nagoya 466-8550, Japan

Juha Kere   Department of Biosciences at Novum and Clinical Research Centre, Karolinska Institutet, 14157 Huddinge, Sweden

Marlies Knipper   Department of Otolaryngology, Tübingen Hearing Research Centre, Molecular Neurobiology, Elfriede-Aulhorn-Strasse 5, D-72076 Tübingen, Germany

Min Goo Lee   Department of Pharmacology, Yonsei University College of Medicine, 134 Sinchon-Dong, Seoul 120-752, Korea

Daniel Markovich   School of Biomedical Sciences, Department of Physiology and Pharmacology, University of Queensland, Brisbane, Queensland 4072, Australia

David B. Mount   Renal Division, Brigham and Women's Hospital and VA Boston Healthcare System, Room 542, Harvard Institutes of Medicine, 4 Blackfan Circle, Boston, MA 02115, USA

Shmuel Muallem   Department of Physiology, Room K4-120, UT South Western Medical Centre, 5323 Harry Hines Blvd, Dallas, TX 75390 9040, USA

Dominik Oliver   Institute of Physiology, University of Freiburg, Hermann-Herder-Straße 7, 79104 Freiburg, Germany

Paul M. Quinton   Department of Pediatrics, University of California San Diego School of Medicine, 9500 Gilman Drive—0831, La Jolla, CA 92093-0831, USA

Michael F. Romero   Department of Physiology & Biophysics, Room #SOM-E563, Case Western University School of Medicine, 2119 Abington Rd, Cleveland, OH 44106-4970, USA

Antonio Rossi   Department of Biochemistry, University of Pavia, Via Taramelli, 3/B, I-27100 Pavia, Italy

Ursula Seidler   Medizinische Hochshule Hannover, Carl-Neuberg-Strasse 1, 30625 Hannover, Germany

Manoocher Soleimani   Division of Nephrology and Hypertension, University of Cincinnati Medical Center, 231 Albert Sabin Way, MSB G259, Cincinnati, OH 45267 0585, USA

Andrew K. Stewart   (*Novartis Foundation Bursar*) RW 773, Molecular Medicine and Renal Units, Beth Israel Deaconess Medical Center, East Campus, 330 Brookline Avenue, Boston, MA 02215, USA

Philip J. Thomas   The University of Texas, Southwestern Medical Center at Dallas, Department of Physiology, Room K4.140A, 5323 Harry Hines Blvd, Dallas, TX 75390-9040, USA

R. James Turner   Gene Therapy and Therapeutics Branch, National Institute of Dental and Craniofacial Research, Building 10, Room 1A01, National Institutes of Health, Bethesda, MD 20892-1190, USA

Susan M. Wall   Renal Division, Department of Medicine, Emory University, Mailstop: 1930/001/1AG (Medicine-Renal), Atlanta, GA 30322, USA

Michael J. Welsh   (*Chair*) Howard Hughes Medical Institute/500 EMRB, Carver College of Medicine, University of Iowa, Iowa City, IO 52242, USA

Chair's introduction

Michael J. Welsh

Howard Hughes Medical Institute/500 EMRB, University of Iowa, Iowa City, IO 52242, USA

In introducing this symposium, I am expecting that this will be an interesting meeting for three reasons. First, I have much to learn. Secondly, this is an interesting family of transporters, because there are multiple members, there is rich diversity, and there are some common themes. Thirdly, this is a small meeting which should permit some excellent discussion.

I am not an expert on the SLC26 family, so what I can say by way of introduction is quite limited. Instead, I'll highlight what I would like to learn from this meeting. First, how do these channels work? Are they electrically conductive? Or are they neutral transporters? If they are conductive, what is their relationship to ion channels? I would like to know how selectivity is determined. As I look at some of the literature, the issues of selectivity should prove very interesting. I would also like to learn something about how this family has evolved. There are multiple members and we may learn something if we look at these transporters through evolution.

The second main question I would like to address is how they contribute to normal physiology, in many different epithelia. What are the common threads and what is unique? If you transplanted one of these transporters into a different place, would it adopt new functions depending on its new home? Or are the functions all intrinsic to the protein?

Third, I hope to learn more about how loss of their function disrupts physiology. What is the pathophysiology associated with loss or mutation? If we understand this better, might it allow us to do something about disease? These are the main things I would like to learn from this meeting.

As I looked through the history of the Novartis Foundation, I came across the following quote from Lord Beveridge, made at the inauguration of the Foundation in 1949. He said, 'This place itself is not a laboratory for mixing compounds, but we do mean to make it a laboratory for mixing scientists'. So I hope we'll mix it up, have a good time, and learn much over the next few days.

Overview of the SLC26 family and associated diseases

Juha Kere

Department of Biosciences at Novum, Karolinska Institutet, 14157 Huddinge, Sweden and Department of Medical Genetics, University of Helsinki, 00014 Helsinki, Finland

Abstract. In the late 1990s the SLC26 family of anion exchangers emerged as the second, structurally distinct gene family capable of similar transport functions as the classical SLC4 or anion exchanger (AE) gene family. The observations leading to the characterization of the SLC26 family were firmly based on research on rare human diseases and aided by comparison to *Caenorhabditis elegans*. *SLC26A1*, or rat sulphate/anion transporter 1 (*Sat1*), was the first gene cloned in mammals, but not characterized in humans until the year 2000. Three rare recessive diseases in humans, namely diastrophic dysplasia (cartilage disorder resulting in growth retardation), congenital chloride diarrhoea (anion exchange disorder of the intestine) and Pendred syndrome (deafness with thyroid disorder) turned out to be caused by the highly related genes *SLC26A2* (first called *DTDST*), *SLC26A3* (first called *CLD* or *DRA*) and *SLC26A4* (first called *PDS*), respectively. Subsequently, others and our laboratory cloned prestin, a cochlear motor protein gene (*SLC26A5*), a putative pancreatic anion transporter (*SLC26A6*), and *SLC26A7–SLC26A11*. Some SLC26 family members show highly specific tissue expression patterns, others are widely expressed. The SLC26 exchangers are capable of transporting, with different affinities, at least the chloride, iodide, sulfate, bicarbonate, hydroxyl, oxalate and formate anions, and have distinct anion specificity profiles.

2006 Epithelial anion transport in health and disease: the role of the SLC26 transporters family. Wiley, Chichester (Novartis Foundation Symposium 273) p 2–18

Transport of small molecules across lipid membranes is a fundamental function of all cellular organisms. Hundreds of proteins with specialized transport capabilities are expressed in different tissues of multicellular organisms, and indeed in different domains of membranes in individual cells. The transporter proteins come in families, with different members sharing structural similarities but often with distinct properties and physiological functions that may or may not be interchangeable. Distinct expression patterns in different tissues also suggest that the corresponding genes have highly specialized regulatory elements, in spite of high similarity of coding sequences. Finally, just a few changes in protein sequences may cause radical differences in the transport properties. The SLC26 family of anion exchangers provides examples of a wide spectrum of all these features, and is largely uncharacterized. This is not surprising, considering that most members of the whole gene and protein family were described only a few years ago. Many transporter proteins were first isolated based on their functional properties, the discovery of the *SLC26* gene family has been driven by human disease gene cloning and thereafter genomic approaches, based on the homology of the gene family and availability of whole genome sequences. At the beginning of this odyssey, only one gene belonging to the *SLC26* gene family was known in mammals, the rat sulphate anion transporter 1 (*Sat1*) gene. The next three *SLC26* family members were identified by positional cloning of rare recessive human disease genes (Hästbacka et al 1994, Höglund et al 1996, Everett et al 1997) even though one of them, *SLC26A3*, had been first cloned as a suggested tumour suppressor gene (Schweinfest et al 1993). One gene was first characterized in gerbil rather than human based on its function, motor activity in cochlear cells of the inner ear (Zheng et al 2000). Finally, all the remaining five currently known *SLC26* genes were identified by a genomic homology-driven approach in human (Lohi et al 2000, 2002) and in parallel, by other approaches (Waldegger et al 2001, Toure et al 2001, Vincourt et al 2002, 2003, Mount & Romero 2004). The nomenclature of this gene family follows the convention of other solute carrier genes, starting with *SLC*, followed by the family number and an *A* separating the individual gene number. The individual members of the *SLC26* family got their number identities in July 2000 (for *SLC26A1* to *A6*) and in January 2001 (for *SLC26A7* to *A11*, based on the full or partial human cDNA sequences AF331521 to AF331526 submitted from our laboratory) after exchange of email messages between the author of this review, Dr Elspeth Bruford of the HUGO Gene Nomenclature Committee, and nomenclature reviewers, including Dr Matthias A. Hediger. The entire human SLC26 gene family with references to the earliest GenBank sequence database entries and diseases associated with them are presented in Table 1. In the following paragraphs, I will briefly describe the discovery of the different *SLC26* genes. I will only discuss the molecular cloning of each gene, even though the existence of such transporters had been demonstrated earlier by functional studies, and for considerations of space, I have omitted most of the literature related to their functions. Much additional information has already been revealed about their specific functional properties and physiological roles, and some of the first mouse knockout models are also available. More detailed reviews of these studies will be presented by other papers in this book.

SLC26A1

The rat liver canalicular sulfate transporter was the first gene of the *SLC26* family to be molecularly characterized in mammals, cloned by Bissig et al (1994) and characterized by Markovich et al (1994). Curiously, the human gene remained

TABLE 1 The human SLC26 family of anion transporters

Gene symbol (aliases)	Chromosome location	Main sites of expression	Diseases, OMIM numbers	Sequence accession	References to molecular cloning and diseases
SLC26A1 (SAT1)	4p16	Liver, kidney, pancreas, brain	Not known	AF297659, AY124771	Lohi et al 2000, Regeer et al 2003
SLC26A2 (DTDST)	5q32-q33.1	Ubiquitous, cartilage	DTD, 222600 ACGIB, 600972 AOG2, 256050 EDM4, 226900	U14528	Hästbacka et al 1994, 1996
SLC26A3 (CLD, DRA)	7q31	Ileum, colon, seminal vesicle, eccrine sweat gland	CLD, 214700	L02785	Schweinfest et al 1993, Höglund et al 1996
SLC26A4 (PDS)	7q31	Thyroid, kidney, cochlea	PDS, 274600 DFNB4, 600791 EVA, 603545	AF030880	Everett et al 1997, Haila et al 1998, Li et al 1998
SLC26A5 (PRES)	7q22.1	Outer hair cells of the cochlea	DFNB61, 604943	AF523354, AY289133	Zheng et al 2000, Liu et al 2003
SLC26A6 (PAT1)	3p21	Kidney, pancreas, skeletal muscle	Not known	AF279265, AF288410	Lohi et al 2000, Waldegger et al 2001
SLC26A7	8q21	Kidney, placenta, testis	Not known	AF331521, AJ413228-AJ413230	Lohi et al 2000, 2002, Vincourt et al 2002
SLC26A8 (TAT1)	6p21	Testis	Not known	AF314959, AF331522	Lohi et al 2000, 2002, Toure et al 2001
SLC26A9	1q32	Lung	Not known	AF314958, AF331525	Lohi et al 2000, 2002
SLC26A10	12q14	Brain, ubiquitous	Not known	AF331523, AL050358	Lohi et al 2000
SLC26A11	17q25	Ubiquitous	Not known	AF331524, AF345195	Lohi et al 2000, Vincourt et al 2003

For disease designations, please see text. Many of the genes were first published as GenBank database entries, many independently by two research groups within a few months. Sequence accession numbers are given for the earliest full or partial cDNA sequences.

uncharacterized until several years later, when Hannes Lohi in our group determined and submitted the human sequence to GenBank in August 2000 (AF297659; Lohi et al 2000). The gene structure was confirmed in June 2002 by Regeer, Lee and Markovich (AY124771; Regeer et al 2003). The mouse gene was cloned and characterized by Lee et al (2003).

Diastrophic dysplasia and SLC26A2

Diastrophic dysplasia (DTD) is an autosomal recessive cartilage disorder, leading to a disproportionate growth disorder. DTD is one of those about 30 rare recessive disorders that have been observed at much higher frequencies in Finland than in most other countries, a phenomenon that has been well explained by genetic founder effects; curiously, congenital chloride diarrhoea (CLD) is another. Johanna Hästbacka, a doctoral student of Dr Albert de la Chapelle, first mapped the gene to chromosome 5q31 and later, working as a postdoc in Dr Eric Lander's laboratory, positionally cloned it (Hästbacka et al 1994; sequence AF345195 submitted to GenBank in September 1994). The gene was named DTD sulfate transporter (*DTDST*) and its homology to the rat *Sat1* gene was immediately noticed, suggesting a pathogenetic mechanism for DTD. Cartilage is dependent on high amounts of sulfate and Hästbacka et al (1994) suggested that impaired function of DTDST may lead to undersulfation of proteoglycans in cartilage matrix. Later, mutations in *SLC26A2* were found also in achondrogenesis type IB (ACGIB), atelosteogenesis type II (AOG2), and multiple epiphyseal dysplasia type 4 (EDM4) (for OMIM numbers, see Table 1). Even though *SLC26A2* mRNA is quite abundantly expressed and SLC26A2 protein has been detected in colon, sweat glands, pancreas and placenta in addition to cartilage, symptoms of recessive *SLC26A2* mutations seem to be confined to problems in cartilage development (Hästbacka et al 1996, Haila et al 2001).

Congenital chloride diarrhoea and SLC26A3

With the aim of identifying potential tumour suppressor genes in colon cancer, Schweinfest et al (1993) discovered a gene that was abundantly expressed in normal colon epithelium, but strongly down-regulated in malignant cells; they named the gene Down-Regulated in Adenoma (*DRA*; L02785). The same year that *DRA* was cloned, we published the mapping of the gene for congenital chloride diarrhoea (CLD) to chromosome 7q22-q31 (Kere et al 1993). The recessively inherited intestinal Cl⁻ absorption disorder CLD is unusually common, though still rare (1 : 20 000) in Finland (Kere et al 1999, Mäkelä et al 2002). The likely mutational homogeneity caused by a genetic founder effect suggested a way to its positional cloning by a similar linkage disequilibrium strategy that Johanna Hästbacka used to clone

DTDST. Pia Höglund undertook the cloning work as her PhD thesis project, and we reported the identification of *DRA* as the CLD gene based on mutations in patients from Finland and Poland (Höglund et al 1996). This discovery suggested a new function for the now emerging family of genes, namely Cl⁻/HCO₃⁻ exchange, implicated by the body of physiological information of the defect in CLD (Holmberg 1986). So far, all families with diagnosed CLD have revealed mutations in *SLC26A3*, and no diseases other than CLD have been associated with *SLC26A3* mutations (Mäkelä et al 2002).

Pendred syndrome and SLC26A4

Pendred syndrome is characterized by congenital sensorineural deafness and goitre. Everett et al (1997) cloned positionally the *SLC26A4* gene and observed that it is both highly homologous to the *SLC26A3* gene and located less than 50 kb away from it on chromosome 7 (the sequence AF030880 was submitted in October 1997). The gene had been identified also by Dr Stephen W. Scherer's group as mentioned in a report of the genomic structure of *SLC26A3* (Haila et al 1998). In addition to Pendred syndrome, another recessive, non-syndromic deafness gene *DFNB4* had been mapped to the same general region, and Li et al (1998) found mutations in *SLC26A4* in these patients. Interestingly, despite their close proximity and structural relationship, *SLC26A4* is expressed in the thyroid gland, kidney and cochlea, whereas *SLC26A3* is abundantly expressed only in the colon, suggesting that the genes possess highly specialized regulatory regions. *SLC26A4* was studied as a candidate gene in autoimmune thyroiditis, including Graves' disease, Hashimoto thyroiditis and primary idiopathic myxedema, and genetic association of microsatellite alleles suggested that variation in *SLC26A4* might modify susceptibility to some forms of autoimmune thyroiditis (Kacem et al 2003).

Hearing and SLC26A5

The gene for prestin, or *SLC26A5*, is the second *SLC26* family gene first cloned in a species other than human. It was identified based on subtractive hybridization and differential screening in outer hair cells of the gerbil cochlea (Zheng et al 2000). As a molecular motor protein, its functional properties are distinct from the other SLC26 family members (Dallos & Fakler 2002). The human gene sequence was deposited to GenBank by Liu and co-workers in June 2002 (AF523354) and by David Mount in May 2003 (AY289133), and cloning of the human *SLC26A5* gene and mutations in non-syndromic recessive deafness in at least two families were then described by Liu et al (2003). Even though both of the genes *SLC26A4* and *A5* cause deafness when mutated, they have different functional properties (Oliver et al 2001).

SLC26A6

The fact that three very different recessive diseases were caused by the related genes *SLC26A2*, *A3* and *A4* led us to look into the possibility that several genes of the same structural family might remain unknown, with possibly new diseases to be explained by such genes. The publication of the genomic sequence of the first multicellular organism, *Caenorhabditis elegans*, allowed us to have a comprehensive look at a genomic level. To our pleasant surprise, *C. elegans* possessed no fewer than seven genes homologous to the three disease genes, and this observation suggested that the gene family might also be much larger in human (Kere et al 1999). We then searched all publicly available human expressed sequence tag and high-throughput genomic sequences for homologues of *SLC26A2* to *A4* and identified fragments of new genes that were distinguished by their genomic positions and that were subsequently named *SLC26A6* to *A11* (Lohi et al 2000). *SLC26A6* was the first new gene that we characterized in detail (AF279265, submitted in June 2000). *SLC26A6* sequence was also submitted to GenBank by Waldegger and co-workers a month later (AF288410; Waldegger et al 2001). Polyclonal antibodies revealed that SLC26A6 protein resided on the luminal membrane of pancreatic ductal cells, prompting us to suggest that it might function as the luminal Cl^-/HCO_3^- exchanger of pancreatic ducts (Lohi et al 2000). Wang et al (2002) determined that in the intestine *SLC26A6* is expressed most abundantly in the duodenum, and elsewhere in small intestine but not in colon, a pattern that is opposite to that of *SLC26A3* expression.

SLC26A7

The remaining human *SLC26* family genes were cloned using standard genome-driven approaches, and the work was greatly aided by the rapidly cumulating genome sequences as well as direct PCR cloning using cDNA from relevant tissues. We submitted to GenBank the coding sequence for *SLC26A7* in December 2000 (AF331521) and it was confirmed by J. Girard in September 2001 (AJ413228). In accordance with initial RT-PCR experiments that revealed its abundant expression in the kidney (Lohi et al 2000), we used human kidney cDNA to amplify overlapping fragments of *SLC26A7*. Initial characterization of its transport properties in a frog oocyte expression system supported the concept that *SLC26A7* has anion transport properties (Lohi et al 2002). Independently, Vincourt et al (2002) cloned *SLC26A7* from endothelial cells of high endothelial venules, confirming also its sequence and most abundant expression in the kidney. Both groups found that *SLC26A7* has alternative polyadenylation signals affecting the 3' UTR, but their physiological significance has so far remained poorly characterized. In addition, *SLC26A7* appears also to have alternative splicing that leads to a change in the C-terminal tail, where the last 11 amino acids of the major isoform may be replaced by an 18 amino acids polypeptide in the minor isoform variant (Vincourt et al 2002).

SLC26A8

The testis-specific *SLC26A8* gene was sequenced independently by three groups. David Mount submitted its sequence to GenBank in October 2000 (AF314959) and Lohi and Kere in December 2000 (AF331522; Lohi et al 2000). *SLC26A8* was also characterized by Toure et al (2001) who used yeast two-hybrid affinity cloning to find proteins binding to the *MgcRacGAP* gene, a new male germ cell specific GTPase activating protein that they had previously characterized in spermatocytes. This family of proteins is known for their pleiotropic properties, including effects on cell motility, adhesion, cell cycle progression and cell division. They determined that the SLC26A8 protein binds through its C-terminal part to the N-terminal domain of MgcRacGAP protein, and that SLC26A7 can act as a sulfate transporter (Toure et al 2001). SLC26A8 and its transport properties were also described by Lohi et al (2002) as part of our systematic approach to characterize new members of the SLC26 family. We predicted the exons of *SLC26A8* from genomic sequences at 6p21 where we had first mapped the gene, verified the exon-intron boundaries by PCR from human testis cDNA, and determined that SLC26A8 protein appeared to transport both chloride and sulfate anions, possibly also oxalate (Lohi et al 2000, 2002). Both groups raised polyclonal antibodies against SLC26A8 protein and determined by immunohistochemistry that SLC26A8 is almost exclusively expressed by spermatocytes and spermatids (Toure et al 2001, Lohi et al 2002).

Because the *Drosophila* RotundRacGAP protein, close homologue of MgcRac-GAP is essential for male fertility in fruit fly, and because *SLC26A8* expression is strictly testis-specific in human, it became plausible to hypothesize that *SLC26A8* might be occasionally mutated in human male infertility. This possibility was further indirectly supported by the occurrence of a male-specific infertility-associated chromosome translocation breakpoint on chromosome 6p21 (Paoloni-Giacobino et al 2000). To explore the possibility that *SLC26A8* might be mutated in some forms of male infertility, Mäkelä et al (2005) sequenced the coding parts of *SLC26A8* in 116 infertile men and detected altogether five amino acid substitutions. However, these polymorphisms appeared also in population controls at similar frequencies. Assuming that male infertility would be caused by a recessive mechanism involving *SLC26A8*, Mäkelä et al (2005) concluded that *SLC26A8* mutations are not common in male infertility caused by primary spermatogenic failure.

SLC26A9

The sequence of the *SLC26A9* gene was submitted to GenBank by David Mount in October 2000 (AF314958) and by Lohi and Kere in December 2000 (AF331525). Its expression pattern and first characteristics were described by Lohi et al (2002). Again, we used the genomic sequence as a basis for exon prediction, verified the exon–intron boundaries with PCR on lung cDNA, and expanded the 5' and 3' sequences by rapid amplification of cDNA ends (RACE). Immunohistochemistry using polyclonal antibodies revealed SLC26A9 protein in both bronchial and alveolar epithelial cells. Expression of SLC26A9 in frog oocytes led to a significant increase of chloride, sulfate and oxalate transport above background (Lohi et al 2002).

SLC26A10

Lohi and Kere identified an unspliced cDNA sequence homologous to the *SLC26* family members and submitted a putative coding sequence to GenBank in December 2000 (AF331523). An mRNA sequence for *SLC26A10* has also been submitted by Koehrer and co-workers (AL050358). More detailed characterization of *SLC26A10* has not yet been published, and it is presently unsettled whether this gene is functional or not.

SLC26A11

A partial sequence for *SLC26A11* was submitted to GenBank by Lohi and Kere in December 2000 (AF331524) and its complete coding sequence by David Mount in February 2001 (AF345195). Subsequently, cloning of *SLC26A11* and initial characterization of its sulfate transport ability were published by Vincourt et al (2003). *SLC26A11* was found expressed by all tissues on a multiple tissue cDNA panel, and its sequence was determined based on an adrenal gland clone (Lohi et al 2000); Vincourt et al (2003) used endothelial venule and kidney cells to verify its sequence. Vincourt et al (2003) suggested that *SLC26A11* might be a candidate gene for congenital deafness (DFNA20) and Usher syndrome (USH1G), but these hypotheses remain untested.

Conclusions

The history of the discovery of the *SLC26* gene family is unique in that it was mostly driven by positional cloning of rare human diseases and by genomic approaches. Nevertheless, it has interested physiologists, because some of these genes may turn out to be responsible for transporter functions long known to exist but remaining unidentified. Remarkably, a PubMed search in May 2005 found nearly 250 citations with the keyword "slc26a*", when the number of citations for the classical anion exchanger family "slc4a*" was about 140. Such a rapid publication burst testifies to the exciting scientific questions that the *SLC26* gene family pose. As an example, we still know very little about the control of tissue-specific expression of these genes.

Acknowledgements

The studies conducted in our laboratory were carefully performed by my former students Pia Höglund, Siru Mäkelä (née Haila), Hannes Lohi and presently, Minna Kujala. I wish to thank them and our many collaborators. We have enjoyed long-term financial support for work in our laboratory by the Sigrid Jusélius Foundation and Academy of Finland. The *SLC26* gene project has been specifically supported by the Ulla Hjelt Fund and Helsinki University Hospital Research funds.

References

Bissig M, Hagenbuch B, Stieger B, Koller T, Meier PJ 1994 Functional expression cloning of the canalicular sulfate transport system of rat hepatocytes. J Biol Chem 269:3017–3021

Dallos P, Fakler B 2002 Prestin, a new type of molecular motor. Nat Rev Mol Cell Biol 3:104–111

Everett LA, Glaser B, Beck JC et al 1997 Pendred syndrome is caused by mutations in a putative sulphate transporter gene (PDS). Nat Genet 17:411–422

Haila S, Höglund P, Scherer SW et al 1998 Genomic structure of the human congenital chloride diarrhea (CLD) gene. Gene 214:87–93

Haila S, Hästbacka J, Böhling T, Karjalainen-Lindsberg M-L, Kere J, Saarialho-Kere U 2001 SLC26A2 (diastrophic dysplasia sulfate transporter) is expressed in developing and mature cartilage but also in other tissues and cell types. J Histochem Cytochem 49:973–982

Hästbacka J, de la Chapelle A, Mahtani MM et al 1994 The diastrophic dysplasia gene encodes a novel sulfate transporter: positional cloning by fine-structure linkage disequilibrium mapping. Cell 78:1073–1087

Hästbacka J, Superti-Furga A, Wilcox WR, Rimoin DL, Cohn DH, Lander ES 1996 Atelosteo-genesis type II is caused by mutations in the diastrophic dysplasia sulfate-transporter gene (DTDST): evidence for a phenotypic series involving three chondrodysplasias. Am J Hum Genet 58:255–262

Höglund P, Haila S, Socha J et al 1996 Mutations in the down-regulated in adenoma (DRA) gene cause congenital chloride diarrhoea. Nat Genet 14:316–319

Holmberg C 1986 Congenital chloride diarrhea. Clin Gastroenterol 3:583–602

Kacem HH, Rebai A, Kaffel N, Masmoudi S, Abid M, Ayadi H 2003 PDS is a new susceptibility gene to autoimmune thyroid diseases: association and linkage study. J Clin Endocr Metabol 88:2274–2280

Kere J, Sistonen P, Holmberg C, de la Chapelle A 1993 The gene for congenital chloride diarrhea maps close to but is distinct from the gene for cystic fibrosis transmembrane conductance regulator. Proc Natl Acad Sci USA 90:10686–10689

Kere J, Lohi H, Höglund P 1999 Genetic disorders of membrane transport III. Congenital chloride diarrhea. Am J Physiol 276:G7–13

Lee A, Beck L, Markovich D 2003 The mouse sulfate anion transporter gene Sat1 (Slc26a1): cloning, tissue distribution, gene structure, functional characterization, and transcriptional regulation thyroid hormone. DNA Cell Biol 22:19–31

Li XC, Everett LA, Lalwani AK et al 1998 A mutation in PDS causes non-syndromic recessive deafness. Nat Genet 18:215–217

Liu XZ, Ouyang XM, Xia XJ et al 2003 Prestin, a cochlear motor protein, is defective in non-syndromic hearing loss. Hum Mol Genet 12:1155–1162

Lohi H, Kujala M, Kerkelä E, Saarialho-Kere U, Kestilä M, Kere J 2000 Mapping of five new putative anion transporter genes in human and characterization of SLC26A6, a candidate for pancreatic anion exchanger. Genomics 70:102–112

Lohi H, Kujala M, Makela S et al 2002 Functional characterization of three novel tissue-specific anion exchangers SLC26A7, -A8, and -A9. J Biol Chem 277:14246–14254

Mäkelä S, Kere J, Holmberg C, Höglund P 2002 SLC26A3 mutations in congenital chloride diarrhea. Hum Mut 20:425–438

Mäkelä S, Eklund R, Lähdetie J, Mikkola M, Hovatta O, Kere J 2005 Mutational analysis of the human SLC26A8 gene: exclusion as a candidate for male infertility due to primary spermatogenic failure. Mol Hum Reprod 11:129–132

Markovich D, Bissig M, Sorribas V, Hagenbuch B, Meier PJ, Murer H 1994 Expression of rat renal sulfate transport systems in Xenopus laevis oocytes. Functional characterization and molecular identification. J Biol Chem 269:3022–3026

Mount DB, Romero MF 2004 The SLC26 gene family of multifunctional anion exchangers. Pflugers Arch 447:710–721

Oliver D, He DZZ, Klocker N et al 2001 Intracellular anions as the voltage sensor of prestin, the outer hair cell motor protein. Science 292:2340–2343

Paoloni-Giacobino A, Kern I, Rumpler Y, Djlelati R, Morris MA, Dahoun SP 2000 Familial t(6;21)(p21.1;p13) translocation associated with male-only sterility. Clin Genet 58:324–328

Regeer RR, Lee A, Markovich D 2003 Characterization of the human sulfate anion transporter (hsat-1) protein and gene (SAT1; SLC26A1). DNA Cell Biol 22:107–117

Schweinfest CW, Henderson KW, Suster S, Kondoh N, Papas TS 1993 Identification of a colon mucosa gene that is down-regulated in colon adenomas and adenocarcinomas. Proc Natl Acad Sci USA 90:4166–4170

Toure A, Morin L, Pineau C, Becq F, Dorseuil O, Gacon G 2001 Tat1, a novel sulfate transporter specifically expressed in human male germ cells and potentially linked to RhoGTPase signaling. J Biol Chem 276:20309–20315

Vincourt JB, Jullien D, Kossida S, Amalric F, Girard JP 2002 Molecular cloning of SLC26A7, a novel member of the SLC26 sulfate/anion transporter family, from high endothelial venules and kidney. Genomics 79:249–256

Vincourt JB, Jullien D, Amalric F, Girard JP 2003 Molecular and functional characterization of SLC26A11, a sodium-independent sulfate transporter from high endothelial venules. FASEB J 17:890–892

Waldegger S, Moschen I, Ramirez A et al 2001 Cloning and characterization of SLC26A6, a novel member of the solute carrier 26 gene family. Genomics 72:43–50. Erratum in: Genomics 2001 77:115

Wang Z, Petrovic S, Mann E, Soleimani M 2002 Identification of an apical Cl/HCO3 exchanger in the small intestine. Am J Physiol Gastrointest Liver Physiol 282:G573–579

Zheng J, Shen W, He DZ, Long KB, Madison LD, Dallos P 2000 Prestin is the motor protein of cochlear outer hair cells. Nature 405:149–155

DISCUSSION

Welsh: As you described the congenital chloride diarrhoea (CLD), it sounds like a complete loss of function, on the basis of the mutations you have and looking at families. With other diseases, is it always a loss of function? Is there ever an autosomal dominant? Is there heterogeneity in the loci for the clinical phenotype?

Kere: I don't think there is any example of a dominant inheritance for these diseases. With regard to heterogeneity for hearing disorders, there are a couple of hundred genes for different hearing defects. The non-syndromic type that goes with the Pendred syndrome is just one of those. Clinically, it is difficult to distinguish among these different hearing disorder types. When we look at the mutation spectrum, as we start with patients we are trying to find those mutations that cause loss of function. Other kinds of mutations might occur in the populations, but they would not cause any problem. With CLD we have an example of this kind of mutation that changes the coding region but occurs as a common polymorphism in populations. This is C307W.

Welsh: In thinking about loss of function, some of these are in sites accessible to biopsy such as the colon and rectum. Is what is found there consistent with what you find, for example, in some of your nonsense mutations and splice site variations where you would expect a decrease in the amount of message?

Kere: We have done very little with these patients. The Finnish patients show normal levels of the message, as expected, because it is just a change in the protein.

Knipper: You mentioned the close neighbourhood of *SLC26A3* and *A4*. They are localized close together on chromosome 7. *A6* is on chromosome 3. Do you think *A3* and *A4* came from gene duplication?

Kere: It would be interesting to look at *A3* and *A4* now that the full genomic sequence is available. In human, their coexistence close to each other with such high homology is indicative of a very recent duplication event.

Knipper: Is there homology with *C. elegans* sequences?

Kere: Yes, indeed, but *C. elegans* has fewer members, with just seven genes.

Everett: *A3* and *A4* split after divergence of mammals from fish. *A3* and *A4* are conserved in mammals. It may be interesting to look at other vertebrates though, for example chicken.

Markovich: David Mount suggested that *A10* is a pseudogene. What is the closest gene to *A10*, and do you agree that it is a pseudogene?

Kere: I have no idea.

Mount: From sequencing expressed sequence tags (ESTs) we gave up on *A10* relatively quickly; in mouse and human there are some ESTs that have stop codons within the putative coding region. To my understanding it is an expressed pseudogene, I haven't seen any evidence to the contrary.

Markovich: What is it closest related to?

Mount: We can't answer that because we weren't able to parse together a coherent open reading frame. Just from the evolutionary perspective, it is interesting that there are 10 genes in *Drosophila*, but except for prestin orthologues and some

A6-like homology, you can't find clear orthologues of the mammalian genes in lower vertebrates. This is in contrast with most other transporters. You can find something close to the *CLC2* gene in *C. elegans* (Rutledge et al 2001); in the cation chloride transporter family there is clearly an *NKCC1*-like gene and a *KCC*-like gene in this species. When did the mammalian-type *SLC26* orthologues appear? What does this tell us about the role of the human *SLC* genes? An intriguing aspect of this family is that there are so many of them and they are not particularly conserved between invertebrates and mammals.

Kere: One of the interesting features that deserves greater study is that even though the structures are so well conserved (the exons are the same size and so on) there must be big differences in the regulatory sequences. I have seen relatively little published work looking at the promoters and what makes some of these transporters so tissue specific and others broadly expressed.

Welsh: The multi-tissue Northern blots you showed are interesting, but tissues are of course made of many different types of cells. Has the location of transporters been examined within different organs? I am particularly interested in the brain, where *A1* was expressed. In the brain I'm thinking about the choroid plexus, of course. But there was quite heavy expression there: is it in neurons, astrocytes or endothelium, for example?

Kere: We don't know about brain. For us, one of the striking features has been this chloride diarrhoea. When we look at the colonic epithelium, it is not expressed in the crypts, which is consistent with the notion that it is downregulated in adenoma. The least differentiated cell types do not express it, but as the colon epithelial cells climb up on the surface epithelium, they start to express this. So there is a cell differentiation stage-specific pattern of expression for some of these transporters, which is highly interesting. The same is true for the testis-specific expression, where it is expressed by the germ cell lineage, but not by all types.

Soleimani: Many of the *SCL26* family members show distinct expression patterns in various cell types both in the kidney and gastrointestinal (GI) tract. As an example, in the kidney, A1 is expressed on the basolateral membrane of the proximal tubule whereas A4 is detected on the apical membrane of certain intercalated cells in cortical collecting duct. A6 is located on the apical membrane of the kidney proximal tubule whereas A7 is detected on the basolateral membrane of alpha intercalated cells in outer medullary collecting. These results indicate that even within the same tissue, SLC26 isoforms show cell-specific expression patterns. The same observation is true in the GI tract, where A6, A7 and A9 are demonstrated in the stomach by tissue blot, but each is targeted to different cell types in the same tissue.

Muallem: The same transporter can either go to the apical membrane or the basolateral membrane.

Aronson: With A7, we have some antibodies that show apical staining in the proximal tubule, as well as the basolateral staining that Mannoocher Soleimani has seen. This is still a preliminary finding.

Muallem: No one knows how these transporters are targeted in different tissues.

Case: We talk about tissue specificity. It reminds me of drugs: when they are invented they are always 'tissue specific', but the more we know about them the looser the specificity becomes. Are we going to find that these channels have a lot wider distribution than we've been led to believe?

Kere: Clearly, some of them are more widely distributed than we first thought. I showed them as they were in the papers originally. Now we know that A6, for example, is present in kidney. Those that still show rather specific expression patterns are chloride diarrhoea, Pendred syndrome, prestin and TAT1 or SLC26A8.

Case: Have thorough searches been made?

Soleimani: The tissue distribution of SLC26 family members in human and experimental animals may be different. The case in point is SLC26A8. The original studies by Juha Kere in human tissues detected A8 only in testis whereas in mouse, A8 expression is widespread. The story with A11 is also similar to A8 with regard to its distribution in human vs. experimental animals.

Mount: It is like a housekeeping gene. We have not as yet achieved convincing functional expression for either the human or mouse orthologue, despite numerous attempts with chloride, sulphate and oxalate uptakes.

Case: Isn't that typical of scientists? If it is everywhere, we don't find it interesting, but if it is in your left toe it suddenly becomes the most important thing around!

Alper: Housekeeping may not have the same meaning in every tissue or cell type.

Seidler: I am interested in the PDZ domain consensus sites that you find in four of the 11. Whenever I look for PDZ domain proteins I find them near the apical region, not the basolateral membrane. One of the four was A7: my understanding what that it is basolaterally localized. I hear there is a discussion about this.

Mount: There are PDZ domain Lin7 homologues in the kidney that are basolateral (Straight et al 2000). As epithelial physiologists we appear to be biased towards thinking that PDZ domains are all apical, but there clearly are basolateral PDZ-domain proteins.

Alper: There are two cloned isoforms of A7 in the original paper. They differ in the C-terminus, and I remember a PDZ recognition motif in only one of them. Therein might be a more conventional explanation.

Chan: With regard to the tissue distribution of the gene family, I want to add that A6 is also found in the uterus, in the endometrium to be exact. There seems to be interaction between this and cystic fibrosis transmembrane regulator protein (CFTR).

Markovich: We published a paper last year (Reeger & Markovich 2004) looking at A1 trafficking. We found it contains a PDZ domain. It is trafficked to the baso-lateral membrane and proximal tubule. If we mutate a dileucine motif in its C-terminus, this stops it getting to the basolateral membrane. When we mutate the PDZ domain it still gets to the basolateral membrane. We think the dileucine motif might be important for trafficking, which was shown for the sodium phosphate co-transporters as well.

Ashmore: I want to jump back to the discussion on evolution. What is the conclu-sion about the nature of the ancestral member of this SLC26 family? What do the dendrograms show?

Welsh: Are these in prokaryotes?

Alper: SLC26 genes are found widely among prokaryotes. Their number only proliferates.

Welsh: What do we know about their function in prokaryotes?

Alper: The first paper on function appeared in December 2004 (Price et al 2004). This was done by a [^{14}C]bicarbonate flux assay. It appears to be a bicarbonate transporter, but it is likely that there will be many substrates.

Kere: In human the gene structures fall into two classes. A1 and A2 are related, with four coding exons, and the remaining genes have 21 exons.

Ashmore: So the conclusion is that A1 and A2 are the earliest members?

Kere: I'm not sure this is true: introns may have been added or lost.

Mount: The interesting question is, when did some of the mammalian ortho-logues emerge? It is not the primordial ancestor that interests me, but where in the middle of evolution did the event happen? This may be different for different branches of the gene family.

Alper: How do you decide that? The lower metazoans have many genes, but they appear to be equidistant from all the mammalian *SLC26* genes.

Mount: In *Xenopus* and *Ciona intestinalis*, there are some that look almost like A2 and A6, but not completely. This is based on blast alignments in large part, and not a rigorous assessment.

Everett: I have actually performed a careful phylogenetic study of all fully sequenced model organisms, so perhaps I can clear this up a little. Basically, I used the most conserved regions of these proteins to perform a TBLASTN search on the full genomic sequence for each organism, rather than using cDNA or other expressed sequences, so that I know I included the full set of SLC26 paralogues for each organism in the analysis. From my results, it's probably most relevant to mention the comparison of the set of mammalian paralogues with those of fish such as *Fugu*. We see that the duplication of some paralogues occurred before the divergence of mammals and fish (i.e. there is direct *Fugu* orthologue for a mam-malian transporter), and some afterwards (i.e. there is no direct orthologue, or there are two fish genes for one mammalian or vice versa). On the other hand, if you go further back, looking at non-vertebrates (for example *C. elegans*, *Drosophila* and *Arabidopsis*), you will find that usually, just one ancestral gene was present (or retained) upon the divergence of each lineage, which has later duplicated such that each organism has its own set of paralogues.

Alper: Juha Kere, I'd like to come back to the point you made about the other allele in some of the patients with recessive disease in whom only a single mutation was found on one allele. Has anyone had the other allele completely sequenced at a genomic level?

Kere: This is in our plan. Conceivably, there may be something in the promoter. There could be something in the 5′ introns. It could be something downstream of the last exon. The sequencing effort required is not that bad. The problem is that it may be a single nucleotide change in one of the introns. There are lots of SNPs in introns, so finding the right one is going to be tough.

Soleimani: You referred to DRA (or A3) as a chloride–bicarbonate exchanger, and indeed some functional studies in certain *in vitro* expression systems show that DRA mediates chloride bicarbonate exchange. However, one can not conclude that DRA is a chloride bicarbonate exchanger just based on the presence of chloride losing diarrhoea and metabolic alkalosis in patients with a mutation in DRA. Even if DRA mediates a base other than bicarbonate in exchange for chloride, still its inactivation, as in the case of congenital chloride diarrhoea, should lead to hypochloraemia and as a consequence metabolic alkalosis.

Seidler: There are several expression studies for the human sequence that clearly show a preference for chloride–bicarbonate exchange over chloride–base exchange. The same is true when the brush border membranes are isolated and chloride–base versus chloride–bicarbonate exchange is looked for. The data show that bicarbonate is transported preferentially.

Soleimani: The point is that this is similar to A6; one can get multiple exchange modes with DRA depending on many factors, including the use of different expression systems or imposition of certain ionic gradients.

Mount: There are two issues. One is that the volume depletion is why you get the hypokalemic alkalosis. The other question becomes one of specificity, which I find the most puzzling in this area.

Alper: Physiologically, they are all operating in the presence of 10–25 mM bicarbonate and sub-micromolar hydroxyl. The buffer capacity says that if they show *in vitro* preference for bicarbonate, then it is bicarbonate exchange.

Aronson: Henry Binder's group (Rajendran & Binder 1994) has shown that chloride–butyrate exchange and the recycling of butyrate can accomplish the same thing. Has this specifically been tested?

Alper: Yes, at least for some.

Mount: We have looked at butyrate uptake and have not had convincing transport for A6 and other SLC26s, with SLC5A8 as a positive control.

Soleimani: I wanted to emphasize that congenital chloride diarrhoea by itself doesn't mean that we are missing chloride–bicarbonate, but rather chloride–base exchange.

Markovich: My understanding of evolution is that genes that cause diseases evolve faster.

Welsh: Is that true?

Markovich: Chris Ponting from Oxford University had a Genome Research paper last year in which he looked at the evolution of genes in mammals (Winter et al 2004). He found that disease-causing genes evolve faster. I would suggest that *SLC26* members that may not be linked to diseases could be older.

Welsh: It seems surprising.

Alper: Disease-causing genes are often those with vital functions in tissues which have no alternate back-up mechanisms that can be relied upon to compensate for the loss-of-function.

Welsh: You did a nice job explaining how we have learned about this family. Most of the information has come from humans, which is in contrast to some other fields. In this regard, it seems like this is a wonderful opportunity to look at people with mutations in, say, CLD, and see whether they have phenotypes in other organs. It is interesting to consider whether there might be propensities for diseases associated with haploinsufficiency.

Kere: Going back to the gut situation, perhaps it is physiologically interesting that even though the DTDST (SLC26A2) and many other members are present in the gut, they don't replace the missing function arising from mutations in the chloride diarrhoea gene.

Soleimani: One specific way to test this issue is to overexpress *A6* in the colon of an animal model of congenital chloride diarrhoea (*A3* deletion) and see whether the diarrhoea stops.

Kere: Then, for the heterozygous state, the chloride diarrhoea gene was originally suggested as a gene down-regulated in colon cancer. There was a study done by Pia Höglund and her husband Dr Aaltonen who is a cancer geneticist, looking at cancer incidence increase among the heterozygous carriers of the mutation. The result showed no relevant increase in any type of cancer (Hemminki et al 1998). The heterozygous state seems to be normal for cancer. We are all carriers of half a dozen or so recessive disease alleles, so this may contribute in combination with other genes to some disease susceptibility. The other gene that we have studied for disease associations in human is *A8*. In human there is a testis-specific expression pattern. In a recent study we picked males with infertility hoping to see mutations in *A8* (Mäkelä et al 2005). The only things that we could pick up were a couple of polymorphisms in a fairly large crowd of infertile men.

Markovich: Are they in the coding region?

Kere: Yes, some were (five lead to amino acid substitutions).

References

Hemminki A, Hoglund P, Pukkala E et al 1998 Intestinal cancer in patients with a germline mutation in the down-regulated in adenoma (DRA) gene. Oncogene 16:681–684

Price GD, Woodger FJ, Badger MR, Howitt SM, Tucker L 2004 Identification of a SulP-type bicarbonate transporter in marine cyanobacteria. Proc Natl Acad Sci USA 101:18228–18233

Makela S, Eklund R, Lahdetie J, Mikkola M, Hovatta O, Kere J 2005 Mutational analysis of the human SLC26A8 gene: exclusion as a candidate for male infertility due to primary spermatogenic failure. Mol Hum Reprod 11:129–132

Regeer RR, Markovich D 2004 A dileucine motif targets the sulfate anion transporter (sat-1) to the basolateral membrane in renal cell lines. Am J Physiol Cell Physiol 287:C365–372

Rutledge E, Bianchi L, Christensen M et al 2001 CLH-3, a ClC-2 anion channel ortholog activated during meiotic maturation in C. elegans oocytes. Curr Biol 11:161–170

Straight SW, Karnak D, Borg JP, Kamberov E, Dare H, Margolis B, Wade JB 2000 mLin-7 is localized to the basolateral surface of renal epithelia via its NH(2) terminus. Am J Physiol Renal Physiol 278:F464–475

Winter EE, Goodstadt L, Ponting CP 2004 Elevated rates of protein secretion, evolution, and disease among tissue-specific genes. Genome Res 14:54–61

Individual characteristics of members of the SLC26 family in vertebrates and their homologues in insects

Marlies Knipper, Thomas Weber*, Harald Winter, Claudia Braig, Jelka Cimerman, Juergen T. Fraenzer and Ulrike Zimmermann

*University of Tübingen, Department of Otolaryngology, Tübingen Hearing Research Center, Molecular Neurobiology, Elfriede-Aulhorn-Strasse 5, D-72076 Tübingen, Germany and *St. Jude Children's Research Hospital, Department of Developmental Neurobiology, 332 North Lauderdale, Memphis, TN 38105–2794, USA*

Abstract. The 10-member *SLC26* gene family encodes anion exchangers of which SLC26A5 appears to be restricted to the outer hair cells of the inner ear. Here, the so-called prestin protein acts as a molecular motor, thought to be responsible for active mechanical amplification in the mammalian cochlea. We introduce special characteristics of SLC26A5 which may have relevance for other members of the family as well. As such, data point to a characteristic transcriptional control mechanism of which thyroid hormone surprisingly takes a role not only as an enhancer of expression, but also as a regulator of the subcellular redistribution of the prestin protein. Of significance for other members of the SLC26 family may be the observation that the failure of the subcellular redistribution of prestin protein prior to the onset of hearing leads to severe deficit of mature prestin function. Data will furthermore be argued in the context that prestin-related SLC26 proteins in the auditory organs of non-mammalian vertebrates and insects are widespread, possibly ancestral constituents of auditory organs and are likely to serve salient roles in mammals and across taxa.

2006 Epithelial anion transport in health and disease: the role of the SLC26 transporters family. Wiley, Chichester (Novartis Foundation Symposium 273) p 19–41

The fifth member of the SLC26 family, SLC26A5 (also known as prestin), was recently identified due to amino acid sequence and gene structure and its high homology within the sulfate transport region (Zheng et al 2000). While no doubt exists anymore about prestin's function as the motor protein of outer hair cells (OHCs) in the inner ear (Liberman et al 2002, Zheng et al 2002), its function as a solute carrier (SLC) with anion transport capability is still a matter of debate (Chambard & Ashmore 2003, Zheng et al 2002). Nevertheless, despite high tissue specificity of the distinct members of the SLC26 family, the homologous structure of all SLC26 members, well conserved across different species, may indicate comparable features of regulation and function. As such, preliminary data on the transcriptional regulation and alteration of subcellular distribution of SLC26A5, as well as its homologues in fish and insects, may have some relevance for other members of the SLC26 family.

Materials and methods

All the material and methods have already been described elsewhere and will be mentioned in brief. Tissue preparation, immunohistochemistry, riboprobe synthesis and *in situ* hybridization were performed as previously described in Knipper et al (1999, 2000) and Weber et al (2003). Zebrafish riboprobe synthesis was performed as described in Weber et al (2003). The quantity of electrically functional prestin molecules present in the OHC membrane was assessed by the magnitude of the voltage-dependent capacitance (C_{nonlin}) resulting from translocation of its voltage sensor (Santos-Sacchi 1991). Voltage-dependent capacitance was measured by using a software-based lock-in technique (phase tracking) as described (Oliver & Fakler 1999). The amount of axial, electromechanical force produced by an isolated OHC against a calibrated mechanical load at its apical end was measured as described previously (Gummer et al 2002).

Results

Considering the intriguing roles of the SLC26 anion exchangers in normal physiology and human pathophysiology, an understanding of the principles of transcriptional regulation is of fundamental interest. SLC26A5 (prestin) (Zheng et al 2000, Oliver et al 2001) exhibits a restricted expression pattern in the lateral wall of OHCs in the cochlea (Belyantseva et al 2000).

Prestin expression in the inner ear

Initially we studied prestin expression during development at the protein and mRNA level. Prestin mRNA is up-regulated during a critical developmental period of the inner ear, when most genetic and acquired hearing deficits are manifested. Prestin mRNA was noted from P2 onwards, first distributed around the nucleus in the cytoplasm of the OHC, shown for the midbasal turn at P7 (Fig. 1), while from approximately P8 onwards mRNA appeared to move towards the apical part of the hair cells, where it was restricted beyond P10, shown for P12 (Fig. 1). Using antibodies directed to the C-terminal part of SLC26A5 (Weber et al 2002) from approximately P3 onwards prestin protein was distributed across the entire OHC membrane (shown for P7, Fig. 1), while coincidentally to the redistribution of mRNA, the prestin protein became redistributed to the lateral membrane of the OHC membrane (shown for P12, Fig. 1).

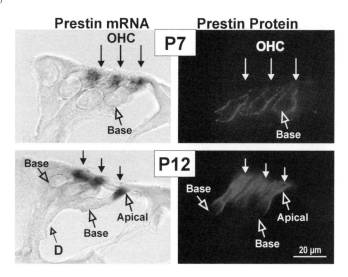

FIG. 1.   Localization of prestin mRNA and protein in cross-sections of rat cochleae at P7 and P12 using *in situ* hybridization and immunohistochemistry. Note the restricted localization of mRNA to the apical part of OHC, whereas prestin protein becomes redistributed from the entire OHC membrane (P7) to the lateral plasma membrane (P12). OHC, outer hair cell; D, Deiters cell.

Transcriptional control of prestin

As described (Weber et al 2002), we identified a putative response element (TRE^{Prest}) for nuclear receptors starting at position −416 relative to the ATG codon (Weber et al 2002) which has a human prestin gene homologue in approximately the same position. To determine the relative position of TRE^{Prest} we identified the transcriptional start site of the prestin gene performing a 5′-rapid amplification of cDNA ends and detected so far unknown fragments, the sequence of which covered a presumptive first and second exon, consistent with the presence of two untranslated exons (exons I and II) upstream of the exon containing the ATG codon (exon III) (Fig. 2A). Using electrophoretic mobility shift assay (EMSA) and reporter gene studies (Weber et al 2002), we learned that TRE^{Prest} specifically binds thyroid hormone receptors (TRs) and that TRα-T3-dependent and TRβ-T3-dependent increase in reporter gene activation could be further enhanced by RXRα and all-*trans*-retinoic acid. These data indicate that thyroid hormone (TH) through TRβ/RXR positively modulates prestin expression, ensuring its timely up-regulation prior to the onset of hearing function (Weber et al 2002).

While it needs to be clarified whether a TH/TR transcriptional modulation is a cell-specific feature of SLC26A5 in OHCs or also acts on other members of the

SLC26 family, data claim that further *trans*-acting proteins act on prestin expression. We screened the 5'-untranslated region of the *SLC26A5* gene in humans and mice including the two untranslated exon I and II for *cis*-acting elements and focused in a first approach on the identified GC and TATA boxes upstream and within exon I. Co-transfection of either SP1 or TFIID, the *trans*-acting proteins of GC or TATA motifs respectively, revealed specific binding in humans and mice for SP1 (Fig. 2B) while no specific binding could be achieved with any of the TATA motifs (data not shown). Various gene constructs within or upstream of exon I displayed multi-fold enhancement of luciferase activity in reporter gene studies (Fig. 2C), indicating a critical role for *cis*-acting elements within this upstream region.

In conclusion, TR/RXR acting on TRE, probably in addition to *cis*-acting elements in the 5'-upstream regions covering exon I and an upstream region from exon I, plays a crucial role in the transcriptional control of prestin in humans and rodents.

Alteration of subcellular distribution

We analysed prestin protein distribution immunohistochemically in the cochlea of hypothyroid animals at P28 using double-labelling of anti-prestin and anti-synaptophysin. An immature prestin protein distribution across the entire OHC membrane was still visible in the absence of TH (Fig. 3A, Hypo P28), which allowed us to examine the role of subcellular prestin distribution in hair cell function.

Motile responses of OHCs, approaching $0.2 \mu m$, as well as prestin protein density can be measured by the analysis of non-linear capacitance (Ashmore 1990, Oliver & Fakler 1999) and the force produced by motile responses is detectable by atomic force microscopy (Gummer et al 2002). In the absence of TH non-linear capacitance (C_{nonlin}) (Fig. 3B) and charge density (data not shown) were still reduced at P15 corresponding to the prestin density found at P8 in normally developing rats (Oliver & Fakler 1999). At a later developmental stage (P28), C_{nonlin} of hypothyroid

FIG. 2. (A) Position of TRE[Prest] within the prestin gene. Based on homologous human genomic sequence, a model was designed proposing the position of TRE[Prest] within the exon–intron map of the prestin gene. Note its position downstream of exons I and II and upstream of exon III. (B) Position of GC- and TATA-elements within the prestin gene. GC- and TATA-elements are noted in the 5'-upstream region of exon I or within the exon I. GC-boxes in humans (B, position 1–6) and mice (B, position 7, 8) but not TATA-boxes neither in humans nor mice (data not shown) bind specifically their correspondent *trans*-active protein SP1 or TFIID. (C) Analysis of 5'-upstream regions of SLC26A5 in reporter gene studies. A presumptive minimal promoter function of the 5'-region upstream of exon II is supported upon differential use of distinct sequences of this region within luciferase reporter gene studies.

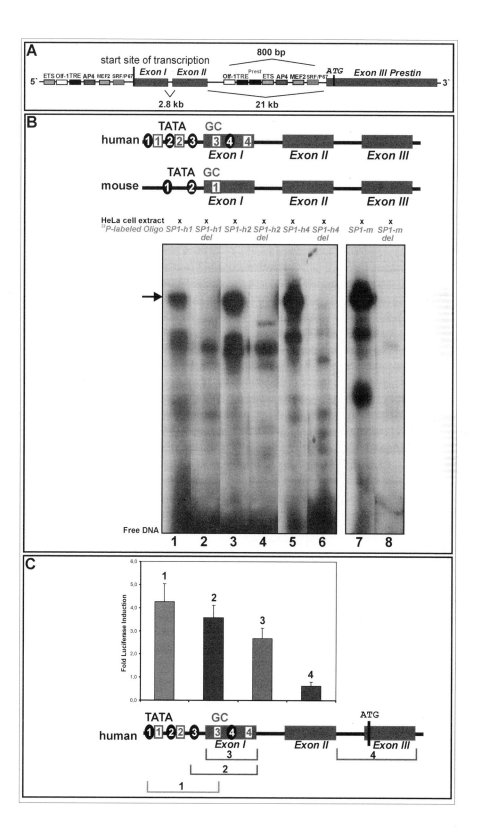

animals recovered to the level found in control animals (Fig. 3B), indicating that the absence of TH caused a retardation of the developmental increase of prestin expression. Actual force measurements carried out with an atomic force microscope on isolated OHCs from hypothyroid rats at P15 and at P28 revealed significantly reduced levels in comparison to controls (Fig. 3C). Various control experiments

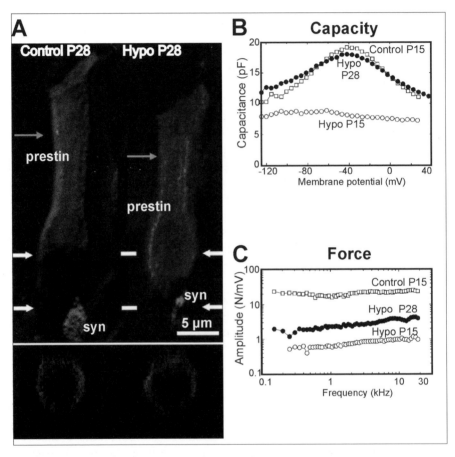

FIG. 3.   (A) Uniform distribution of prestin in the absence of TH. Prestin and synaptophysin (syn) immunofluorescence in OHCs from medial cochlea turns of rat at P28, for a control animal (Control) and for a hypothyroid animal (Hypo). (B) Delay of voltage-dependent charge transfer in the absence of TH. Non-linear capacitance measured in OHCs from a control animal at P15 (open squares) and from hypothyroid animals at P15 (open circles) and P28 (solid circles). (C) Reduced electromechanical force in the absence of TH. Amplitude of the (isometric) electromechanical force as a function of stimulus frequency, for isolated OHCs from a control animal at P15 (open squares) and from hypothyroid animals at P15 (open circles) and P28–31 (solid circles).

indicated furthermore that the reduction in force production under hypothyroid conditions was not due to structural parameters (Zimmermann et al 2006, in preparation).

In conclusion, TH affects the resonance behaviour of the basilar membrane of the inner ear by influencing the subcellular distribution of prestin protein in OHCs and thereby the generation of mechanical force.

Prestin homologues in non-mammalian vertebrates and insects

When we questioned evolutionary novelty of prestin motor in mammalian OHCs we were surprised to get positive hybridization with the rat prestin riboprobe in the auditory organ of mosquitoes, the Johnston's organ, which consists of thousands of radially arranged multi-cellular mechanoreceptor units (Fig. 4A).

By using the deduced amino acid sequence 465–704 of the rat prestin gene for BLAST algorithms of the NCBI, Ensembl and FlyBase genome browsers (Weber et al 2003) alignments were performed and extended to the other members of the SLC26 family (A1–A9). As described (Weber et al 2003) Zebrafish I gene product was revealed as most closely related to prestin (SLC26A5) of rat, mice and gerbil (Fig. 4*B*, Zebrafish I). In *Drosophila*, the FlyBase BDGB BLAST server identified a predicted gene with highest similarity to rodent prestin, CG5485 (Weber et al 2003). Zebrafish I and *Drosophila* CG5485 were cloned and again specific riboprobes showed hybridization with the chordotonal sensilla of Johnston's organ (Fig. 5*A*), and in the region of the auditory organs respectively (Fig. 5B) (Zebrafish I Genbank submission accession number AY278118).

In conclusion, data point to the existence of SLC26A5 homologues in zebrafish and *Drosophila* which are expressed in sensory organs, challenging the question to their function as either anion transporters and/or active amplifiers.

Discussion

Transcriptional regulation

Our data suggest TH/TR as a first transcriptional regulator of one of the members of the SLC26 family, the inner ear-specific SLC26A5 (prestin) (Weber et al 2002). Thus, TH/TR might regulate prestin expression in a cell-specific manner together with other as yet unknown *trans*-acting elements to define its cell-specific temporally precise expression (Weber et al 2002).

SLC26A6 proteins show best sequence similarity to SLC26A5 (Lohi et al 2000). Two alternative 5′ non-coding regions and transcriptional start sites with Kozak sequences have been predicted for SLC26 members with start codons either in exon 1 or exon 2 (Lohi et al 2000). However, information about upstream sequences displaying promoter function within the SLC26 family is rare. Cloning of the

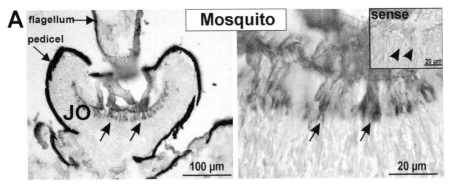

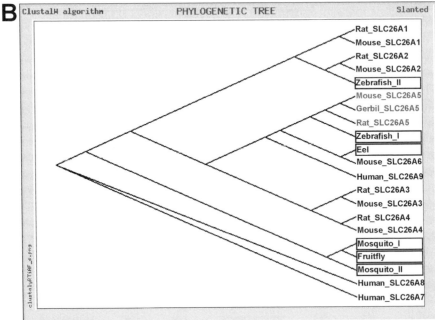

FIG. 4. (A) mRNA expression in cross-sections of the mosquito *Toxorhynchites brevipalpis* using *in situ* hybridization with rat prestin-specific riboprobes. In cross-sections of *T. brevipalpis* antennae, hybridization signals occur in the second antennal segment (pedicel), in the chordotonal sensilla, the auditory mechanosensory units of Johnston's organ (JO). No signals were observed with the sense probe (inset). (B) Phylogenetic tree. The relationship between prestin homologous genes from different species and from a selection of representative members of the SLC26 family. For the generation of the phylogenetic tree the alignment from Fig. 2A in Weber et al (2003) was used.

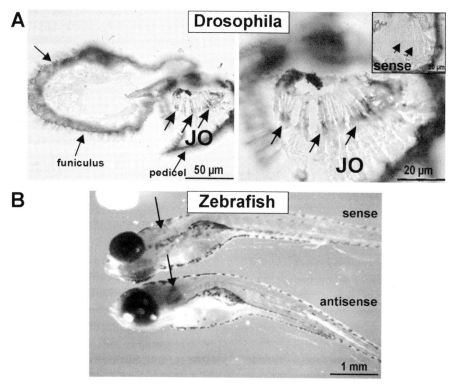

FIG. 5.   (A) mRNA expression in cross-sections of the fruit fly *Drosophila melanogaster* using *in situ* hybridization with *Drosophila* CG5485-specific riboprobes. Reactive signals were detected in cells within the Johnston's organ in the second segment of antennae (pedicel) of *Drosophila melanogaster* and (B) in the utriculus, sacculus and lagena of the hearing organ of the zebrafish.

human prestin cDNA revealed a gene with 21 exons including two untranslated exons (Liu et al 2003), confirming our recent findings (Weber et al 2002). The first two exons were suggested to be important for prestin expression (Zheng et al 2003). Here, we report a functional role of TREs within the first 600 bp of the homologous intron of the human prestin gene and describe specific binding of GC elements but not TATA within a 5′-untranslated region downstream and within exon I. Further studies are required to verify the function of other *cis*-acting elements within the 5′-upstream region of SLC26A5 and challenge its comparison with other members of the SLC26 family.

Regulation of subcellular distribution

The coordinated transport of ions and water across intact epithelia requires selective sorting of receptors, ion channels and transporters to apical or basolateral cell surfaces. PDZ (**P**SD-95/**D**isc-large/**Z**O-1) domains exhibit protein–protein interaction domains that play an essential role for determining cell polarity or membrane targeting (Songyang et al 1997). From all members of the SLC26 family, SLC26A6 is the first described to exhibit PDZ-interaction pathways and its organization in membrane micro-domains together with PDZ domain proteins such as NHERF and E3KARP (Lohi et al 2003).

SLC26A5 does not contain a PDZ-motif in its C-terminal part, and data presented here strongly suggest a mechanism of subcellular distribution of SLC26A5 different from that of other members of the family (Zimmermann et al 2006, in preparation). Indeed, the fact that a correct subcellular distribution is not achieved in the absence of TH, despite normal bell-shaped non-linear capacity curves with equal $V_{1/2}$ values, indicates that at least structural maturation or distinct prestin protein phosphorylation steps, known to affect membrane targeting as well as $V_{1/2}$ values of non-linear capacitance (Deak et al 2005) do not play a role for the regulation of (polarized) subcellular distribution of SLC26A5.

Expression in non-mammalian vertebrates and insects

Beside SLC26A5, no further members of the SLC26 family have been identified in mechanosensitive organs in mammals, non-mammalian vertebrates or insects. This study establishes the presence of SLC26 members for the auditory sense organs of non-mammalian vertebrates, i.e. teleost fish, and insects, documenting that SLC26 proteins occur in a vast variety of ears (Weber et al 2003). Although the ears of insects and vertebrates have evolved independently, their auditory mechanosensors (chordotonal sensilla and hair cells, respectively) may share a common evolutionary origin (Gillespie & Walker 2001). Such a scenario is supported by the fact that homologous genes control the development of these sensors (Hassan & Bellen 2000) and may explain the presence of closely related SLC26 members in insect and vertebrate ears. While non-linear amplification has been described for tetrapods and insects (Gopfert & Robert 2001, 2003), fish are not known to improve audition by active mechanical amplification, although physiological evidence exists that their hair bundles can move in response to electrical stimulation (Rusch & Thurm 1990). Molecular and functional studies are in progress to clarify the evolutionary origin of motile function of SLC26A5 orthologues.

References

Ashmore JF 1990 Forward and reverse transduction in the mammalian cochlea. Neurosci Res 12 (suppl):S39–50

Belyantseva IA, Adler HJ, Curi R, Frolenkov GI, Kachar B 2000 Expression and localization of prestin and the sugar transporter GLUT-5 during development of electromotility in cochlear outer hair cells. J Neurosci 20:RC116

Chambard JM, Ashmore JF 2003 Sugar transport by members of the SLC26 superfamily of anion-bicarbonate exchangers. J Physiol 550:667–677

Deak L, Zheng J, Orem A et al 2005 Effects of cyclic nucleotides on prestin's function. J Physiol 563:483–496

Gillespie PG, Walker RG 2001 Molecular basis of mechanosensory transduction. Nature 413:194–202

Gopfert MC, Robert D 2001 Active auditory mechanics in mosquitoes. Proc R Soc Lond B Biol Sci 268:333–339

Gopfert MC, Robert D 2003 Motion generation by Drosophila mechanosensory neurons. Proc Natl Acad Sci USA 100:5514–5519

Gummer AW, Meyer J, Frank G, Scherer MP, Preyer S 2002 Mechanical transduction in outer hair cells. Audiol Neurootol 7:13–16

Hassan BA, Bellen HJ 2000 Doing the MATH: is the mouse a good model for fly development? Genes Dev 14:1852–1865

Knipper M, Gestwa L, Ten Cate WJ et al 1999 Distinct thyroid hormone-dependent expression of TrKB and p75NGFR in nonneuronal cells during the critical TH-dependent period of the cochlea. J Neurobiol 38:338–356

Knipper M, Zinn C, Maier H et al 2000 Thyroid hormone deficiency before the onset of hearing causes irreversible damage to peripheral and central auditory systems. J Neurophysiol 83:3101–3112

Liberman MC, Gao J, He DZ et al 2002 Prestin is required for electromotility of the outer hair cell and for the cochlear amplifier. Nature 419:300–304

Liu XZ, Ouyang XM, Xia XJ et al 2003 Prestin, a cochlear motor protein, is defective in non-syndromic hearing loss. Hum Mol Genet 12:1155–1162

Lohi H, Kujala M, Kerkela E et al 2000 Mapping of five new putative anion transporter genes in human and characterization of SLC26A6, a candidate gene for pancreatic anion exchanger. Genomics 70:102–112

Lohi H, Lamprecht G, Markovich D et al 2003 Isoforms of SLC26A6 mediate anion transport and have functional PDZ interaction domains. Am J Physiol Cell Physiol 284:C769–779

Oliver D, Fakler B 1999 Expression density and functional characteristics of the outer hair cell motor protein are regulated during postnatal development in rat. J Physiol 519 Pt 3:791–800

Oliver D, He DZ, Klocker N et al 2001 Intracellular anions as the voltage sensor of prestin, the outer hair cell motor protein. Science 292:2340–2343

Rusch A, Thurm U 1990 Spontaneous and electrically induced movements of ampullary kinocilia and stereovilli. Hear Res 48:247–263

Santos-Sacchi J 1991 Reversible inhibition of voltage-dependent outer hair cell motility and capacitance. J Neurosci 11:3096–3110

Songyang Z, Fanning AS, Fu C et al 1997 Recognition of unique carboxyl-terminal motifs by distinct PDZ domains. Science 275:73–77

Weber T, Zimmermann U, Winter H et al 2002 Thyroid hormone is a critical determinant for the regulation of the cochlear motor protein prestin. Proc Natl Acad Sci USA 99:2901–2906

Weber T, Gopfert MC, Winter H et al 2003 Expression of prestin-homologous solute carrier (SLC26) in auditory organs of nonmammalian vertebrates and insects. Proc Natl Acad Sci USA 100:7690–7695

Zheng J, Shen W, He DZ et al 2000 Prestin is the motor protein of cochlear outer hair cells. Nature 405:149–155

Zheng J, Madison LD, Oliver D, Fakler B, Dallos P 2002 Prestin, the motor protein of outer hair cells. Audiol Neurootol 7:9–12

Zheng J, Long KB, Matsuda KB et al 2003 Genomic characterization and expression of mouse prestin, the motor protein of outer hair cells. Mamm Genome 14:87–96

Zimmermann U, Frank G, Oliver D et al 2006 Expression of prestin and cellular force generation by auditory outer hair cells are differentially affected by thyroid hormone depletion during development. In preparation

DISCUSSION

Welsh: How does prestin work? Do we know?

Knipper: I can only say a little bit; Dominic Oliver will say more in his paper. Currently, the concept is that a voltage-dependent conformational change of prestin is discussed as the molecular basis for motile outer hair cell (OHC) responses.

Muallem: Do you have any information about the splice variants? Are there any differences among them?

Knipper: The data are preliminary. We need to have the proper function information if at all possible. We are currently looking for a presumptive role of PDZ domains, and we are planning to test whether the splice variants, when coexpressed, change the charge behaviour.

Muallem: So you don't know that it is even functional?

Knipper: Not at the moment. The nicest thing would be if it has no impact on the motility, but it has an impact on the force.

Alper: Is the boundary between the lateral and basal membranes, which is defined in the mature state by the redistribution of prestin, defined by the Deiter's cell contact? And is it the Deiter's cell interface that is the influence driving prestin out?

Knipper: You have good information about the OHC, because most people here don't even know what and where Deiter's cells are in relation to outer hair cells. The Deiter's cells are the supporting cells apposing outer hair. Indeed, there seems to be a striking temporal parallel between the change in the subcellular distribution of SLC26A5 from a wide one to a restricted membrane distribution, and the enlargement of the contact zone of Deiter's cell and OHCs. While we do not have any functional evidence for this, we have to consider such a concept.

Alper: What happens to the Deiter's cells in hypothyroid state? What happens developmentally?

Knipper: They look nice. There is an inward rectifying K^+ channel in Deiter's cells that is strongly expressed in the apical part, just opposite the KCNQ4 ion channel in the base of OHCs that leads to the repolarization of the cell. In the absence of thyroid hormone (TH) we have current evidence that it looks not so much different. On the level of ion channels, we do not have anatomical or functional evidence that Deiter's cells are affected during hypothyroidism.

Alper: How about cytoskeletal candidate protein redistribution paralleling prestin distribution?

Knipper: No one knows.

Lee: How does TH change the distribution of prestin variants?

Knipper: We performed studies with RT-PCR and found a profound reduction of the splice variants in the absence of TH. This fits with our model. But it is not a total absence.

Lee: In the case of Pendred syndrome, there is an association with hypothyroidism. Is prestin somehow responsible for the deafness in Pendred syndrome?

Knipper: I don't think anyone has looked for prestin in Pendred knockouts. The typical feature or phenotype is the loss of the endocochlear potential as a first sign in the PDS knockout. In other knockouts that we work in we have looked at phenotypes where we have a loss of endocochlear potential. We noticed that in these mice the prestin expression appears normal. Thus the prestin expression is extremely robust. We still can observe it when other outer hair cell ion channels have gone already, indicating a low protein turnover rate which is not that much influenced by metabolic energy.

Everett: I think the fact that Pendred syndrome has features of both inner ear and thyroid disease leads to a lot of confusion. People have long thought that the deafness is thyroid hormone related but I do not believe that hypothyroidism plays any part in the deafness of Pendred syndrome. If you look at people or mice with mutations in thyroid peroxidase, which causes a more severe iodide organification defect and overt congenital hypothyroidism, they don't have a severe hearing defect. I am sure that the deafness in Pendred syndrome has nothing to do with any mild hypothyroidism. The hair cell defects seen in hypothyroidism are very different from the defects of endolymphatic hydrops and swelling of the inner ear in the Pendred knockout. It is just a coincidence that there is also a defect in the thyroid in Pendred syndrome.

Kere: One of the things we haven't discussed yet is the topology of the proteins on the membrane. You showed a model with 10 transmembrane domains. We have tended to favour the 12-transmembrane domain model for the other members.

Knipper: This might be a misunderstanding. We also favour a model with 12 transmembrane domains. I showed a picture which I found in the recent review by Dallos (Dallos & Fakler 2002) in which they analysed the glycosylation/phosphorylation steps. Due to their results a further experimental basis for a 12 transmembrane domain topology was generated.

Turner: There are a number of theoretical methods for predicting membrane protein topology, and these proteins are a mess! Especially at the N-termini where many of the predictive algorithms give quite different results.

Kere: We have some data to show that the C-terminal tail is inside the cell for a few SLC26 proteins.

Knipper: We can confirm this for the A5 members.

Everett: We took all the SLC26 transporters known at the time and analysed them together using a topology program, to give a more robust conclusion that, for all these transporters collectively, the C-terminus is in the cell. This is just a computational prediction though; there is better wet-bench based evidence to support this.

Turner: A number of the recently crystallized membrane proteins have been shown to have hydrophobic regions that don't completely cross the membrane. I think all bets are off for now.

Kere: Does anyone have evidence for an odd number of transmembrane domains?

Ashmore: Historically, prestin was predicted to have 11 domains, and then this was revised to 12. I am surprised that you say that the mess is at the N-terminal rather than the C-terminal end. Or do all of these prediction programs make a mess of both ends?

Turner: There is mess at the C-terminal end as well. But at least there are some clearly defined hydrophobic peaks. The predictive algorithms tend to correspond very well when there is a sharp peak. There are also a couple of not such sharp peaks that might be either one or two membrane-spanning domains.

Welsh: I take it from this discussion no one has mapped the topology.

Mount: And we assume that the topology is going to be the same for all the members, which is probably an oversimplification. In some programs A1 seems to have an extra transmembrane domain early on in the N-terminus. I think there will be common transmembrane domains, but there may be other regions in the protein where there may be some divergence.

Kere: The splice variation produces different numbers.

Muallem: There is probably something in the number 12. It is not only cystic fibrosis transmembrane regulator protein (CFTR), but also almost all the other epithelial chloride channels that have 12 transmembrane domains.

Alper: Except for 18 or 14 for CLC. How do you define precisely what is transmembrane, even with the crystal structure?

Welsh: The structure of CLC threw predictions into disarray.

Alper: I think you can be confident about 12 for the ABC family.

Muallem: When you look at the CLC structure, you can kind of count 12 transmembrane domains, even though there are some additional helices that protrude into the membrane.

Case: Is this effect of TH on the hair cell something seen with other transporters, or is this a unique situation?

Knipper: Yes, we see an effect of TH not only on transporters such as prestin, but also on ion channels, that are expressed during the critical time period prior to hearing function in hair cells and spiral ganglia neurons. Before this time and after this time TH does not display any effect on the inner ear according to our studies. During this critical period, however, TH indeed appears to affect particular ion channels and transporters, those proteins, which are up-regulated during the most final differentiation of hair cells and spiral ganglia neurons. We also presume that this is the period when in the body, the lung, heart, bones and intestines other final differentiation processes occur under the control of TH, including differentiation of functionally relevant ion channels. Thus TH may have a most critical role for synchronous running of final differentiation processes in multiple organs and cells.

Mount: We were able to PCR a fair amount of A5 from brain. There is some expression in heart as well. Have you looked with your antibody outside of the inner ear?

Knipper: Yes. We cannot find anything.

Oliver: We were also able to PCR A5 from brain of rat and mouse.

Mount: So you don't see prominent expression in rare cell types.

Oliver: No.

Knipper: The promoter of prestin was used to drive Cre recombinase expression at a time when promoter sequences were still presumed to be upstream of exon III. The result was a extraordinarily low expression also in inner hair cell in addition to outer hair cells (Tian et al 2004). Thus it can be concluded that there is a very low expression in the inner hair cell. We however, could never detect prestin protein at the inner hair cell level.

Ashmore: Even so, it is tempting to think that prestin is present throughout subsystems of the brain, and not just the auditory and ascending auditory pathways. It would be interesting to know specifically where it is in the brain.

Alper: If so, do you think it would serve the same electromechanical function?

Ashmore: I don't know.

Knipper: It would be interesting if we knew how this protein functions, if we do not have a dense expression pattern. Despite the fact that a variable density of the protein in transfected cells may nevertheless result in conformational change leading to detectable movement or better non-linear capacity, we might not necessarily pick it up if prestin isn't expressed apart from in OHC.

Welsh: How dense is its expression?

Knipper: It is extraordinarily dense. This is comparable to what has been described for A1 in erythrocytes.

Ashmore: There are 10 million copies per cell. About 40% of the surface membrane is prestin.

Soleimani: Are there other chloride transporters in OHC?

Knipper: We have strong evidence that SLCA2 that is expressed over the entire OHC membrane is partially overlapping with prestin.

Soleimani: What about NKCC1 or others? The reason for this question is that we did some experiments on prestin in collaboration with Dr Dallos. We couldn't detect any chloride–bicarbonate exchange mediated by prestin, but he later showed that if cells were depleted of chloride, the motility that is generated by prestin stops, even though prestin does not transport chloride. These results suggest that intracellular chloride is essential for the activity of prestin. Is it possible to inhibit another chloride transporter and therefore deplete intracellular chloride and as a result inhibit prestin?

Oliver: A similar idea was proposed by Joe Santos-Sacchi (Rybalchenko & Santos Sacchi 2003). He showed that there is a chloride conductance in the lateral membrane of the outer hair cell, which seems to be stretch sensitive. It is not clear what molecule provides this conductance. Chloride flux through this conductance may modulate the intracellular chloride concentration and modify the function of prestin.

Soleimani: What is the effect of furosemide on this chloride conductance? This question can be rephrased into whether prestin is the final pathway for a good number of diseases that cause deafness.

Oliver: Furosemide impairs inner ear function mainly by abrogating the endocochlear potential, the electrical driving force that is crucial for the electrical activity of the hair cells. This then also blocks OHC motility. There is chloride transport in other parts of the inner ear too, namely in the stria vascularis and the supporting cells, which is absolutely critical for inner ear function without directly pertaining to prestin function. Transport in the stria vascularis generates the endocochlear potential and involves the Na-K-2Cl co-transporter (NKCC1/Scl12a2) which explains the furosemide sensitivity (Delpire et al 2003).

Knipper: Every protein that would change the turgor of the outer hair cell would lead to deafness. From this point of view, we can hardly understand why prestin has become so famous. For the hair cell, the consequences are the same.

Welsh: Are people with prestin mutations completely deaf?

Knipper: Liu et al (2003) found a mutation in a splice acceptor site to be responsible for deafness. This is one of the first human diseases related to prestin.

Oliver: Carriers in whom both alleles are affected are completely deaf (Liu et al 2003), but this might also have to do with the degeneration of hair cells. In the prestin knockout mouse, hair cells also tend to degenerate (Liberman et al 2002).

Welsh: You would predict, therefore, that if the hair cell didn't degenerate but you took away prestin, people would still hear, although the acuity might decrease.

Oliver: Yes, I predict that there would be a drop-off in sensitivity of 50–60 db. This is actually what was observed in cochlear regions of prestin knockout animals where the hair cells were fully intact.

Markovich: Prestin seems to be the odd gene out in this family. It doesn't seem to have any anion transport. Why? Has it lost certain sequences?

Oliver: I don't know whether or not it has anion transport activity. As far as I know, people haven't looked very carefully.

Soleimani: We tested both mouse and human prestin, generous gifts from Peter Dallos. However we did not detect any chloride–bicarbonate exchange by either clone.

Knipper: So the answer is no?

Soleimani: Dr Karniski examined radiolabelled chloride uptake in oocytes expressing prestin. My understanding was prestin did not show any measurable chloride uptake (personal communication via Dr Dallos).

Ashmore: I still think it transports, but at a very low rate!

Soleimani: Perhaps the activity of prestin requires the right environment in terms of substrates, osmolarity, pH or the expression system (oocytes, mammalian cells, etc.).

Markovich: So it could be an anion exchanger?

Thomas: Along these lines, we heard about hypothyroidism and the fact that those patients don't exhibit this localization in OHC and, yet, there is charge movement without mechanical function. Let's just presume that under these expression levels you get some paracrystalline array. My guess is that no one is testing function under those conditions where we know in the native cell it is actually performing its mechanical function.

Welsh: You may get the same kind of array in an oocyte. I don't know whether it has a particular characteristic that you can recognize, for example, by scanning EM.

Knipper: No one has done a crystal array.

Thomas: At the higher density, is it functional?

Oliver: Basically, we don't get these high densities in our expression systems.

Alper: In transiently transfected cells we can get about a million copies per cell.

Oliver: Looking at the charge transfer density normalized to the linear membrane capacitance (which is a measure for the membrane area), we are about a factor of 10 lower than in OHCs.

Mount: Aquaporin 4 gives these orthogonal arrays in the proximal tubule of the kidney, which are lost in knockout mice (Verbavatz et al 1997, Yang et al 1996). You would think that it might be a characteristic of the protein itself, to make some kind of crystalline structure in the membrane, which may be dependent on the concentration.

Knipper: This is the reason why we have to get it out. What is the reason for this high expression of prestin in the membrane, a feature clearly different from other SLC26 members? This comes back to the question of whether prestin might have a physiological function as a transporter. Both high expression and loss of transporter function point to SLC26A5 as an exception among the other members. The question raised by this is whether at the time of the divergence of other vertebrates from fish it was necessary to invent something auditory-specific, since there is a protein which has only this particular characteristic and function in outer hair cells of the inner ear. If this is the case, we also have the solution for the striking cell-specific expression pattern of the other SLC26 members. The auditory hearing organ can be traced back at the molecular level to a common origin as might other distinct organs as well. Thus, prestin might indeed be an exception. It had 100 000 years to develop into a motile protein with molecular characteristics different from the other members.

Aronson: At least in principle, you don't need a large structural difference. It is a subtle kinetic change to let it do a transport hemi-cycle. You say there is an inside chloride-dependent conformational change; if you just had an anion exchanger with a very slow off-rate, it could have a reorientation step and a low rate of anion exchange. It really is an anion exchanger, just a very sluggish one, so it mainly has conformational changes with a low flux rate.

Muallem: Before we can answer these questions, someone needs to take the hair cell itself and measure something in it. It wouldn't be too difficult to measure pH and current. The cell milieu might provide some kind of regulation in addition to the structure.

Mount: A broader point is that we are biased that these are anion exchangers. In the CLC literature Chris Miller (Accardi & Miller 2004) and others (Picollo & Pusch 2005, Scheel et al 2005) have shown that the CLC chloride 'channels' can function as exchangers, exchanging chloride with a proton. The level of homology between SLC26 paralogues is not that high. It is not surprising that some of these are channel-like, and some of them have different stoichiometries. It doesn't bother me that it is not an exchanger; it may be just some variant of a global transport characteristic which all of them have at varying levels. There is this tendency to think that they are all anion exchangers. If anything comes out of this meeting, it will be that this is probably not the case.

Muallem: I agree with you, obviously! Not only that, but also as we learn more about them we know very little about what the true substrate specificity is. I would not be surprised to find out that this transporter is very good at transporting some other molecule.

Mount: I am not happy with negative results, such that I no longer have faith that if I can't find something transported by a given member of this family that it isn't transported. There are phosphorylation-dependent effects of chloride that we may be forgetting about, such that the conditions we use might not lead to successful transport. For example, intracellular chloride activates or inactivates the various modes of cation-chloride co-transport (Lytle & McManus 2002).

Alper: In this respect, with this question of conformational change without transport and conformational change with slippage to make for a leaky kind of transport, it is of note that that GFP-tagged prestin still localizes normally and gives the same electromechanical non-linear capacitance under conditions of hypothyroidism. With the non-linear capacitance you have protein which on a single molecule basis is moving at 50000 Hz or so. Could you see that with your GFP-tagged prestin? Are there optical tools that would allow you to resolve that movement, especially in the case where that movement is uncoupled from your measured whole cell contractile force?

Knipper: This brings us back to a basic question concerning *in vitro* expression systems. Can we mimic hair cells in a culture system? In a culture system we can produce a situation where the SLC26A5 can produce motility with a voltage charge comparable to transfected cells, but we can never produce force.

Mount: Can you define the difference between the capacitance changes and the force?

Knipper: If we isolate hair cells we mimic the *in vivo* situation, because people put the hair cells in a kind of chamber that mimics somehow cell turgor. The hair cells are soaked in a pipette and thereby only motile responses and force can be investigated.

Oliver: We should distinguish between the microscopic force generated by the single molecule and the macroscopic force that is measurable from the whole cell. The second will depend on structural things like the cytoskeleton and so on, but most probably not the microscopic force produced by the conformational change of a single molecule.

Mount: So in the hypothyroid state in the outer hair cells you see capacitance changes but you don't see force.

Knipper: Yes, this is measured in *in vivo* situations. It is reduced force. We need to know more about the normal *in vivo* situation, including details of phosphorylation and glycosylation. The specificity of the ion channel may be influenced by these kinds of biochemical and structural changes. In *in vitro* expression systems these are different from the *in vivo* system. Therefore I presume we have a gap to fill out. We cannot mimic the *in vivo* system if we do not know how much influence the biochemical changes have on a specific motile cell response. If I give you a 3D culture system where the cochlea remains intact, and someone tries to convince you that you could mimic the *in vivo* system, I would say they are wrong. This is because in the *in vitro* situation some ion channels are missing which are important for normal outer hair cell function. This is often disregarded. For example, KCNQ4 is missing in any *in vitro* situation, emphasizing the artificial nature of an *in vitro* situation. I presume that in every *in vitro* expression system that others are looking at, the situation is similar. Within hours, and even minutes, distinct ion channels go down and change. We can't disregard this fact. The phenotype is different, and we need to bear this in mind.

Welsh: What happens to people who have hypothyroidism? The changes you saw were quite dramatic. Do you see changes consistent with hypothyroidism?

Knipper: The kind of deafness I described was as a result of a hypothyroid state during embryonal development and not in adults. We learned from animal experiments that TH affects the ear only during a limited window, which in humans would be between the second and sixth embryonal months. We furthermore learned that every period of TH deficiency during this critical time period, even if only transient, leads to deafness, and thus would lead to different degrees of hearing loss in newborn children. You would not be able to diagnose the child as hypothyroid and would not be able to link the different degrees of deafness to TH deficiency. In adults experiencing a deficiency of thyroid hormone, the hearing would not be expected to be dramatically effected.

Welsh: You can measure the movement of the outer hair cells. Does that change?

Ashmore: There are also distortion product oto-acoustic emissions (DPOAEs) which are an indication of outer hair cell function.

Knipper: There seems to be some controversial evidence as to whether DPOAEs change when adults suffer from hypothyroidism. I asked medical doctors working with hypothyroid patients, and got various controversial answers. Some said that over a long period of hypothyroidism the patients have a hearing deficit, but with compensation therapy this is reversed.

Quinton: Does anything occur with ageing?

Knipper: We work on two relevant ageing animal models, Fischer rats and gerbils. These get profound human-like hearing deficits, comparable to what is known in humans. We have one phenotype that has smaller efferents, and when we measure the DPOAE, concomitant with the hearing loss there is a loss of active cochlear mechanics in both of these animals, even though their hair cells are present and prestin is present. In age-dependent hearing loss in humans, the mechanism of loss of active cochlear mechanics is also controversial. So for humans it is an open question, but for animal models we have some evidence that the active cochlear mechanics may be slightly affected.

Quinton: Is there any difference in expression?

Knipper: We don't see any.

Ashmore: In the cochlea there is no turnover of cells. There may be turnover of protein, but this is inadequately documented. You are born with a fixed number of cells in your cochlear so you need to look after them.

Kere: What do we know about the mutation spectrum in humans?

Knipper: The only study currently published is by Liu et al (2003) who saw a mutation in the splice acceptor site.

Kere: Are there any coding short nucleotide polymorphisms (SNPs) in public databases?

Knipper: People in the hearing centre in Tuebingen and probably also in other places are currently looking for it.

Kere: Has anyone screened deaf individuals?

Knipper: This is ongoing; we have a database of 2000 people in our hearing research centre.

Alper: What is known about the lack of replacement of cochlear hair cells in $Rb^{-/-}$ or $Rb^{+/-}$ mice? There is new work about the regeneration of hair cells with the retinoblastoma gene product. Is prestin working similarly in the different tonotopic sections of the cochlear, with equal abundance?

Chan: There is literature on the regeneration of outer hair cells after damage by antibiotics, indicating that these cells do have the ability to regenerate. Whether these cells can regenerate could affect the interpretation of some of the data presented.

Knipper: In our hearing research centre Professor Gummer works on motile responses in isolated OHCs. The message is that if you take OHCs from either the apical or basal part, they appear to exhibit the same voltage charge responses.

Ashmore: There is a battle going on at the moment between researchers in the field as to whether amplification in the ear is to do with prestin, or—perish the thought—nothing to do with prestin but rather something to do with the hairy end of the cell.

Oliver: There is one paper that describes that the density of the charge movement via prestin increases towards higher frequencies (Santos-Sacchi et al 1998).

Knipper: Jonathan Ashmore, isn't it true that force production is dramatically influenced by the basilar membrane structure? Along the tonotopic axis, the anatomical structure determines the extent of motile responses? In the apical part the larger outer hair cells are not embedded in such a fixed structural environment as is the case in the basal cochlear turns.

Ashmore: The complication in the cochlear mechanics arises because at the high-frequency end of the basilar membrane the structure gets stiffer. Therefore you need larger forces to compensate for this increased stiffness and presumably an increase in associated viscosity. At the low frequency end the problem is not so extreme. This is why the observation that the density of prestin increases at the high frequency end is working in the right direction. The current debates in the field surround whether it is helping enough.

Welsh: You mentioned that you see A5 in mosquito and *Drosophila*. It looked like it was in the proboscis.

Knipper: In the mosquito I showed the SLC25A5 homologue expression in the thousands of radially oriented sensory receptor units in the chordotonalorgan, which is the hearing organ.

Welsh: Where else do you see it?

Knipper: The brain. There are ongoing studies for detection of prestin expression outside the ear in *Drosophila*.

Welsh: Has anyone else looked at analogues of this in non-vertebrates?

Knipper: I presume one of the challenging questions is has this kind of homologue in the inner ear undergone a special evolution or not? In other organs it might be an ion channel, but in the inner ear it could be a protein which somehow has progressed and developed to become finally a motile protein. The striking question thus is, does the homologue move or not? Does it show non-linear capacitance or not? These are critical questions. Does A6, which is closely related, have the same properties?

References

Accardi A, Miller C 2004 Secondary active transport mediated by a prokaryotic homologue of ClC Cl⁻ channels. Nature 427:803–807

Dallos P, Fakler B 2002 Prestin, a new type of motor protein. Nat Rev Mol Cell Biol 3:104–111

Delpire E, Lu J, England R, Dull C, Thorne T 2003 Deafness and imbalance associated with inactivation of the secretory Na-K-2Cl co-transporter. Nat Genet 22:192–195

Liberman MC, Gao J, He DZ, Wu X, Jia S, Zuo J 2002 Prestin is required for electromotility of the outer hair cell and for the cochlear amplifier. Nature 419:300–304

Lytle C, McManus T 2002 Coordinate modulation of Na-K-2Cl cotransport and K-Cl cotransport by cell volume and chloride. Am J Physiol Cell Physiol 283: C1422–1431

Liu XZ, Ouyang XM, Xia XJ et al 2003 Prestin, a cochlear motor protein, is defective in non-syndromic hearing loss. Hum Mol Genet 12:1155–1162

Picollo A, Pusch M 2005 Chloride/proton antiporter activity of mammalian CLC proteins ClC-4 and ClC-5. Nature 436:420–423

Rybalchenko V, Santos-Sacchi J 2003 Cl⁻ flux through a non-selective, stretch-sensitive conductance influences the outer hair cell motor of the guinea-pig. J Physiol 547:873–891

Santos-Sacchi J, Kakehata S, Kikuchi T, Katori Y, Takasaka T 1998 Density of motility-related charge in the outer hair cell of the guinea pig is inversely related to best frequency. Neurosci Lett 256:155–158

Scheel O, Zdebik AA, Lourdel S, Jentsch TJ 2005 Voltage-dependent electrogenic chloride/proton exchange by endosomal CLC proteins. Nature 436:424–427

Tian Y, Li M, Fritzsch B, Zuo J 2004 Creation of a transgenic mouse for hair-cell gene targeting by using a modified bacterial artificial chromosome containing prestin. Develop Dynamics 230:199–203

Verbavatz JM, Ma T, Gobin R, Verkman AS 1997 Absence of orthogonal arrays in kidney, brain and muscle from transgenic knockout mice lacking water channel aquaporin-4. J Cell Sci 110:2855–2860

Yang B, Brown D, Verkman AS 1996 The mercurial insensitive water channel (AQP-4) forms orthogonal arrays in stably transfected Chinese hamster ovary cells. J Biol Chem 271: 4577–4580

Sulfate transport by SLC26 transporters

Daniel Markovich

Department of Physiology and Pharmacology, School of Biomedical Sciences, University of Queensland, Brisbane, St. Lucia, QLD 4072, Australia

Abstract. Sulfate is the fourth most abundant anion in human plasma that is essential for numerous physiological functions, including biotransformation of xenobiotics, steroids, non-steroidal anti-inflammatory drugs (NSAIDs), adrenergic stimulants/blockers and analgesics. Sulfate is also required for activation of many endogenous compounds (heparin, heparan sulfate, dermatan sulfate, bile acids) and utilized in the metabolism of neurotransmitters. Sulfation of structural components, including glycosaminoglycans and cerebroside sulfate, is essential for the maintenance of normal structure and function of tissues. Due to its hydrophilic nature, sulfate cannot readily cross the lipid bilayer of cells, thus plasma membrane proteins, known as sulfate transporters, are required for the movement of sulfate into/out of cells. Sulfate transporters can be divided into two distinct groups: Na$^+$-dependent sulfate transporters belonging to the SLC13 gene family and Na$^+$-independent sulfate transporters (antiporters, exchangers) belonging to SLC26 gene family. There are 11 members of the SLC26 family (including Sat1, DTDST, CLD, pendrin, prestin, cfex) whose structures and functions have been only partially characterized. In this presentation, the current information on the structures and functions of the sulfate transporters in the SLC26 gene family will be described and the issue that certain members of this family are unable to transport sulfate, will be addressed.

2006 Epithelial anion transport in health and disease: the role of the SLC26 transporters family. Wiley, Chichester (Novartis Foundation Symposium 273) p 42–58

SLC26 genes encode sulfate transporters

SLC26A1 (Sat1)

The first member of the SLC26 gene family was identified by expression cloning using a *Xenopus* oocyte system to screen a rat liver cDNA library by radiotracer sulfate uptake (Bissig et al 1994). A single functional clone was isolated, which became known as Sat1 (<u>s</u>ulfate <u>a</u>nion <u>t</u>ransporter <u>1</u>). Functional characterization in *Xenopus* oocytes using radiotracer sulfate measurements, revealed Sat1 to encode a Na$^+$-independent sulfate transporter (see Table 1), with a high affinity ($K_m = 136 \mu M$) for sulfate, that was strongly inhibited by a stilbene derivative DIDS (4,4′-diisothiocyanato-2,2′-disulfonate) ($IC_{50 \, DIDS} = 28 \mu M$) and oxalate, but not by succinate or cholate (Bissig et al 1994). These properties correlated closely with the functional activities of the sulfate/bicarbonate exchanger in liver canalicular

TABLE 1 Sulfate transport by SLC26 transporters

HGNC Name	Gene/Protein Name	Tissue distribution	Transports Sulfate	DIDS Sensitive[2]
SLC26A1	Sat1	widespread	Yes[1]	Yes
SLC26A2	DTDST	ubiquitous	Yes	Yes
SLC26A3	DRA, CLD	widespread	Yes	Yes
SLC26A4	PDS, Pendrin	inner ear, kidney, thyroid	No	Yes
SLC26A5	Prestin	widespread	ND	
SLC26A6	PAT1, CFEX	widespread	Yes	Yes
SLC26A7	—	kidney	Yes	Yes
SLC26A8	Tat1	sperm, brain	Yes	Yes
SLC26A9	—	lung	Yes	Yes
SLC26A10	pseudogene	ND	ND	ND
SLC26A11	—	widespread	Yes	Yes

HGNC, Human Genome Nomenclature Committee; ND, not determined; widespread, four or more tissues.

[1]K_m for sulfate has been determined at 0.1–0.3 mM.

[2]Sulfate transport is significantly inhibited by anion exchange inhibitor DIDS, a stilbene-derivative.

membrane vesicles (Meier et al 1987). The coding region of the rat *Sat1* (*rSat1*) cDNA predicts a protein of 703 amino acids with a calculated molecular mass of 75.4 kDa. Hydrophobicity analysis of rSat1 protein suggests 12 putative transmembrane domains (TMDs)(Markovich 2001). Northern blot analysis detected a single mRNA (3.8 kb) transcript for Sat1 very strongly in liver and kidney, with weaker signals in skeletal muscle and brain (Bissig et al 1994). The abundance of *Sat1* mRNA in the kidney was identical to its expression in the liver (the tissue from which it was cloned), prompting us to examine the role of Sat1 in the kidney. When rat kidney mRNA was injected into *Xenopus* oocytes it induced a Na$^+$-independent sulfate transport activity which was inhibitable by DIDS, probenecid and phenol red (Markovich et al 1994). The degree of inhibition by these compounds was closely correlated with the inhibition pattern of the renal basolateral membrane (BLM) sulfate anion exchanger (David & Ullrich 1992), as well as with Sat1 cRNA-induced activity in *Xenopus* oocytes (Bissig et al 1994). Furthermore, using a hybrid depletion strategy, Sat1 antisense oligonucleotides led to a complete abolition of the kidney mRNA-induced Na$^+$-independent sulfate transport activity (in *Xenopus* oocytes), confirming that *Sat1* encodes the Na$^+$-independent sulfate transporter, whose function most closely correlates with the proximal tubular BLM sulfate/ bicarbonate anion exchanger. This finding was later confirmed by immunocyto-chemistry using Sat1-specific peptide polyclonal antibodies showing the location of the Sat1 protein in the kidney to be restricted to the BLM of kidney proximal tubules (Karniski et al 1998).

We have also identified Sat1 expression in the brain (Lee et al 1999). Injection of rat brain mRNA into *Xenopus* oocytes induced a saturable Na$^+$-independent sulfate uptake, which was inhibited by DIDS (IC$_{50\text{ DIDS}}$ = 150 ± 30 μM) and oxalate, with properties identical to Sat1 anion exchange activity. The involvement of a Sat1-like transporter responsible for the sulfate transport in rat brain was confirmed by hybrid depletion using Sat1 antisense oligonucleotides, which completely abolished the brain mRNA-induced sulfate transport activity in *Xenopus* oocytes (Lee et al 1999). Furthermore, *in situ* hybridization using Sat1 probes detected *Sat1* mRNA expression most prominently in the hippocampus and cerebellum, in particular in the neuronal-rich cell layer, suggesting that Sat1 may play an important role in sulfate transport in neuronal or glial cells (Lee et al 1999).

Recently, this laboratory isolated the mouse (Lee et al 2003) and human (Regeer et al 2003) *Sat1* orthologues (named *mSat1* and *hSAT1*, respectively) of the rat *Sat1* cDNA, using a homology screening approach. We also identified their gene structures, determined their protein function in *Xenopus* oocytes and characterised promoter activities in mammalian renal cell lines. *mSat1* encodes a 704 amino acid protein (75.4 kDa) with 12 putative transmembrane domains (TMDs) that induces sulfate, oxalate and chloride transport in *Xenopus* oocytes (Lee et al 2003). By RT-PCR, *mSat1* mRNA expression was found to be strongest in the kidney and liver, with lower levels observed in caecum, calvaria, brain, heart and skeletal muscle. Two distinct transcripts were expressed in kidney and liver due to alternative utilization of the first intron, corresponding to an internal portion of the 5′-untranslated region. The mouse *Sat1* gene (*Slc26a1*) is approximately ~6 kb in length and consists of 4 exons, with an optional intron 1. The *Sat1* promoter is ~52% G + C rich and contains a number of well-characterized *cis*-acting elements, including sequences resembling responsive elements to thyroid hormone (T$_3$REs) and vitamin D (VDREs). We demonstrated that *Sat1* promoter driven basal transcription in renal opossum kidney (OK) cells was stimulated by tri-iodothyronine (T$_3$), but not by 1,25-(OH)$_2$ vitamin D$_3$ (vitamin D). Site-directed mutagenesis of an imperfect T$_3$RE sequence at −454 bp led to a loss of *Sat1* promoter inducibility by T$_3$. This suggests that position −454 in the *Sat1* promoter contains a functional T$_3$RE that is responsible for transcriptional activation of *Sat1* by T$_3$.

By radiotracer measurements in *Xenopus* oocytes, mSat1 led to a ~20-fold induction in Na$^+$-independent sulfate transport, a ~sixfold induction in oxalate and chloride transport, but was unable to induce any formate, L-leucine or succinate uptakes, when compared to water injected (control) oocytes. This suggests that mSat1 transport was restricted to the anions SO$_4^{2-}$, oxalate and chloride. As shown for rSat1 (Satoh et al 1998), mSat1-induced SO$_4^{2-}$ transport occurred only in the presence of extracellular chloride, suggesting its requirement for chloride. Typical Michaelis-Menten saturation kinetics were observed for mSat1-induced SO_4^{2-} uptake, with a calculated $K_m = 0.31 \pm 0.05$ mM and $V_{max} = 143.90 \pm 5.32$ pmol/h for SO_4^{2-} interaction. These values are in close agreement with rSat1-induced SO_4^{2-} transport kinetics ($K_m = 0.14$ mM) (Bissig et al 1994). mSat1-induced SO_4^{2-} transport was significantly inhibited by molybdate, selenate, tungstate, DIDS, thiosulfate, phenol red and probenecid, whereas citrate and glucose had no effect. These data suggest that mSat1 encodes a functional SO_4^{2-}/chloride/oxalate anion exchanger, whose activity is blocked by anion exchange inhibitors DIDS and phenol red, tetra-oxyanions (selenate, molybdate and tungstate) and thiosulfate.

Using a similar strategy, we characterized the human SAT1 (hSAT1) transporter. hSAT1 encodes a 701 amino acid protein (75 kDa), with 12 putative TMDs, that induces sulfate, chloride and oxalate transport in *Xenopus* oocytes (Regeer et al 2003). By RT-PCR, *hSAT1* mRNA expression was found most abundantly in the kidney and liver, with lower levels in the pancreas, testis, brain, small intestine, colon and lung. The human *SAT1* gene (*SLC26A1*) is comprised of four exons stretching approximately 6 kb in length, with an alternative splice site formed from an optional exon (exon II). The *SAT1* promoter is ~60% G + C rich, contains a number of well-characterized *cis*-acting elements, but lacks any canonical TATA- or CAAT-boxes. The *SAT1* 5′ flanking region led to basal promoter activity in renal OK and LLC-PK1 cells. The use of *SAT1* 5′ flanking region truncations has shown the first 135 bp to be sufficient for basal promoter activity in both cell lines. Unlike in the mouse *Sat1* promoter, neither tri-iodothyronine (T_3), nor 1,25-$(OH)_2$ vitamin D_3 (vitamin D) could *trans*-stimulate *SAT1* promoter activity, despite the presence of sequences in its promoter resembling responsive elements to thyroid hormone (T_3REs) and vitamin D (VDREs). Mutation of the activator protein 1 (AP-1) site at position −52 in the *SAT1* promoter led to loss of transcriptional activity, suggesting the requirement of AP-1 for *SAT1* basal transcription.

By radiotracer measurements in *Xenopus* oocytes, hSAT1 led to ~40-fold induction in Na^+-independent sulfate uptake, ~fivefold induction in chloride uptake and ~sixfold induction in oxalate uptake, but was unable to induce formate uptake, when compared to the water injected (control) oocytes. This is in agreement with the transport activities for rSat1 and mSat1, both of which transport sulfate, oxalate and chloride, but not formate (Bissig et al 1994, Lee et al 2003). Typical Michaelis-Menten saturation kinetics were observed for hSAT1-induced sulfate uptake in *Xenopus* oocytes, with a calculated K_m for SO_4^{2-} of 0.19 ± 0.066 mM and a V_{max} of 52.35 ± 3.30 pmol/oocyte/hour (Lee et al 2003). Various compounds were tested as possible inhibitors of hSAT1-induced sulfate uptake in *Xenopus* oocytes: the stilbene-derivative DIDS, phenol red, thiosulfate, molybdate, tungstate and selenate all significantly inhibited hSAT1-induced SO_4^{2-} transport, whereas citrate and glucose had no effect.

mSat1 and hSat1 proteins share 94% and 77% amino acid identities, respectively, with the rSat1, suggesting that they encode orthologues of the same protein and belong to the same gene family (SLC26). Interestingly, there are two consensus regions which are highly conserved between the Sat1 proteins, as well as other members of the SLC26 gene family. These are the 'phosphopantetheine attachment site' (PROSITE PS00012) motif at residues 415–430 of rSat1 (QTQLSSV-VSAAVVLLV) and the 'sulfate transporter signature' (PROSITE PS01130) motif at residues 98–119 of both rSat1 and mSat1 (PIYSLYTSFFANLIYFLMGTSR). Furthermore, there is a 'sulfate transporter and anti-sigma antagonist' (STAS) domain found in the C-terminal cytoplasmic domain of the Sat1 protein, which appears conserved in all members of the SLC26 members (Mount & Romero 2004). The functional significance of these motifs is yet unknown.

SLC26A2 (DTDST)

By positional cloning, a gene was identified on human chromosome 5q to be linked to an autosomal recessive osteochondrodysplasia called diastrophic dysplasia (DTD; OMIM 222600), with clinical features including dwarfism, spinal deformation and abnormalities of the joints (Hastbacka et al 1994). The human DTD gene (SLC26A2) is comprised of four exons and three introns spanning over 40 kb. DTD shares 48% and 33% amino acid identities with rSat1 (Slc26a1) and DRA (Slc26a3), respectively (Hastbacka et al 1994). Due to its high sequence similarity with rSat1, the exact function of DTD was sought. Primary skin fibroblasts obtained from normal, carrier and DTD patients (with mutations in the DTD gene) were cultured and assayed for sulfate uptake (Hastbacka et al 1994). Cells from normal and carrier samples showed saturable sulfate transport, whereas cells from the DTD patients showed a greatly diminished sulfate uptake (Hastbacka et al 1994), suggesting that the normal (wild-type) DTD gene encodes a functional sulfate transporter, which loses its transport activity when mutated (as in DTD patients). Thus the cDNA encoding the DTD gene became known as the diastrophic dysplasia sulfate transporter (DTDST) (Hastbacka et al 1994). Translation of DTDST cDNA produced a protein of 739 amino acids, with a predicted molecular mass of 82 kDa, containing 12 putative TMDs (Hastbacka et al 1994). DTDST mRNA expression was ubiquitous, detected by Northern blot analysis in all tissues of the body (Hastbacka et al 1994). At least 30 different mutations have been identified in the human DTDST gene (Dawson & Markovich 2005), which are responsible for a variety of recessively inherited chondrodysplasias, including diastrophic dysplasia (DTD), atelosteogenesis type 2 (AO2) and a lethal condition known as achondrogenesis 1B (ACG1B) (Hastbacka et al 1994). For a recent review on DTDST mutations and their contributions to these pathogenic states, see Dawson & Markovich (2005).

Subsequently, the rat *DTDST* gene and cDNA (*rDtdst*) were cloned and its function was characterised in *Xenopus* oocytes (Satoh et al 1998). rDtdst induced Na^+-independent sulfate transport, which was inhibited by chloride, unlike rSat1 which required extracellular chloride for sulfate transport (Satoh et al 1998). Furthermore, rDtdst-induced sulfate uptake was inhibited by thiosulfate, oxalate and DIDS, with a *cis*-inhibition pattern identical to rSat1 and DRA. No information is presently available on sulfate transport kinetics of DTDST. The mouse *Dtdst* (*mDtdst*) cDNA (initially named *st-ob* for sulfate transporter in osteoblasts) was isolated from mouse fibroblastic cells C3H10T1/2 (Kobayashi et al 1997). The mDtdst protein shares 80% amino acid identity with the human DTDST protein. Tissue distribution showed the mDtdst mRNA strongly expressed in the thymus, testis, calvaria and the osteoblastic MC3T3-E1 cells, having a reduced expression in the undifferentiated C3H10T1/2 cells. The expression of *mDtdst* mRNA in C3H10T1/2 cells was shown to be increased by transforming growth factor $\beta1$ (TGF$\beta1$), retinoic acid and dexamethasone, as well as BMP2 (Kobayashi et al 1997). BMP2 also led to a twofold increase in sulfate uptake in C3H10T1/2 cells when compared to untreated (control) cells, suggesting that *st-ob* encodes a functional sulfate transporter, whose activity was up-regulated by BMP2. Since osteoblasts actively take up sulfate to synthesize proteoglycans (being major components of the extracellular matrix of bone and cartilage), the functional activity of mDtdst protein as a sulfate transporter may be important for osteoblastic differentiation.

SLC26A3 (DRA/CLD)

By subtractive hybridization using cDNA libraries from normal colon and adenocarcinoma tissues, a human cDNA which was down-regulated in adenomas (*DRA*) was isolated (Schweinfest et al 1993). Human *DRA* (*hDRA*) mRNA was found to be exclusively expressed in normal colon tissues, with its expression significantly decreased in colonic adenomas (polyps) and adenocarcinomas. *hDRA* mRNA was detected throughout the intestinal tract (duodenum, ileum, caecum, distal colon), but not in the oesophagus or stomach (Byeon et al 1996). The hDRA gene (*SLC26A3*) was mapped to human chromosome 7q22-q31.1 (Taguchi et al 1994). The predicted topology of this polypeptide was a 84.5 kDa protein, having either 10, 12 or 14 TMDs, with charged clusters of amino acids at its N- and C-termini, potential nuclear targeting motifs, an acidic transcriptional activation domain and a homeobox domain (Schweinfest et al 1993). These putative domains in the hDRA protein suggest it could interact with a transcription factor, which would be consistent with hDRA having a role in tissue-specific gene expression and/or as a candidate tumour suppressor gene. By use of the baculovirus expression system, hDRA was expressed in insect Sf9 cells where it led to greater than a threefold increase in sulfate uptake when compared to control cells (Byeon et al 1998). In *Xenopus* oocytes, hDRA-induced Na^+-independent sulfate, chloride and oxalate transport, which was DIDS sensitive (Silberg et al 1995, Moseley et al 1999, Lohi et al 2003), suggesting that *hDRA* encodes a functional sulfate/chloride/oxalate anion exchanger. No information is presently available on the transport kinetics of DRA. The hDRA gene (*SLC26A3*) was found to be defective in patients with congenital chloride diarrhoea (CLD; OMIM 214700) syndrome (Moseley et al 1999), a recessively inherited defect of intestinal chloride/bicarbonate exchange. CLD is a potentially fatal diarrhoea characterized by a high chloride content (Holmberg 1986, Kere et al 1999). To date, over 30 different mutations have been identified in CLD patients, from various ethnic populations (Kere et al 1999). For a recent review on hDRA mutations and their contributions to pathogenic states, see Dawson & Markovich (2005). *DRA* was implicated as a positional and functional candidate for CLD (Höglund et al 1995, 1996). Due to its link with congenital chloride diarrhoea, *hDRA* became known as the CLD gene. The CLD gene spans ≈39 kb and is comprised of 21 exons (Haila et al 1998). Analysis of the putative CLD promoter region (570 bp) showed putative TATA and CCAAT boxes and multiple transcription factor binding sites for AP-1 and GATA-1 (Haila et al 1998). However, the functional significance of these sites is unknown. The mouse *DRA* (*mDRA*) cDNA orthologue was cloned using a combination of RT-PCR and RACE techniques (Melvin et al 1999). When expressed in human embryonic kidney HEK293 cells, mDRA protein conferred Na^+-independent, electroneutral chloride/bicarbonate exchange activity. Northern blot analysis showed mDRA mRNA expressed at high levels in caecum and colon, and at lower levels in small intestine (Melvin et al 1999).

SLC26A4 (pendrin) and SLC26A5 (prestin)

Due to significant structural similarities with members of the SLC26 gene family, the pendrin (*SLC26A4*) (Everett et al 1997) and prestin (*SLC26A5*) (Zheng et al 2000) genes were initially thought to encode sulfate transporters. However, pendrin was unable to induce any radiotracer sulfate uptake, but rather led to the transport of iodide, chloride (Scott et al 1999) and formate (Scott & Karniski 2000) when expressed in *Xenopus* oocytes and insect Sf9 cells, suggesting that pendrin had anion preferences for chloride, iodide and formate, rather than sulfate. Similarly, upon its isolation, prestin was suggested to encode a motor protein of the cochlear outer hair cells (Zheng et al 2000) and no evidence to date has been presented that it functions as a sulfate transporter. This broke down the 'central dogma' that all SLC26 proteins encode functional sulfate transporters and due to this incongruence, the significance of these proteins to this discussion is inappropriate, but will form the focus of other articles in this volume.

SLC26A6-A10

Using bioinformatics/genomic approaches, the last remaining six members of the SLC26 gene family (SLC26A6-A11) were recently identified by screening databases using sequence information of the first five SLC26 members (Lohi et al 2000, Vincourt et al 2003). Each gene encoding SLC26A6-A11 was found to reside on a different human chromosome, with a unique tissue distribution (Lohi et al 2000, Mount & Romero 2004). Functional expression of SLC26A6, A7, A8 and A9 individually in *Xenopus* oocytes demonstrated significant radiotracer uptake of sulfate, chloride, oxalate and formate, which was inhibited by DIDS (Lohi et al 2002, Jiang et al 2002). SLC26A10 was found to be expressed in the brain (Lohi et al 2000). However, its function is yet unknown and has been suggested to be an expressed pseudogene (Mount & Romero 2004). The *SLC26A11* cDNA was isolated from human high endothelial venule endothelial cells (HEVECs) and when expressed in *Xenopus* oocytes led to Na^+-independent sulfate transport that was DIDS-sensitive (Vincourt et al 2003). No information is presently available on the transport kinetics of SLC26A6–A11 proteins.

Summary

The SLC26 gene family represents a group of structurally related proteins that encode anion exchangers, with two exceptions, SLC26A5 and SLC26A10. Most members of this family can mediate sulfate transport, with one exception, SLC26A4, which prefers other anions. The localizations of the SLC26 proteins in various tissues, highlights the importance of sulfate in mammalian physiology. However, the overall functional significance of these transporters and the fact that they can transport other anions in addition to sulfate, still needs to be further characterized. Future studies with these unique proteins, will be aimed at accurately determining their substrate preferences and transport kinetics, following the example of the first member (Sat1; SLC26A1) being the most characterized. Similarly, the significance of the STAS and the 'sulfate transporter signature' domains conserved in all members of this family, will yield more insights on the structure/function relationships of these proteins. Finally, the big picture of how these proteins contribute to sulfate homeostasis and the pathophysiological consequences of their absence in our genomes can only be determined when knockout animals of these transporters will be generated and their phenotype(s) characterized in the near future.

Acknowledgements

This work was supported in part by the Australian Research Council, the National Health and Medical Research Council of Australia and the University of Queensland Foundation.

References

Bissig M, Hagenbuch B, Stieger B, Koller T, Meier PJ 1994 Functional expression cloning of the canalicular sulfate transport system of rat hepatocytes. J Biol Chem 269:3017–3021

Byeon MK, Westerman MA, Maroulakou IG et al 1996 The down-regulated in adenoma (DRA) gene encodes an intestine-specific membrane glycoprotein. Oncogene 12:387–396

Byeon MK, Frankel A, Papas TS et al 1998 Human DRA functions as a sulfate transporter in Sf9 insect cells. Protein Expr Purif 12:67–74

David C, Ullrich KJ 1992 Substrate specificity of the luminal Na^+-dependent sulphate transport system in the proximal renal tubule as compared to the contraluminal sulphate exchange system. Pflugers Archiv 421:455–465

Dawson PA, Markovich D 2005 Pathogenetics of the Human SLC26 Transporters. Curr Med Chem 12:385–396

Everett LA, Glaser B, Beck JC et al 1997 Pendred syndrome is caused by mutations in a putative sulphate transporter gene (PDS). Nat Genet 17:411–422

Haila S, Höglund P, Scherer SW et al 1998 Genomic structure of the human congenital chloride diarrhea (CLD) gene. Gene 214:87–93

Hastbacka J, de la Chapelle A, Mahtani MM et al 1994 The diastrophic dysplasia gene encodes a novel sulfate transporter: positional cloning by fine-structure linkage disequilibrium mapping. Cell 78:1073–1087

Holmberg C 1986 Congenital chloride diarrhoea. Clin Gastroenterol 15:583–602

Höglund P, Sistonen P, Norio R et al 1995 Fine mapping of the congenital chloride diarrhea gene by linkage disequilibrium. Am J Hum Genet 57:95–102

Höglund P, Haila S, Scherer SW et al 1996 Positional candidate genes for congenital chloride diarrhea suggested by high-resolution physical mapping in chromosome region 7q31. Genome Res 6:202–210

Jiang Z, Grichtchenko II, Boron WF, Aronson PS 2002 Specificity of anion exchange mediated by mouse Slc26a6. J Biol Chem 277:33963–33967

Karniski LP, Lotscher M, Fucentese M et al 1998 Immunolocalization of sat-1 sulfate/oxalate/bicarbonate anion exchanger in the rat kidney. Am J Physiol 275:F79–87

Kere J, Lohi H, Höglund P 1999 Genetic disorders of membrane transport III. Congenital chloride diarrhea. Am J Physiol 276:G7–13

Kobayashi T, Sugimoto T, Saijoh K, Fujii M, Chihara K 1997 Cloning and characterization of the 5'-flanking region of the mouse diastrophic dysplasia sulfate transporter gene. Biochem Biophys Res Commun 238:738–743

Lee A, Beck L, Brown RJ, Markovich D 1999 Identification of a mammalian brain sulfate transporter. Biochem Biophys Res Commun 263:123–129

Lee A, Beck L, Markovich D 2003 The mouse sulfate anion transporter gene Sat1 (Slc26a1): cloning, tissue distribution, gene structure, functional characterization, and transcriptional regulation thyroid hormone. DNA & Cell Biol 22:19–31

Lohi H, Kujala M, Kerkela E et al 2000 Mapping of five new putative anion transporter genes in human and characterization of SLC26A6, A candidate gene for pancreatic anion exchanger. Genomics 70:102–112

Lohi H, Kujala M, Makela S et al 2002 Functional characterization of three novel tissue-specific anion exchangers SLC26A7, -A8, and -A9. J Biol Chem 277:14246–14254

Lohi H, Lamprecht G, Markovich D et al 2003 Isoforms of SLC26A6 mediate anion transport and have functional PDZ interaction domains. Am J Physiol Cell Physiol 284:C769–779

Markovich D 2001 The physiological roles and regulation of mammalian sulfate transporters. Physiol Rev 81:1499–1534

Markovich D, Bissig M, Sorribas V et al 1994 Expression of rat renal sulfate transport systems in Xenopus laevis oocytes. Functional characterization and molecular identification. J Biol Chem 269:3022–3026

Meier P, Valantinas J, Hugentobler G, Rahm I 1987 Bicarbonate sulfate exchange in canalicular rat liver plasma membrane vesicles. Am J Physiol 253:461–468

Melvin JE, Park K, Richardson L et al 1999 Mouse down-regulated in adenoma (DRA) is an intestinal Cl(−)/HCO(3)(−) exchanger and is up-regulated in colon of mice lacking the NHE3 Na(+)/H(+) exchanger. J Biol Chem 274:22855–22861

Moseley RH, Höglund P, Wu GD et al 1999 Downregulated in adenoma gene encodes a chloride transporter defective in congenital chloride diarrhea. Am J Physiol 276:G185–192

Mount DB, Romero MF 2004 The SLC26 gene family of multifunctional anion exchangers. Pflugers Arch 447:710–721

Regeer RR, Lee A, Markovich D 2003 Characterization of the human sulfate anion transporter (hsat-1) protein and gene (SAT1; SLC26A1). DNA & Cell Biol 22:107–117

Satoh H, Susaki M, Shukunami C et al 1998 Functional analysis of diastrophic dysplasia sulfate transporter. Its involvement in growth regulation of chondrocytes mediated by sulfated proteoglycans. J Biol Chem 273:12307–12315

Schweinfest C, Henderson K, Suster S, Kondoh N, Papas T 1993 Identification of a colon mucosa gene that is down-regulated in colon adenomas and adenocarcinomas. Proc Natl Acad Sci USA 90:4166–4170

Scott DA, Karniski LP 2000 Human pendrin expressed in Xenopus laevis oocytes mediates chloride/formate exchange. Am J Physiol Cell Physiol 278:C207–211

Scott DA, Wang R, Kreman TM, Sheffield VC, Karnishki LP 1999 The Pendred syndrome gene encodes a chloride-iodide transport protein. Nat Genet 21:440–443

Silberg DG, Wang W, Moseley RH, Traber PT 1995 The down regulated in adenoma (dra) gene encodes an intestine-specific membrane sulfate transport protein. J Biol Chem 270:11897–11902

Taguchi T, Testa RT, Papas TS, Schweinfest C 1994 Localization of a candidate colon tumor suppressor gene (DRA) to 7q22-q31.1 by fluorescence in situ hybridization. Genomics 20:146–147

Vincourt JB, Jullien D, Amalric F, Girard JP 2003 Molecular and functional characterization of SLC26A11, a sodium-independent sulfate transporter from high endothelial venules. FASEB J 17:890–892

Zheng J, Shen W, He DZ et al 2000 Prestin is the motor protein of cochlear outer hair cells. Nature 405:149–155

DISCUSSION

Welsh: What would happen if you knocked A2 out and in its place expressed A4 at the same site? Would there be any phenotype?

Markovich: It hasn't been tried.

Muallem: If you look at these transporters, their properties are completely different: none of them work the same. We don't know, of course, how all of them function. In our hands A3 transports sulfate about as well as gluconate, which is close to nil. A6 seems to behave the same. For A7, there isn't much sulfate transport by this transporter. For all of these that you are describing as sulfate transporters, the assay that you are using is based on isotope fluxes. You are incubating oocytes for about an hour to get something. You end up with some counts, but I'm not sure these are physiologically significant fluxes.

Welsh: Shall we make a bet about the consequences of replacing one transporter with another? How about a bottle of wine?

Muallem: Look at the diseases resulting from loss of the specific transporters: they are very different.

Welsh: I will bet that you could make a knockout A2 or A6, express a different one at the same location, and rescue the disease. The results might prove informative.

Soleimani: One has to take into consideration the membrane targeting (apical vs. basolateral) as well as the electrogenicity of these anion exchangers when designing experiments of this nature.

Muallem: That is my whole point. The transport properties of A6 and A3 are completely different. If you look at pendrin, its substrate specificity is very different from that of A3 and A6. If we look closely the differences are very prominent.

Welsh: I don't argue with that. But are the differences physiologically important? And might the site of expression affect their physiological function?

Soleimani: There is another aspect that is worth mentioning. All these tissues have their own unique milieu. We need to think about the effect of the indigenous milieu on the transport activity. A3 (DRA) is expressed in colon, where there is abundant concentration of ammonium, somewhere in excess of 25–50 mM, which is generated locally and absorbed. No other tissue has a comparable level of ammonium. I am not sure that anyone has looked at whether ammonium has the same effect on A3 as it does on A6 in chloride–chloride, chloride–bicarbonate or chloride–oxalate exchange mode. Even the milieu of proximal tubule differs versus duodenum where A6 is expressed. It seems that milieu is another unknown variable.

Alper: We have looked at ammonium, and it differs by species among the same orthologues.

Soleimani: I am not sure that one could substitute with the other one. We don't know the impact of indigenous milieu on the function of each of these genes, if they are going to be expressed in another tissue.

Muallem: I am addressing a problem that is a lot more basic than that: what are the transport properties and the physiological substrates of these proteins? The more we look at them, they appear very different. It is like asking the Ca^{2+}-activated chloride channel to substitute for cystic fibrosis transmembrane regulator protein (CFTR). It is no coincidence that it doesn't because these proteins do more than one thing, and the transport properties are different enough that even when they both function as chloride channels we can still not substitute one for the other. In this case we are dealing with transporters that have the same function of Cl⁻ channel activity.

Welsh: I stand duly chastized!

Muallem: It is a good experiment, though: rescue experiments may tell you something about the other functions of the protein or the degree to which you can spare Cl^-/HCO_3^- exchange by DRA or sulfate transporters by A1, for example.

Seidler: I think Daniel Markovich has data that show that the transport properties of sulfate, for example, will be very different on the basis of the individual protein, when he showed the competition experiments. He stimulates sulfate uptake of Sat1 by adding extracellular chloride. The speed of sulfate transport will be very different based on the individual protein expressed in the same cell type. On top of that, different organ contexts will have very different conditions. One transporter expressed in one organ might not transport sulfate at all, although it has some sulfate transport capability. The next one might have a higher rate.

Markovich: The mice who have some of these transporters knocked out could be useful here. You could use them and look at whether you get changes in expression of mRNAs for these different transporters.

Romero: The only thing that mRNA expression patterns for other transporters in knockout animals will tell you is that in the tissues where those genes are normally expressed, whether the expression of the transporter genes has changed? What Mike was proposing is if you were to take one of these homologue transporters and put it into the tissue where it was knocked out, would it replace function? So for the SLC26A6 knockout, if you put DRA into that same tissue, or you take A6 and put it into the colon under a colon-specific promoter, this would be the more appropriate experiment. But a gene array of the knockouts will just tell you the systemic response of the absence of this one gene.

Soleimani: That is an excellent point. Dr Gary Shull examined the differentially regulated genes in the small intestine of the NHE3 knockout mouse by DNA microarray. The NHE3 knockout mice are dehydrated as a result of decreased absorption of salt and fluid in the small intestine. He found changes in the expression of certain specific genes involved with the regulation of salt and water absorption. This suggests compensatory alteration in the expression of genes that could affect the absorption of NaCl in this animal model.

Seidler: I disagree. I think the NHE3 knockout model was a good example of what goes on in terms of adaptation in the human body to loss of proximal intestinal Na⁺ uptake. There is a very strong transcriptional up-regulation of the epithelial sodium channel, as well as the colonic H^+/K^+ ATPase, under the regulatory influence of elevated aldosterone concentration (Schultheis et al 1998). Similar findings have been seen in patients with chronic diarrhoea. It appears that the organism can, or cannot, adapt to the changes one creates by knocking out a particular gene, based on whether those changes somehow occur in a physiological or pathophysiological situation.

Romero: It needs to be under the control of the right promoter to respond in the right way.

Mount: We find sulfate to be a much more discriminatory substrate than oxalate in the oocyte. We find human DRA/A3 can mediate significant chloride transport. It is very robust, at the level of mouse A6. We reproducibly cannot find any sulfate transport with the appropriate positive controls. With the mouse orthologue of A7 and A9, we get very good modest chloride uptake for A7 but no sulfate transport whatsoever. This is where I get frustrated: it is the absence of positive data.

Markovich: You are using a different species: your mouse clones.

Mount: I am saying that for human DRA we get no sulfate transport. It may be a latitude–longitude effect, because across the street Seth Alper gets similar results!

Markovich: Previous studies by Silberg et al (1995) have done that.

Mount: Not for human A3.

Kere: For DRA it has been shown by others that SLC26A3 transports sulfate and we have reproduced this together with Richard Moseley and Gary Wu (Moseley et al 1999).

Mount: The Moseley paper had a twofold increase in uptake, versus the several hundred-fold increase we see for chloride transport by SLC26A3. It went from 1 pM/oocyte per hour sulfate uptake to two or three. I grew up in a field of cation chloride transport, where *Xenopus* oocytes are pretty much the only place we can express NKCC2, for example. We became sensitized to the fact that there is tremendous variability among oocyte groups. On a given day we get water uptakes of 200–500 pM per oocyte in chloride. Statistically, if we use anything less than 15–20 oocytes per group we see differences between injected groups. If we use 25–30 oocytes, for human A3 we do not see sulfate transport. It is the same for Sat1.

Alper: It is the same for us. Mouse A3 mediates very low levels of sulfate transport in our hands, but human A3-mediated sulfate transport is below our threshold of detection.

Mount: Part of this is because of species differences. But we see this reproducibly for human A3 where we have that difference. We also don't see chloride uptake for mouse Sat1. In a classical anion exchanger, if we have a substrate and do radioactive uptake, and then add cold substrate and see *cis*-inhibition, this is consistent with anion exchange. We see some sulfate uptake in A1 that is markedly potentiated by any other extracellular monovalent anion (Xie et al 2002). I don't know how to reconcile this with anion exchange, or the lack of effect of basolateral chloride on sulfate exchange in the proximal tubule. We are not all finding the same things.

Alper: Could you review the proximal tubule data on basolateral sulfate. Has this experiment been done?

Aronson: It has been done in basolateral vesicles, which is probably the cleanest system, because you can have no chloride. You get perfectly fine bicarbonate–sulfate exchange (Pritchard & Renfro 1983, Kuo & Aronson 1988).

Alper: What happens when you are looking in the presence of some chloride?

Aronson: When we looked we didn't see any inhibition by chloride (Kuo & Aronson 1988). That doesn't mean that chloride doesn't do something in an intact living cell if it has some type of indirect effect.

Mount: Also, this gets back to the issue that we are assuming that these are all simple anion exchangers. There may be some chloride conductance that we don't see. I don't think A1 is a sulfate–chloride exchanger. I find this hard to reconcile with this *cis* activation.

Aronson: There may be a finite rate of sulfate transport through several of these transporters when studied in expression systems. The issue is the relative affinities of the anions compared to the physiological concentrations. Extracellular chloride is 100 times greater than sulfate. Even if they had equal affinity, the outside site would mainly be occupied by chloride. If you take a tracer and look at the uptake in the absence of competing anions, this doesn't tell you what the function would be under physiological conditions. For example, A3 may be capable of transporting sulfate but may not do so under physiological conditions.

Muallem: This is the power of the disease. This is why we know that A2 is a sulfate transporter. But we don't know *why* it is a sulfate transporter.

Mount: It is a potent chloride and oxalate transporter, as well.

Soleimani: Dr Muallem, your analogy is only applicable to disease states. There are at least six other members of this family that are not associated with any disease. What would be the best way to obtain that information? It seems that it has to be based on *in vitro* expression studies and then on studies in mice lacking those genes.

Alper: You are also following the squeaky wheel hypothesis. The phenotype of the A2 deficiency is achondrodysplasia. There is a real deficiency in sulfated proteoglycan biosynthesis as a result. There may be other phenotypes. Our attention is drawn to the major phenotype. When you evoke phenotypes with some other provocative stimulus, you may uncover other phenotypes.

Muallem: That is true. The fact that A2 is so widespread means that it would be surprising if it doesn't have other phenotypes. I was waiting to see the lung phenotype in this case to see whether there was any problem in synthesis or sulfation of mucins. It is a sulfate transporter, whatever else it is. It may be able to transport other anions, but our job is to find out why it is a sulfate transporter, how it does it and why this is a problem in the disease. In this case there is a clear-cut answer:

that is what it does. The same thing is true for DRA: it is a chloride transporter. Pendrin is the same.

Romero: What keeps being repeated is that the context of these cells and transporters is critical. In the colon the environment is wildly different from that seen by other cells in the body. The same is true of chondrocytes, which have a different environment from cells of the proximal tubule. Notwithstanding just the ion concentrations, we all know that there are different ways that these transporters are predicted to be regulated. We have to remember that these are three-dimensional proteins that have three-dimensional surfaces. It is the three-dimensional structure and what is around that is going to determine what these proteins do in the cells that they are in. Even if you could do the gene replacement experiment with the right promoter, you would then have to fix all the other factors. You have to get it talking to the right intracellular molecules and interacting with other proteins, such as CFTR and NHERF, to get it to respond the way the native proteins work. This is assuming that the ion gradients, K_m and V_{max} are all the same. Electrochemical potentials matter.

Muallem: Remember *Drosophila*. Rescues teach us a lot, even if they are only partial. The only problem with the experiment is that it is too expensive.

Alper: I'd like to return to this *cis* requirement of chloride for Sat1-mediated sulfate transport. Has anyone tried to define the concentration dependence of that requirement and the anion selectivity?

Mount: We published that other monovalent ions could substitute for chloride (Xie et al 2002). Other halides could also activate it. Lactate did as well. All the monovalent ions we looked at would *cis* activate, whereas oxalate and cold sulfate dramatically *cis* inhibited.

Alper: Did *cis*-gluconate serve to potentiate sulfate uptake?

Mount: All our uptakes are done in gluconate.

Ashmore: Could you explain the phenomenon for non-transporter people?

Mount: This was doing a hot sulfate uptake in the presence or absence of chloride. The Japanese group (Satoh et al 1998) found very little if any sulfate uptake. We get statistically significant and reproducible sulfate transport activity for A1 in the absence of chloride. But we get a significant activation in the presence of chloride. It is more or less the same at both 10 and 100 mM. It is completely different from the way any of the others behave. We also looked at chloride uptake in the presence of extracellular sulfate and didn't see any. We looked at oxalate uptake in the presence of extracellular chloride and got a more modest effect, but we did see an activation. Both divalent substrates that in our hands are reproducible for A1 seem to be potentiated by extracellular chloride.

Seidler: Did you clamp the intracellular chloride concentration?

Mount: No

Seidler: Could the extracellular chloride be serving as an intracellular exchange substrate? In mammalian cells there is a rapid leakage of chloride into cells.

Mount: The oocytes are pretty full of chloride.

Muallem: The oocytes are highly impermeable to Cl^- and lose very little Cl^- after 2 h incubation in a Cl^--free medium. We had a difficult time depleting oocytes of Cl^-.

Soleimani: Is it possible that Sat1 or A1 can also mediate chloride bicarbonate exchange? If so, then this changes the intracellular pH of the oocyte which can secondarily activate the sulfate uptake.

Knipper: Did I understand correctly that there is no structural correlate between those SLC26 channels that have a preference for either mono of bivalent anions?

Mount: For the one where we can compare the three groups, human A3, we don't find any sulfate transport whereas Daniel does. I find that oxalate is a more permissive anion: some of the things that don't transport sulfate do transport oxalate. I don't know that it is going to be a straightforward issue of divalent charge.

Knipper: A brief comment on the knock-in and knockout experiments. I presume it would work if you delete the gene and put it under a cell-specific promoter, leading to deletion restrictively in that distinct organ. Then all the problems which we discussed with the cellular context should be solved.

Romero: No, because the transport rates and the kinetics of the various transporters vary dramatically for these SLC26 transporters. The regulation of these transporters also varies.

Knipper: But if you put it in the correct genetic context it should work.

Romero: If, for example, the transporter doesn't have the ability to be regulated by PKC or PKA, and the regulation is at that level, then this 'proper genetic context' won't work.

References

Kuo SM, Aronson PS 1988 Oxalate transport via the sulfate/HCO3 exchanger in rabbit renal basolateral membrane vesicles. J Biol Chem 263:9710–9717

Moseley RH, Hoglund P, Wu GD et al 1999 Downregulated in adenoma gene encodes a chloride transporter defective in congenital chloride diarrhea. Am J Physiol 276:G185–192

Pritchard JB, Renfro JL 1983 Renal sulfate transport at the basolateral membrane is mediated by anion exchange. Proc Natl Acad Sci USA 80:2603–2607

Satoh H, Susaki M, Shukunami C, Iyama K, Negoro T, Hiraki Y 1998 Functional analysis of diastrophic dysplasia sulfate transporter. Its involvement in growth regulation of chondrocytes mediated by sulfated proteoglycans. J Biol Chem 273:12307–12315

Schultheis PJ, Clarke LL, Meneton P et al 1998 Renal and intestinal absorptive defects in mice lacking the NHE3 Na+/H+ exchanger. Nat Genet 19:282–285

Silberg DG, Wang W, Moseley RH, Traber PG 1995 The down regulated in adenoma (dra) gene encodes an intestine-specific membrane sulfate transport protein. J Biol Chem 270:11897–11902

Xie Q, Welch R, Mercado A, Romero MF, Mount DB 2002 Molecular characterization of the murine Slc26a6 anion exchanger, functional comparison to Slc26a1. Am J Physiol 283: F826–838

Sugar transport by members of the SLC26 superfamily

Jonathan Ashmore and Jean-Marie Chambard*

*Department of Physiology, University College London, Gower Street, WC1E 6BT London, UK and the UCL Ear Institute, 332 Gray's Inn Road, London WC1X 8EE, and *Assay Development & Compound Profiling, GlaxoSmithKline, Gunnels Wood Road, Stevenage, Hertfordshire SG1 2NY, UK*

Abstract. Two members of the SLC26 superfamily of transport proteins are expressed in the mammalian inner ear. These are pendrin (SLC26A4) and prestin (SLC26A5). Mutations in either lead to deafness but for different reasons. Prestin forms the molecular underpinning of the mechanical feedback step in cochlear amplification. The outer hair cell (OHC) motor basolateral membrane contains a high density of prestin (more than 10^7 copies per cell). OHCs can thus act as a laboratory to study this transporter. Before prestin was identified by a subtractive cDNA strategy, it was recognized that OHCs selectively transport fructose. Cylindrical OHCs swell reversibly and therefore shorten on exposure to iso-osmotic replacement of bath glucose with fructose. The uptake is saturable with $K_m = 16$ mM. We studied such transport in a heterologous expression system. Prestin expression levels, assayed by electrophysiology, were at least 30 times lower than in OHCs. Radioactive flux measurements show that fructose was transported into transfected cells. Measuring fluxes by cell swelling suggested that about 300 water molecules are co-transported per fructose, probably through the same salicylate-sensitive pore. Pendrin shares the same properties. The phenomenon reveals features of the extracellular surface of SLC26 while charge translocation reveals anion binding features of intracellular surface.

2006 Epithelial anion transport in health and disease: the role of the SLC26 transporters family. Wiley, Chichester (Novartis Foundation Symposium 273) p 59–73

Two members of the SLC26 family have featured in the hearing literature. These are SLC26A4 (pendrin) and SLC26A5 (prestin) both of which will be the topics of other chapters in this volume. Pendrin is expressed in the transport epithelium, the stria vascularis of the cochlea, and mutations lead to developmental abnormalities of the inner ear and deafness. Prestin is expressed at very high levels in the outer hair cells of the cochlea but for more subtle reasons. Mutations in prestin also lead to deafness (Liberman et al 2002). The exceptionally high expression density of this protein in this class of hair cells was a key feature leading to its identification from subtractive library for genes expressed preferentially in outer hair cells (OHCs) (Zheng et al 2000).

The mammalian cochlea, the organ of hearing, contains a variety of different cell types, all of which are precisely regulated in number and integral to its correct functioning. The sensory hair cells have been the subject of intense study, not least because the cellular structures are very striking but also because alterations in hair cell function lead to deafness. There are two distinctive hair cell classes arranged along the length of the cochlea, inner (IHCs) and outer (OHCs) hair cells but of these, only OHCs exhibit a phenomenon known as electromotility. It has been argued that electromotility underpins auditory sensitivity (see e.g. Ashmore et al 2000). Over 40 genes have been identified which when mutated lead to deafness and of these about half affect some feature of the hair cell function.

Why look for sugar transport in SLC26 family members?

In seeking hair cell-specific markers, it was reported a decade ago that antibodies raised against GLUT5, the fructose transporter (SLC2A5) found in gut, also labelled cells in the cochlea (Nakazawa et al 1995). The immunochemical label was strongest on the basolateral membrane of OHCs and it was suggested then that sugar transport could be there to maintain the metabolic demands of these cells. With hindsight, it seems likely that the GLUT5 antibody was cross-reacting with other components of the basolateral membrane. There is indeed weak sequence similarity between GLUT5 and SLC26A5 in the C-terminal region against which the antibody was raised. Nevertheless, to a physiologist this was a challenge and we tested the idea that sugars could be transported across the basolateral membrane. We found that fructose was indeed preferentially transported into hair cells (Géléoc et al 1999).

The experiments are simple. They involve exposing isolated OHCs to iso-osmotic solution changes, replacing the glucose in the cell bathing solution for fructose. Cells swelled in volume by up to about 10% when 30 mM fructose replaced the normal external 30 mM glucose. We interpreted this observation as water entering together with the sugar to maintain osmotic balance. We also found that the initial swelling rates depended on the fructose concentration in a saturable manner, fitted with a $K_m = 16$ mM, and a Hill coefficient of 1. The changes depended on the type of sugar (fructose was preferentially transported over galactose, mannose, sucrose and glucose).

OHCs exhibit a gating charge movement when the membrane potential is stepped. The charge movement is closely related to the OHC electromotile length change both in time course and voltage dependence. The associated voltage dependent capacitance of the OHC membrane thus becomes an electrophysiological fingerprint for the length change. The motility can be driven experimentally at frequencies approaching 80 kHz. This is too fast to be described by water movements across the membrane or even by conformational changes as in ionic channels. It can be explained by mechanical area changes of the OHC motor in the plane of the membrane, but without any associated bulk flow. Lateral conformational changes can be fast. The speed of the motor suggested to us that the motor protein had to be a transporter, not an ion channel (Gale & Ashmore 1997). Prestin, SLC26A5, found 3 years later, has many properties of this transporter except that it appears to be a quite inefficient anion exchanger (Oliver et al 2001).

Sugar transport in an expression system

Thus OHCs are cells where electrophysiology and transport mechanisms meet. The advantage of using OHCs is that the prestin density is very high (about 10^7 copies per cell), and so small currents passing through prestin can be measured. Our experiments with OHCs show that these cells could be expressing further membrane transporters. By turning to an expression system, it can be shown that SLC26A5 takes up sugars when transfected into several distinct cell lines.

Direct measurements of radioactively labelled fructose transport into transfected COS-7 cells have been carried out. COS-7 cells were used because approximately 70% of these cells become transfected. We used [^{14}C]fructose as a probe, loading the cells by replacing glucose with fructose. These experiments show that fructose is taken up into prestin-transfected, but not untransfected, cells. A standard volume of cells transfected with GFP and prestin yielded a count (in dpm, disintegrations per minute) of 35907 ± 4146 dpm ($n = 3$), whereas as control for background, cells transfected with GFP alone yielded a count of 6083 ± 424 dpm ($n = 3$). Thus fructose uptake increased by approximately sixfold in cells expressing prestin.

It is easier, although indirect, to measure transport by imaging the cell swelling. As with OHCs, it is straightforward to show that transfected cells undergo reversible volume changes when exposed to iso-osmotic glucose/fructose exchange. By carrying out the electrophysiology on the cells, quantitative measurements of the copy number, at least of SLC26A5 where there is quantifiable charge movement, can be obtained.

Human embryonic kidney (HEK-293), COS-7 and Chinese hamster ovary (CHO) -K1 cells were used for these experiments. They were transfected directly using 'LipofectAmine' with either GFP, GFP-prestin (a gift from B. Fakler), GFP-pendrin (a gift from B. Trembath) and as a control for the expression, by a plasmid coding for the 6 transmembrane KCNQ4 potassium channel (a gift from A. Tinker). The C-terminal GFP-tagged rat prestin construct was expressed in HEK-293, COS-7 and CHO cells, with prestin-associated fluorescence appearing predominantly in the membrane. When the GFP was expressed alone it was found in the cytoplasm and not in the membrane. There are no native fructose transporters in these cells, and this was checked by carrying out RT-PCR experiments.

For a spherical cell, the relation between cell diameter D, cell area A, and cell volume V is:

$$\frac{\delta V}{V} = \frac{3}{2}\frac{\delta A}{A} = 3\frac{\delta L}{D}$$

so that in principle cell volume changes can be determined from dynamic cell diameter changes. We determined D using special purpose software to an accuracy of better than 1%. Changes can also be measured from the cross sectional area A, although the image processing is slightly more elaborate. Figure 1 shows the change in cell diameter of a typical prestin-expressing HEK-293 cell when the external glucose was replaced iso-osmotically with fructose. For a nearly spherical cell about $20\,\mu$m in diameter, the cell increased by 0.5–$0.8\,\mu$m in diameter, or equivalently by $2.9 \pm 1.7\%$ (mean $\pm$ SD, $n = 20$).

The observations are consistent with selective transport of fructose. A similar transport was observed using a C-terminal GFP-tagged pendrin (*SLC26A4*) construct. Cells transfected with GFP alone did not respond to sugars. However, the

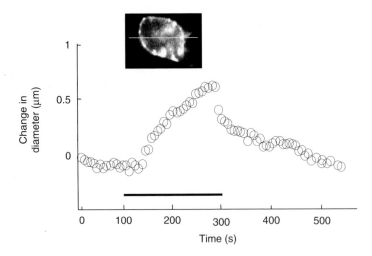

FIG. 1.   Diameter change during sugar application on HEK-293 cells expressing the rat prestin. The exchange of glucose for fructose is indicated by the line. When HEK-293 cells were transfected with the GFP construct alone, there was no change in diameter or area when 30 mM fructose was substituted for 30 mM glucose in the external media (data not shown). In order to eliminate the possibility that transport occurred via a pathway activated by expression of membrane proteins, a membrane protein not related to the SLC26 family, a potassium channel, was used as a control expressed protein. In such cells there was no cell swelling, even though independent whole cell recording showed that this protein was functionally present in the membrane.

fluorescence intensity in cells co-transfected with GFP and pendrin was measured in the cytoplasm before, during and after sugar application. The reduction of the cytoplasmic GFP signal was entirely consistent with water moving in to establish osmotic balance and diluting the fluorescent signal.

There are two additional advantages of using imaging methods rather than radioactive fluxes. First the temporal behaviour of the system can be measured easily. The initial rate of the *volume* change of the cell (V_i) could be fitted by a Michaelis-Menten kinetics for a saturable uptake mechanism

$$V_i = V_{max}[S]/(K_m+[S])$$

with V_{max} the maximum rate, $[S]$ the sugar concentration, and K_m the half-saturating concentration. The data are consistent with fructose being transported by prestin with an apparent $K_m = 24$ mM. V_{max} was $0.04\%/s^{-1}$. These values show that these cells reached a steady state diameter in approximately $2.9/0.04 = 75$ s after sugar exposure. For comparison, the length of OHCs (where $K_m = 16$ mM for fructose) reached a steady state in approximately 5 s. There was no evidence for a baseline substantial exchange of glucose across the cell membrane as no change in area was detectable when 30 mM sucrose was replaced by 30 mM glucose in the bathing medium. Replacing 30 mM glucose with 30 mM mannose produced a ninefold increase in cell volume. Thus mannose transport occurs in HEK-293 cells even though there is still a substrate specificity for fructose.

The second advantage of this system is that the copy number of prestin can be determined by electrophysiology. Voltage-dependent capacitance changes, although small in transfected cells, are measurable by using a lock-in amplifier (Neher & Marty 1982, Gale & Ashmore 1997). From these measurements, the number of copies of prestin protein per cell can be estimated from the maximum non-linear capacitance (C_{max}) and given by the number N of charges moved through the membrane field

$$N = \frac{4\alpha C_{max}}{e}$$

where $\alpha = (ze/k_B\ T)^{-1}$ is the slope factor of the voltage dependence of charge transfer (here k_B is Boltzmann's constant, T is absolute temperature and ze is fractional electronic charge moving through the field). Such measurements suggest that $N = 250\,000$, significantly less (by about 50 times) than found in OHCs.

Blocking sugar transport

Salicylate blocks charge movement in OHCs (Tunstall et al 1995). It is thought to behave as a competitive inhibitor of chloride binding in the prestin molecule (Oliver et al 2001) and also inhibits fructose uptake in OHCs (Ashmore et al 2000).

To test the effect of salicylate on prestin-transfected HEK-293 cells, iso-osmotic replacement of glucose with fructose was carried out in the presence of 10 mM salicylate. The cell medium was replaced by 30 mM glucose solution containing 10 mM salicylate and allowed to stand for 5 minutes with the result that there was no longer a response to 30 mM fructose. Thus salicylate blocked the fructose-induced cell volume change and may indicate that the anion movement and fructose share a common pathway.

Estimation of the sugar transport rate per prestin molecule

The initial rate of cell volume change when exposed to 30 mM fructose was

$$(d/dt)(\delta V/V) = 3 \times 0.03/100 = 0.001 s^{-1}$$

This value is about two orders of magnitude less than found in adult OHCs (Ashmore et al 2000). Since the experiment used iso-osmotic replacement of 30 mM of a non-transported sugar (glucose) with a transported sugar (fructose), we can calculate the number of fructose molecules transported on the assumption that the fructose entering the cell neither concentrated nor diluted the cell contents. For a typical HEK-293 cell (volume = 4 pl), the initial transport rate was

$$0.001 \times 4 \times 10^{-12} \times 0.320 = 1.3 \times 10^{-15} \text{ moles.s}^{-1}$$

or 3080 fructose/prestin.s^{-1}. The maximum transport rate is thus about 6000 fructose molecules per prestin per second.

Water co-transport with fructose

Coupled water transport has been described for other systems (e.g. Zeuthen et al 2001). We carried out experiments which show that water was carried with fructose through the prestin transport pathway.

Figure 2 shows the reversible swelling of a prestin-transfected cell produced by a hypo-osmotic solution. Transfected cells were bathed in standard glucose containing solution which was then made hypo-osmotic by removing the glucose. After 200 s the cells had increased in volume by 5.2 ± 1.0% but, unlike sugar replacement experiments, only showed a partial volume recovery on return to 30 mM glucose. Prolonged application of the hypo-osmotic solution produced a final volume increase (after 480 s, dashed lines) of 10 ± 1.6% ($n = 3$), which is consistent with a final equilibrium determined by osmotic balance.

The initial slope of these data ($3.0 \pm 0.5 \times 10^{-4}$ s^{-1}, $n = 4$) can be used to determine an apparent water permeability, L_p. The value was $2.0 \pm 0.3 \times 10^{-4}$ cm.s^{-1} (when $\Delta\pi = 30$ mOsm/kg), consistent with water diffusion through lipid bilayers but not through an aquaporin in the membrane (Verkman & Mitra 2000). The

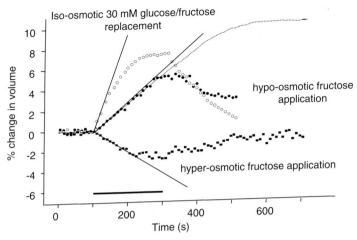

FIG. 2. Water movement in HEK-293 cells transfected with rat prestin. Cell swelling caused by iso-osmotic 30 mM glucose/fructose replacement (open circles) and 30 mOsm/Kg hypo-osmotic glucose solution application (solid circles). Exchange indicated by line. With prolonged hypo-osmotic glucose solution exposure (dotted line) the cell reached a steady state. Cell shrinking caused by hyper-osmotic fructose solution application (solid squares). Cell equilibrated in 20 mM fructose solution and then exposed to 10 mOsm/Kg hyper-osmotic solution containing 30 mM fructose. Although the cell volume decreased, the water permeability (L_p) is estimated to be the same as in the iso-osmotic fructose replacement.

volume increase produced by fructose was 3.3 times greater than by hypo-osmotic stimulation. The final steady volume was 108% of that with the control glucose solution, and could be reversed at almost the same rate. The data are consistent with a rapid movement of fructose across the cell membrane bringing water with it to balance osmotic pressure.

It is possible to calculate the volume flow of water through prestin. It is for a typical HEK-293 cell,

$$4\text{ pl} \times 0.001\text{ s}/250\,000 = 16\text{ zl H}_2\text{O.prestin}^{-1}.\text{s}^{-1}$$

where 1 zl = 10^{-21} l. This is a number close to that found for water movement through SGLT1 in oocytes ($3.2\text{ pl transporter}^{-1}.\text{s}^{-1}$; Loo et al 1999) where a different method of estimating transporter number was used. Thus the number of water molecules transported with each fructose is

$$16\text{ zl} \times N/V_\text{w,} = 180\text{ H}_2\text{O.fructose}^{-1}.\text{s}^{-1}$$

(where N is Avogadro's number and V_w = 8 ml/mol the partial molar volume of water).

Sugar transport by other SLC26 transporters

We tested pendrin (SLC26A4) for sugar transport by similar experiments but using a GFP tagged pendrin construct in HEK-293 cells. Transfected cells also produced a reversible swelling. On replacement of glucose with fructose and mannose, pendrin-transfected cells swelled by 7.6 ± 4.8% ($n = 12$) and 5.4 ± 3.7% ($n = 4$) respectively. This was more than that found with prestin-transfected cells. It does show however that pendrin is also capable of using either mannose or fructose as a substrate for transport. We have not investigated other members of the SLC26 family.

Significance of sugar transport by SLC26 members

These experiments indicate that members of the SLC26 family may be broader spectrum transporters than originally thought. Both prestin and pendrin are expressed in the cochlea. Prestin's main role is that of the motor for OHC electromotility whereas pendrin (SLC26A4) is present in the stria vascularis of the cochlea and may function as a transporter regulating endolymphatic concentrations during development. Fructose transport by prestin does not necessarily imply that fructose is used as the substrate in the cochlea, but it does indicate that with these two transporters a wide spectrum of substrates, both charged and uncharged, are transported. Either transporter could be involved in osmoregulation of the cells, a role for prestin important in OHCs. Likewise, pendrin may well act in the osmoregulation of the fluids of the cochlear endolymph. There is not evidence that fructose is present in measurable quantities in the cochlear fluids whereas glucose is present at milllimolar concentrations.

The transport rate for mannose for both prestin and pendrin in expression systems is close to the rate for fructose. In OHCs these two sugars differ in their ability to produce shape changes (Géléoc et al 1999). The present data suggest that the interaction domain of the sugar in SLC26A5 needs to be identified. It is still possible that the transport of sugars depends on a protein complex of protein with several other as yet unidentified sub-units. In this case each sub-unit could act as a cofactor for the motor protein, sensitive to the intracellular anion concentration and in such multimeric structures a central pore could always allow transport of non-charged molecules.

Some indications as to why SLC26 transporters also transport sugars may be suggested by the recent structural modelling of the archetypal sugar transporter GLUT1/SLC2A1 (Salas-Burgos et al 2004). The similarity in hydrophobicity plots between the SLC2 and SLC26 is striking and could indicate that there are a number of structural motifs in common between these families, both of which contain 12 transmembrane α helices. Indeed, modelling considerations suggest that the specificity of the glucose transporters may reside in the shape of the extracellular vestibule which can bind the smaller five-membered ring of the ketose sugars but not the six-membered ring of glucose. Modelling also suggests that, in SLC2A1, the cytoplasmic vestibule appears to be considerably larger than the exofacial pocket and may form the structure in which the charge movement is possible.

Finally, it has been observed that sugar dependent swelling rates in OHCs depend on membrane voltage (Géléoc et al 1999, Belyantseva et al 2000). It seems most probable that this effect is a consequence of the voltage dependent transport cycle of SLC26A5. It is more problematic to investigate in detail this question in transfected cells as the expression level is low although in principle the experiment is feasible. The transport in such systems would be predicted to be voltage dependent. Nevertheless, the movement of solutes and water controlled by membrane potential could be a route by which OHCs control their internal pressure and hence the forces which they exert in cochlear mechanics. Thus this member of the SLC26 family fulfils multiple roles in hearing: it is a motor, it is a mechanism through which the cell controls its internal pH and finally through the transport of a hexose sugar, it controls the osmotic balance of a hair cell.

Acknowledgements

We thank H. Dorricott for technical assistance and G. Géléoc and L. Harbott who carried out the early experiments on outer hair cell transport. B. Fakler, R. Trembath and A. Tinker kindly gave us gifts of constructs. We also thank B. Perez-Mansilla and G. Thomas for their help with radioactive transport experiments and D. Muallem for discussion about mechanisms of prestin transport. This work was supported by the Wellcome Trust.

References

Ashmore JF, Géléoc GS, Harbott L 2000 Molecular mechanisms of sound amplification in the mammalian cochlea. Proc Natl Acad Sci USA 97:11759–11764

Belyantseva IA, Frolenkov GI, Wade JB, Mammano F, Kachar B 2000 Water permeability of cochlear outer hair cells: characterisation and relationship to electromotility. J Neurosci 20:8996–9003

Gale JE, Ashmore JF 1997 An intrinsic frequency limit to the cochlear amplifier. Nature 389:63–66

Géléoc GS, Casalotti SO, Forge A, Ashmore JF 1999 A sugar transporter as a candidate for the outer hair cell motor. Nat Neurosci 2:713–719

Liberman MC, Gao J, He DZZ, Wu X, Jia S, Zuo J 2002 Prestin is required for electromotility of the outer hair cell and for the cochlear amplifier. Nature 419:300–304

Loo DDF, Hirayama BA, Meinild A-K, Chandy G, Zeuthen T, Wright EM 1999 Passive water and ion transport by cotransporters. J Physiol 518:195–202

Nakazawa K, Spicer SS, Schulte BA 1995 Postnatal expression of the facilitated glucose transporter GLUT 5 in gerbil outer hair cells. Hear Res 82:93–99

Neher E, Marty A 1982 Discrete changes of cell membrane capacitance observed under conditions of enhanced secretion in bovine adrenal chromaffin cells. Proc Natl Acad Sci USA 79:6712–6716

Oliver D, He DZ, Klocker N et al 2001 Intracellular anions as the voltage sensor of prestin, the outer hair cell motor protein. Science 292:2340–2343

Salas-Burgos A, Iserovich P, Zuniga F, Vera JC, Fischbarg J 2004 Predicting the three-dimensional structure of the human facilitative glucose transporter Glut1 by a novel evolutionary homology strategy: insights on the molecular mechanism of substrate migration, and binding sites for glucose and inhibitory molecules. Biophys J 87:2990–2999

Tunstall MJ, Gale JE, Ashmore JF 1995 Action of salicylate on membrane capacitance of outer hair cells from the guinea-pig cochlea. J Physiol 485:739–752

Verkman AS, Mitra AK 2000 Structure and function of aquaporin water channels. Am J Physiol Renal Physiol 278:F13–28

Zeuthen T, Meinild AK, Loo DD, Wright EM, Klaerke DA 2001 Isotonic transport by the Na+-glucose cotransporter SGLT1 from humans and rabbit. J Physiol 531:631–644

Zheng J, Shen W, He DZ, Long KB, Madison LD, Dallos P 2000 Prestin is the motor protein of cochlear outer hair cells. Nature 405:149–155

DISCUSSION

Muallem: Is there actually a Cl⁻ flux associated with sugar transport?

Ashmore: This is on our list of things to study. We have an electrophysiological way of measuring Cl⁻. We think there is evidence for a Cl⁻ current that is modulated by bicarbonate. We haven't yet done the experiment where we exchange external glucose for fructose. This would be the other way round: looking for a Cl⁻ current under the influence of fructose.

Muallem: This means that the Cl⁻ is needed to move the fructose. Is it some kind of coupling?

Ashmore: If pushed, I would say that Cl⁻ and fructose are sharing the same pathway through that complex. Does one exclude the other? I can't say.

Muallem: It is difficult for me to understand how you can move 150 molecules of water by the movement of one fructose molecule.

Ashmore: It is because the space around the fructose molecule is large.

Muallem: But can you fit 150 molecules of water around it?

Quinton: The water doesn't necessarily have to go through that pathway. It can go anywhere across the membrane. The number '150' that he gives for water molecules is equivalent to the number of water molecules (about 200) moved iso-osmotically per molecule of solute.

Alper: Ernie Wright proposes that the water indeed traverses the solute transporter polypeptide.

Quinton: I know that is what Ernie wants, but are you taking the same tack that this is an aquaporin?

Ashmore: No. On the basis of the hydraulic permeability of the cells, it appears that there aren't any aquaporins in the basolateral membrane of these cells. The water is not going through an aquaporin.

Muallem: You just showed us that you can swell the cells by putting them in hypotonic media.

Ashmore: This is with HEK cells, not hair cells. But we can also swell hair cells with hypotonic media.

Muallem: It is interesting that the fructose doesn't come faster, which means that whatever water movement you have, it has to be mediated by some water channels which are activated by fructose. The fact that you are getting swelling that is much faster by adding the fructose than that obtained by osmotic shock alone, means that the water must be moving by some route that is associated with fructose transport.

Ashmore: Why can't it be moving through A5?

Muallem: I am not saying that it can't; I am trying to conceptualize how 150 water molecules move for one fructose molecule.

Gray: It could be that the sugar is opening up a conductive pathway for ions.

Muallem: I want to know how you can take one fructose molecule and get movement of 150 water molecules.

Gray: What if fructose was opening up the structure of the protein?

Muallem: It would become a water channel; that is fine. I don't have a problem with that.

Turner: If you calculate the volume of 150 water molecules is it reasonable compared with the size of these proteins?

Ashmore: It is minute, isn't it?

Thomas: I want to talk further about scale here. If we think about the expression of 250 000 molecules per HEK293 cell, we can make a conservative estimate that this represents 5% of the total membrane. Are you dealing with a 6% change in surface area? If so, these are dramatic changes in the size/conformation of this protein.

Romero: You are comparing apples and oranges. The estimate of 250 000 was based on the density in hair cells.

Ashmore: No, this is the number in HEK cells. I am told by people who know that this is a typical expression level for most proteins with transient transfection.

Turner: Let's say it is less than 5% of the total membrane surface area. You are getting a 6% change in the membrane surface area.

Welsh: No, he is getting a 6% change in volume. This is a much smaller change in the area.

Turner: All I am trying to say is that the cartoons may under-represent the kinds of conformational changes.

Mount: Have you looked at Glut5-transfected HEK cells? Do they differ in behaviour from prestin-transfected cells? This would help you separate the dose effect from the prestin-specific effect.

Ashmore: No, we haven't looked.

Lee: You showed that fructose can be transported by the prestin. Do you think this is related to the mode of transducing activity of prestin?

Ashmore: I don't think so. I think prestin is behaving as a conduit for fructose plus the water. There is a voltage dependence to the uptake of fructose which could be explained by when you depolarize the cells you somehow modify the shape of that pore. This is consistent with the idea that the pore closes down when it is depolarized.

Knipper: If you think that prestin takes the job of the fructose transporter, what is the job of the glucose-5 transporter? Do you know anything about the knockout of Glut5?

Ashmore: All we know is what the immunohistochemistry tells us. The jury is still out as to whether there is Glut5, or whether it is bad immunohistochemistry which explains the Glut5 presence there. It led us down a rather strange route that essentially got us sitting in this meeting here. There has been talk of a Glut5 knockout, but I haven't seen anyone doing it.

Soleimani: Have you examined the effect of amiloride or amiloride analogues in the experiments testing the effects of fructose?

Ashmore: We only tried DIDS on the prestin-transfected cells, not on the pendrin-transfected cells. We haven't tried amiloride.

Alper: What are the metabolic effects of such high fructose concentrations in terms of changing NADP/NADP balance? Could you be triggering such downstream effects by activating a pentose shunt?

Ashmore: Yes. The fact that we get the uptake graded and described by a simple Michaelis–Menten equation sort of argues against that suggestion, but not completely. We thought of using this as an assay for fructose uptake at one stage. Each time the hexokinase acts on the fructose a proton is liberated, so pH could be used as an assay. We were certainly worried about the possibility that additional osmolytes were being generated when the fructose was taken up.

Alper: Is there a physiological source of fructose in the inner ear that would suggest that this has something to do with the physiological function of the outer hair cells (OHC)?

Ashmore: I don't think it has any physiological function; it is just a way of probing what the prestin complex can do. There is certainly glucose present surrounding the basolateral membrane, but not fructose. You don't suddenly start hearing better if you eat an apple!

Aronson: In terms of the water pathway that is prestin independent, and whether the enhanced water permeability is parallel or co-transport, co-transport would predict that there should be coupling of the water flux and fructose flux, so if you made an osmotic gradient when you were measuring [^{14}C]fructose uptake, you should actually see water-driven fructose transport. This wouldn't be true with the parallel pathway. Have you ever tried this?

Ashmore: No.

Welsh: Prestins are regulated as the hairs on the OHCs move. Is this by the voltage?

Ashmore: The sequence of events is that mechanical distortions of the cochlea deflect the stereocilia of hair cells and this opens ion channels at the tips. Current flows in and changes the membrane potential and then the conformational change in prestin is driven by potential. This explanation brings in other electrophysiological problems which researchers are still fighting over.

Welsh: If you change the volume of that OHC, does that then modify what is seen in terms of the movement of the cell? Does it change stiffness, for example?

Ashmore: The actual voltage dependence of that anion movement depends on the lateral tension in the membrane of the OHC. There is a coupling. The biological piezoelectricity of the membrane prestin sheet goes both ways: voltage produces area changes in the membrane and the area changes produce changes in the membrane potential. The molecular basis for this might be that one is changing the structure of the vestibule.

Romero: The flip way of thinking about this is that if you do a voltage clamp experiment where the voltage is controlled, can you control the amount of fructose that is moved in and out, either in the hair or HEK cells? The prediction would be that you should be able to have a voltage-dependent fructose transport, if in fact you are causing this conformational change that you are supposing leads then to water conductivity.

Ashmore: The only experiment that addresses this question is one where we looked at the fructose-induced swelling at different membrane potentials. As we depolarized the cell more there was less fructose uptake—if you hold at +20 mV you get a lower initial rate of change of swelling than if you hold at −50 mV. The interpretation is that the fructose uptake decreases as the cell is depolarized.

Romero: What are the relevant voltage ranges for the hair cell?

Ashmore: The cell normally sits at −70 mV. This is way out of the range of potentials where we saw an effect on fructose transport.

Romero: Under the normal transduction pathway, you open up the channels and see membrane voltage changes, and this sets about a cascade. Presumably these changes have been measured by tweaking of the stereocilia and other things. What is the range that an OHC membrane might expect to see?

Ashmore: Everything in the cochlea takes place on a very small scale. The displacement changes that are critical for hearing are measured in nanometers; the changes in membrane potential may be no more than a few millivolts.

Alper: After your first paper on the fructose appeared, one of the experiments we did was to ask whether fructose could block pendrin-mediated isotopic chloride transport. In our hands there was no effect of 30 mM fructose, mannitol or glucose.

Ashmore: Did you see fructose uptake?

Alper: We didn't have labelled fructose. We asked whether the excess sugar would alter the isotopic chloride movement. It didn't.

Mount: We tried [^{14}C]fructose uptake. Without prestin as a positive control we didn't see it for A6 or a *Xenopus* orthologue of pendrin. But we did this with our standard uptake, using gluconate as our chloride substitution.

Muallem: It is not clear to me why we are completely dismissing this mechanism as having no physiological significance. It doesn't have to be fructose, but the fact that you can see such rapid changes in cell volume might be very important as a mechanism for regulating the volume of the OHCs.

Ashmore: I agree. There are several different time scales involved in hearing. There is a microsecond timescale, which is the one involved in fast amplification, and there is one on a 10 s to 1000 s timescale important for overall gain setting which could well be mediated by movements of solutes such as this. It is clearly important that the cell maintains an internal turgor pressure so that when prestin does generate a force in the basolateral membrane it is transmitted all around on the cell.

Mount: Aren't these cells also anatomically circumscribed? How much can they swell on an anatomical basis? I always thought they were held.

Ashmore: They are held tightly in the organ of Corti. The thinking is that you need to generate the forces along the axis of the cell. This is the force that pushes down onto the basilar membrane.

Welsh: It could go either direction. Neighbouring cells could move up and down also.

Ashmore: The efficiency is less in that situation.

Mount: If you had aquaporins in there, wouldn't this impair the ability of the cells to do that?

Ashmore: Yes, you would end up with a less stiff OHC.

Mount: You don't want something floppy. This kind of mechanism where you don't have aquaporins but instead solute-driven volume changes that are related to prestin are advantageous.

Ashmore: There are also other mechanisms that regulate the stiffness of OHCs. There are clear indications that there are signal transduction pathways that control cytoskeletal stiffness. The internal structure of these cells is pretty complicated. There is an actin–spectrin spring, an endoplasmic reticulum that runs up and down the side, and there are mitochondria everywhere. One could provide a semiclassical model where pumping water is the most important stiffness regulator, but this is clearly not the only cell biology involved.

Muallem: Unless I've got this wrong, it was striking that the water movement with prestin is much faster than that possible with aquaporins. HEK293 cells do have aquaporins, and you are still getting a rate that is faster than swelling without fructose.

Ashmore: You increase the rate by increasing the number of prestin molecules. This is exactly what the hair cell has done.

Romero: How fast is the red cell going through the inner medulla and out of the inner medulla?

Alper: Our experience with 293 cell volume subjected to a 60 milliosmolar shift is that the initial osmotic volume change appears complete within tens of seconds.

Turner: You are thinking about recovery from osmotic shock, which is on a second time scale. The initial shrinkage takes place very quickly.

Alper: It is limited by the time of the bath change.

Romero: In the vesicle studies there are insanely fast water transfer rates (4–18 × 10^{-3} cm/s).

Muallem: But in intact cells, when they are exposed to a hypotonic solution, the process of swelling takes about half a minute. It is definitely not instantaneous.

Case: But what about red cells on their journey through the renal circulation?

Romero: That's for a red cell, which has fewer aquaporins per surface area than prestin molecules per hair cell.

Muallem: Red cells are rather unique, but average secretory cells don't swell and sink that fast. Their volume regulates over minutes, but the swelling and shrinkage takes about 30 s. Cotton et al (1989) calculated carefully the time of water movement and came to the same conclusion. What you are showing us is close to 50–100 times faster.

Ashmore: The fructose experiments have the cells swelling on a timescale of 10 s.

Muallem: If another molecule is the physiological substrate, not fructose, which can do what fructose is doing, coupled with the level of prestin in the hair cell, we have a mechanism to swell and shrink the cells in less than a second. This is much faster than we are used to seeing.

Reference

Cotton CU, Weinstein AM, Reuss L 1989 Osmotic water permeability of Necturus gallbladder epithelium. J Gen Physiol 93:649–679

SLC26A3 and congenital chloride diarrhoea

Pia Höglund

Hospital for Children and Adolescents, University of Helsinki, FIN-00029 HUS, Helsinki, Finland

Abstract. Congenital chloride diarrhoea (CLD, OMIM214700) is a rare genetic disease caused by mutations in a plasma membrane protein, the solute-linked carrier family 26 member A3 (SLC26A3) protein, which encodes for an epithelial anion exchanger for Cl^- and HCO_3^-. The main clinical symptom is a lifetime watery diarrhoea with a high Cl^- content and low pH, causing dehydration and hypochloremic metabolic alkalosis. CLD may be fatal, if not adequately treated by substitution of NaCl, KCl and fluid lost in the faeces. Long-term prognosis is generally favourable, but complications such as renal disease, inflammatory bowel disease, hyperuricemia, inguinal hernias, spermatoceles and male subfertility are possible. The role of dysfunctional SLC26A3 in the pathogenesis of these complications is poorly known. Altogether 30 different mutations of the *SLC26A3* gene are currently known among patients with CLD; the most common of them being the three founder mutations present among Finns (V317del), Polish (I675-676ins) and Arabic (G187X) populations. Individual variation in the clinical picture of CLD is common, but not known to associate with the genotype.

2005 Epithelial anion transport in health and disease: the role of the SLC26 transporters family. Wiley, Chichester (Novartis Foundation Symposium 273) p 74–90

Clinical picture of CLD is characterized by a lifetime, severe diarrhoea

Congenital chloride diarrhoea (CLD, OMIM 214700) is a rare recessively inherited disease with approximately 270 known cases worldwide (reviewed in Mäkelä et al 2002). CLD is caused by a loss/reduced function of a chloride–bicarbonate exchange mechanism located in the ileal and colonic epithelium. As a result, massive loss of Cl^- in the stools occurs, together with metabolic alkalosis because of a failure in the excretion of HCO_3^- into the intestinal lumen. Acidic bowel content subsequently also limits the absorption of Na^+ through a decreased function of a Na^+/H^+ exchanger (Holmberg et al 1975, 1977a). Defective reabsorption of both Cl^- and Na^+ is followed by water, manifesting as voluminous watery diarrhoea.

Diarrhoea begins *in utero*, leading to intestinal dilation and polyhydramnios, and often premature birth. Neonates with CLD usually present with abdominal distention, absence of meconium and electrolyte abnormalities, including hyponatremic, hypochloremic metabolic alkalosis and often hyperbilirubinemia (Holmberg et al 1977a, Holmberg 1986). The diagnosis can be confirmed by measuring faecal concentration of Cl⁻, which always exceeds 90 mmol/l in patients with corrected water and electrolyte balance (Holmberg et al 1975). DNA analyses are also available. The diagnosis can be delayed, because the watery diarrhoea may be mistaken for urine.

The treatment consists of supplementation of NaCl, KCl and fluid lost in the faeces to maintain normal serum levels of Cl⁻, Na⁺, K⁺ and bicarbonate (Holmberg 1986). Adequate replacement therapy from birth is life-saving and allows normal growth and developmental milestones. Maintenance of normal fluid- and electrolyte-balance is also likely to prevent contraction-induced renal impairment and mortality during acute gastroenteritis (Holmberg et al 1977a). If untreated, dehydration is often fatal and hyponatremic attacks may lead to mental and psychomotor impairment. Those who survive the first few critical months and remain undiagnosed may adopt a less aggressive, chronic course of the disease presenting with retarded growth, and later renal impairment (Holmberg et al 1977a,b) and gout (Nuki et al 1991).

The replacement therapy, if correctly administered, has no effect on the diarrhoea. Omeprazol has been shown to reduce stool volume and Cl⁻ content in one patient (Aichbichler et al 1997), whereas limited effectiveness have been reported in others (Höglund et al 2001a). A recently published case report suggested that a short-chain fatty acid (SCFA) butyrate might have a beneficial effect on CLD-related diarrhoea (Canani et al 2004). Most experience is available of cholestyramin, shown to reduce the amount of diarrhoea for at least a few weeks (Holmberg 1986). In future, efforts to find therapies that correct defective SLC26A3 directly will hopefully result in an introduction of an easy, permanent and safe treatment to achieve symptomatic relief, as well.

With a combination of optimal substitution therapy and aggressive treatment of acute illnesses, the survival of these patients has dramatically improved. A recent Finnish survey for the long-term prognosis among a large cohort of adolescent and adult patients revealed that when diagnosed early and adequately treated, the long-term prognosis of CLD is excellent, and an increasing number of adult patients live close to normal life despite CLD. Since the treatment does not reduce diarrhoea, soiling and enuresis are common in early life, but usually subside later, and a great majority (92%) of adult patients rate their quality of life as excellent or good. Renal complications have resulted in end-stage renal disease in a few patients, usually associated with suboptimal treatment and follow-up. Other associations have been inflammatory bowel diseases (rare), hyperuricemia (common), inguinal hernias and enamel hypoplasia of the teeth. Spermatoceles and male subfertility were described as novel and relatively common manifestations of CLD (Hihnala et al 2006, Höglund et al 2006).

Incidence of CLD: three clusters and solitary cases throughout the world

CLD is particularly frequent in Finland, affecting approximately 1/40000 people (Holmberg 1986, Höglund et al 1998a). In Poland, the disease is fivefold less frequent affecting only 1 per 200000 live births (Höglund et al 1998a), higher incidences between 1 in 5000 and 1 in 3200 live births have been reported among Arabic people consanguineous. Thus two thirds of all known CLD cases come from these three geographical areas, the rest being solitary cases originating throughout the world. Some of the isolated cases are likely to be missed, since diarrhoea is a very common symptom and an important contributor to the mortality of infants especially in developing countries.

The *SLC26A3* gene

Human genetics studies have shown that CLD is caused by mutations in a transmembranic solute linked carrier gene family 26 (SLC26) member A3 gene, *SLC26A3* (OMIM 126650), previously known as DRA (for down-regulated in adenomas; Schweinfest et al 1993) or CLD (Höglund et al 1996). The gene has been located on chromosome 7q31 and encodes a single 3.7 kb transcript spanning 39 kb of genomic DNA arranged into 21 exons. It resides only 49 kb away from the *SLC26A4* gene (pendrin), which also shows closest homology with *SLC26A3* (44% identity). The protein is a 764 amino acid hydrophobic glycoprotein with a molecular weight of approximately 82 kDa. Its predicted membrane topology is similar to other members of the *SLC26* family, the exact number of transmembrane segments are yet to be characterized, but varies from 10 to 14 times, so that both the N- and C-terminal domains remain intracellularly.

Expression profile and functional properties of the SLC26A3 protein

Immunohistochemical and *in situ* hybridization studies have demonstrated SLC26A3 expression in the apical membranic and apical cytoplasm of the ileal and colonic surface epithelium (Haila et al 2000), in the brush-border membrane of duodenal epithelium (Jacob et al 2002), in the epithelia of seminal vesicles and sweat glands (Haila et al 2000), in mouse heart myocardium (Alvarez et al 2004), and in cultured cells derived from the pancreatic duct (Greeley et al 2001) and trachea (Wheat et al 2000).

Much has been learned about functional properties of the SLC26A3 protein since the initial report indicated SLC26A3 as a transmembranic sulfate transporter (Silberg et al 1995). Functional properties and anion selectivity of both human and mouse SLC26A3, which share 75% amino acid identity, have been widely studied, e.g. in *Xenopus laevis* oocytes. Current models support the idea that SLC26A3 functions mainly as an exchanger for both Cl^-/HCO_3^- and perhaps Cl^-/OH^-, whereas substrate specificity to oxalate, butyrate and sulfate remain of low or close to undetectable magnitude (Moseley et al 1999, Melvin et al 1999, Ko et al 2002, Chernova et al 2003, Alvarez et al 2004). This is in accordance with the major phenotypical manifestation of CLD, characterized by a loss of chloride in acidic stools due to SLC26A3 deficiency. Detailed discussion about functional properties and regulation of SLC26A3 and other members of the SLC26 transporter family is presented elsewhere.

Spectrum of mutations in the *SLC26A3* gene in patients with CLD

Altogether 30 different mutations in the *SLC26A3* gene have been identified among more than 100 individuals with CLD originating from several different geographical areas and having a wide spectrum of ethnic backgrounds (Table 1; Etani et al 1998, Höglund et al 1996, 1998a,b, 2001a,b, Mäkelä et al 2002). Of all the mutations found, approximately 1/3 are microdeletions, 1/3 single nucleotide missense and nonsense substitutions, 20% are splicing mutations and 10% micro-insertions. The remaining two mutations are a more complex deletion combined with an insertion (2104-2105delGGins29bp), and a 3.5 kb genomic deletion removing exons 7 and 8 (Höglund et al 2001b). Thus far, no whole gene deletions, promoter mutations or de novo mutations have been identified.

Altogether four polymorphic nucleotide changes have been characterized in the coding region of *SLC26A3*, two of them are silent single nucleotide polymorphisms and two are polymorphic both at the DNA and protein level (Höglund et al 2001b). The overall paucity of polymorphisms on the coding sequence of *SLC26A3* may reflect an overall high sensitivity of SLC26A3 to any sequence changes.

Genetic epidemiology

Distinct founder mutations have been demonstrated in all three regions with higher than average incidence of CLD. The V317del-mutation is enriched in the Finnish population due to genetic founder effect, and to date, 98% of all CLD-associated chromosomes carry this specific mutation. In Arab countries (Saudi Arabia and Kuwait), the consanguinity has led to the enrichment of the G187X mutation in some tribes, and the mutation is found in 94% of disease-associated chromosomes. In Poland, the major insertion mutation (I675-676ins) segregates in 47% of the CLD-chromosomes unmasking a set of the rarer/solitary mutations as compound heterozygotes that would result in disease very rarely alone, and also emerges as true homozygotes, which together have given rise to an apparently high incidence of CLD in Poland (Höglund et al 1998a).

TABLE 1 Summary of all mutations reported in the *SLC26A3* gene worldwide

Mutation Name	Exon	Nucleotide change	Effect on coding Sequence	Codon	Protein product	Patient origin	Comment
MISSENSE							
G120S	EX 4	358G > A	Glycine → Serine at 120	120	Conserved residue	Poland, Sweden, Norway	recurrent
H124L	EX 4	371A > T	Histidine → Leucine at 124	124	Conserved residue	Poland, Sweden (Polish ancestry)	Polish minor mut.
P131R	EX 5	392C > G	Proline → Arginine at 131	131	partially conserved	U.S.A.	
S206P	EX 6	616T > C	Serine → Proline at 206	206	Conserved residue	Holland (Morocco ancestry)	
D468V	EX 12	1403A > T	Aspartic acid → Valine at 468	468	Conserved residue	Poland	
L496R	EX 13	1487T > G	Leucine → Arginine at 496	496	mouse: loss of function	Hong Kong	
I544N	EX 15	1631T > A	Isoleucine → Asparagine at 544	544	Loss of function	Vietnam	
NONSENSE							
G187X	EX 5	559G > T	Glycine → STOP at 187	187	Truncated	Saudi Arabia, Kuwait, U.K.	Arabic founder mutation in 90% of CLD-associated chr.
Y305X	EX 8	915C > A	Tyrosine → STOP at 305	305	Truncated	Poland	
Q436X	EX 11	1306C > T	Glutamine → STOP at 436	436	Truncated	Holland	Polish minor mut.
W462X	EX 12	1386G > A	Tryptophan → STOP at 462	462	Truncated	U.K.	
DELETION							
145–157del13bp	EX 3		Frameshift	49	Truncated	Belgium (Togo ancestry)	
344delT	EX 4		Frameshift	115	Truncated	Poland	Polish minor mutation
3.5kb deletion	IN 6–8		Loss of exons 7 and 8, frameshift		Deleted	Japan	
V317del	EX 8		In-frame loss of a Valine at 317	317	mouse: loss of function	Finland, Sweden (Finnish ancestry)	Finnish founder mutation in 97% of CLD-associated chr.
1342–1343delTT	EX 12		Frameshift	448	Truncated	Japan	

Mutation	Location	Nucleotide change	Effect	Codon	Consequence	Origin	Notes
1516delC	EX 14		Frameshift	505	Truncated	Poland	
1548–1551delAACC	EX 14		Frameshift	516	Truncated	Poland	
Y527del	EX 14		In-frame loss of a Tyrosine at 527	527	Deleted	Poland	
1609delA	EX 15		Frameshift	537	Truncated	Canada	
2116delA	EX 19		Frameshift	706	Loss of function	Finland	
INSERTION							
177–178insC	EX 3		Frameshift	60	Truncated	U.S.A.	
268–269insAA	EX 3		Frameshift	90	Truncated	Hong Kong	
1675–676insATC	EX 18		In-frame addition of an Isoleucine	676	mouse: loss of function	Poland	Polish founder mutation in 47% of CLD-associated chr.
REPLACEMENT							
2104–2105del GGins29bp	EX 19	replacement of GG at 2104–2105 with 29bp insertion	In-frame substitution G702T and addition of nine amino acids	703		Norway	
SPLICE DEFECT							
IVS5-2A > G	IN 5		Destruction of the intron acceptor site AG			Canada	
IVS5-1G > T	IN 5		Destruction of the intron acceptor site AG			U.S.A.	
IVS11-1G > A	IN 11		Destruction of the intron acceptor site AG			Poland	
IVS12-1G > C	IN 12		Destruction of the intron acceptor site AG			Germany (Palestine ancestry)	
IVS13-2delA	IN 13		Destruction of the intron acceptor site AG			Kuwait	
POLYMORPHISMS							
C/W307	EX 8	921T/G	Cysteine/Tryptophan at 307	307	function normal	Finland, U.S.A, Sweden	
1299G/A	EX 11	1299G/A	No change	433	Normal	U.S.A., Kuwait, U.K., Japan, Sweden	
1314C/T	EX 12	1314C/T	No change	438	Normal	Kuwait	
R554Q	EX 15	1661G > A	Arginine/Glutamine at 554	554	Nonconserved	Japan	

Functional consequences of mutations in the *SLC26A3* gene and relevance for the phenotype

When looking at the distribution of all the mutations along the coding region of the *SLC26A3* gene, there resides a putative mutation hotspot region between nucleotides 1299 and 1661 harbouring more than half of all sequence changes demonstrated to date (Fig. 1). This region encodes the proximal part of the intra-cellular tail of the predicted SLC26A3 membrane protein and overlaps with the sulfate transporter anti-sigma factor antagonist (STAS) domain, homologous to bacterial anti-sigma antagonists. Biochemical consequences of both naturally occurring sequence changes (e.g. I544N and 2116delA) and artificial mutations studied in detail in *Xenopus* oocyte systems suggest that preservation of the STAS domain is a prerequisite for the normal function, regulation and interactive proper-ties of SLC26A3 (Chernova et al 2003, Ko et al 2004).

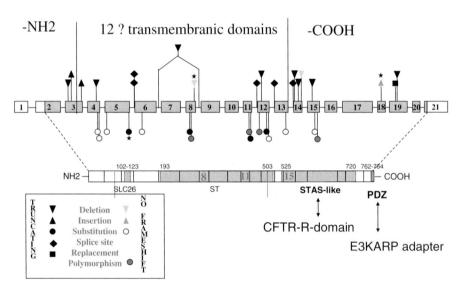

FIG. 1.   A schematic presentation of the *SLC26A3* gene showing its genomic structure, with all the known CLD-associated mutations and functionally important domains. The *SLC26A3* gene consists of 21 exons (drawn to scale), of which 20 are coding (marked in grey). Mutations that are shown above the graph are deletions, insertions and splice site mutations, whereas those below are all single nucleotide substitutions. Truncating mutations are marked with black and non-truncating with white symbols, polymorphisms are marked with gray. The three founder mutations have been marked with asterisks. A mutation hotspot region between exons 11 and 15 resides partially on the STAS-like domain, whereas no point mutations have been found to reside on the PDZ-domain at the very C-terminus of the protein.

Artificial truncations of both the early N-terminal (52 aa) or up to 44 C-terminal amino acids from the putatively cytoplasmic, hydrophilic domain have been shown to decrease but not abolish transport function (Chernova et al 2003), which would also be consistent with our finding that no CLD-associated missense mutations have been found in the very proximal N-terminus of the SLC26A3, nor in the most distal C-terminus of the SLC26A3 (Höglund et al 1998b, Mäkelä et al 2002).

Another potentially important region with regulatory properties is a consensus binding site for the type I PDZ domain. It resides in the very C-terminus of the SLC26A3 protein (Lamprecht et al 2000) and has a potential to recruit SLC26A3 through regulatory proteins, such as E3KARP (NHE3 protein kinase A regulatory protein) with other transporters, such as NHE3 and CFTR (cystic fibrosis transmembrane regulator protein). These large protein complexes are thought to function in close interaction with each other both in the plasma membranic microenvironment and also may form cross-links with the underlying actin cytoskeleton. For example, linked functions of SLC26A3 and CFTR have been suggested to be mediated partially through this pathway (Ko et al 2004), and the same pathway is also likely to connect SLC26A3 and NHE3 (Lamprecht et al 2002).

It is, however, of interest that no diverse or more severe clinical manifestations of CLD are observed in patients, who lack both copies of their SLC26A3-associated PDZ-motifs due to homozygosity for truncating mutations, the G187X 'Arabic' founder mutation being the most common (Höglund et al 1998a). Some compensation is likely to occur, e.g. through an increase in amounts of interactive proteins with concomitant use of alternative binding sites, as proposed for CFTR. High level of expression of SLC26A3 in a cell system has been found to allow coimmunoprecipitation with a mutant CFTR that lacks the PDZ motif (Ko et al 2002).

To date, no clear evidence for a genotype-phenotype variation has been presented, whereas it is not uncommon that even affected siblings with identical mutations present with somewhat diverging phenotypes. More important contributors to the phenotype may be variation in the treatment and individual compensatory mechanisms, such as dietary salt consumption (Höglund et al 2001a). A recent identification of functionally important domains provides novel insights for further studies of possible phenotypic subseries.

SLC26A3 functions as an exchanger for Cl^-/HCO_3^- with distinct physiological roles in different tissues

The basic function of SLC26A3, exchange of intracellular HCO_3^- for extracellular Cl^-, has been found to be modulated through the surrounding biological microenvironment, providing diverse and intriguing roles for SLC26A3 in normal physiology.

Intestine

Under physiological conditions prevailing in the apical membrane of the ileal and colonic surface epithelium, SLC26A3 is likely to mediate the major reuptake of luminal chloride. Joint function of SLC26A3 and a Na^+/H^+ exchanger 3 (NHE3) has a central role in the transepithelial absorption of NaCl (Holmberg et al 1975, Melvin et al 1999, Moseley et al 1999, Chernova et al 2003, Lamprecht et al 2002). Accordingly, defective absorption of ileal and colonic Na^+ in addition to Cl^- is a constant feature in CLD, which can be demonstrated by showing high faecal sodium and chloride content in these patients (Holmberg et al 1975).

In the upper intestine, the brush border membrane of duodenal epithelial cells present with secretive rather than absorptive functions, thus the physiological role of SLC26A3 in duodenum has been proposed to be involved in the secretion of HCO_3^-, which has a central role in the neutralization of intermittent pulses of gastric acid (Jacob et al 2002). In patients with CLD, no clinical consequences, such as duodenal ulcers have been seen, thus loss of putative SLC26A3-mediated neutralizing functions may not have clinical importance in the upper intestine.

Pancreas, sweat gland and lung

The importance of expression of the SLC26A3 protein (or its mouse homologue) in cultured cells derived from pancreas (Greeley et al 2001) in sweat gland (Haila et al 2000) and in lung tissues (Wheat et al 2000) is emphasized by the observation that many of these tissues are most severely affected in cystic fibrosis. Indeed, interaction between CFTR and SLC26A3 is a target of active research, and several molecular mechanisms and novel functional properties (e.g. electrogenic transport activities in ductal systems that concentrate bicarbonate) for SLC26A3 have recently been proposed (Wheat 2000, Greeley et al 2001, Ko et al 2002, 2004, Chernova et al 2003). Detailed discussion about these studies is presented elsewhere. Clinically it is evident that patients with SLC26A3 deficiency do not present with symptomatic pancreatic insufficiency, and children with CLD grow and gain weight normally, when on adequate salt substitution (Holmberg 1986). Thus it seems that the function of SLC26A3 both in the pancreas and in the upper intestine may not be of major importance, or, if defective, can be highly compensated through the action of other transporters.

Similarly, a role for SLC26A3 in sweat formation has been proposed. Using immunohistochemistry, Haila et al (2000) demonstrated that the SLC26A3 is expressed in the coiled secretory part of eccrine sweat glands. The final composition of sweat is modulated through a dual action of coiled secretory part and straight ductular portion of sweat gland, thus the final composition of sweat is dependent on e.g. sweat rate in addition to characteristics of primary sweat (Sato 1993). We have found solitary patients with V317del/V317del genotype and CLD, whose sweat test values are above 60 mmol/L, but the significance of this finding is unknown (Hihnala et al 2006). Minor or occasional (e.g. during excessive perspiration) loss of salt through sweat glands may be possible in CLD, reflected also by the observation that some adult patients with CLD report to benefit from temporary addition of salt substitution after physical activity (Hihnala et al 2006).

Male reproductive tract

In contrast to many other tissues with shown expression of SLC26A3, the male reproductive tract appears to be clearly affected by SLC26A3 deficiency. After the observation that adult male patients with CLD have very few offspring (Hihnala et al 2006), the putative role of SLC26A3 in male reproduction has been evaluated by investigating fertility potential among Finnish adult males with CLD who are homozygous for the V317del mutation (Höglund et al 2006). By history, none of them had naturally conceived children. Their sperm parameters were highly abnormal, and a higher content of chloride and lower pH were found in their seminal plasma when compared with fertile-proven control individuals. In addition, more than 1/3 of these males had spermatoceles. Reproductive hormones were normal and spermatogenesis in testicular biopsies appeared normal. Dysfunction of SLC26A3 in several tissues in the male reproductive tract with shown expression of SLC26A3 (seminal vesicle, efferent duct of the testis) was thus proposed to be responsible for altered water, pH and electrolyte homeostasis in these tissues, with subsequent disruption in normal sperm maturation (Höglund et al 2006).

These findings are important for diagnosis and adequate treatment of infertility in males with CLD. *In vitro* fertilization is likely to give most infertile men with CLD the chance of biological fatherhood. Furthermore, these findings may provide novel insights into cystic fibrosis (CF)-related infertility. Although infertility in males with CF is mainly due to an opposite defect, lack of water in the male genital tract, some of the features, such as low seminal plasma pH is common to both males with CF or CLD. As a recent study implicated a role for defective regulation of SLC26A3 in the CF, it is possible that at least the observed dysfunction of seminal vesicles found in males with CF could be mediated, through a loss-of-action of SLC26A3.

Other tissues

A recent study by Alvarez et al (2004) demonstrated that mouse *Slc2a3* mRNA is abundantly expressed in adult mouse myocardium. In the plasma membrane of cardiomyocytes, SLC26A3 might act as a potential acid-loader together with the isoform SLC26A6, an important function for both normal pH regulation, recovery from ischaemic acidosis and in the development of cardiac hypertrophy (Alvarez et al 2004). A putative role for SLC26A3 in human heart is yet to be clarified. To date, cardiac complications observed among patients with CLD are rare, and related to hypokalemia.

Relevance to IBD and intestinal cancer

Observation of an altered expression of the *SLC26A3* mRNA in the colonic epithelium in inflammatory bowel disease has raised a question, whether dysfunction of the colonic SLC26A3 might be associated with inflammatory bowel disease (IBD)-related diarrhoea (Lohi et al 2002). On the other hand, a slight increase in the figures of incidence of IBD among adult patients with CLD has been observed (Hihnala et al 2006). So far, identification of a role, if any, for dysfunctional SLC26A3 in the pathogenesis of IBD-related diarrhoea is uncertain, and other features, such as acidic intestinal content, subsequent alteration in the intestinal microbial flora and increased flux of bile acids into the colon, may all be factors that generate and maintain chronic inflammation in those few patients with CLD and IBD.

Even more arguable is a possible role of SLC26A3 in the formation of malignancies. The undeniable finding is that the expression of SLC26A3 is lost during neoplastic evolution, initially interpreted to be due to tumour suppressor properties (Schweinfest et al 1993). Indeed, patients and carriers of the V317del-mutation have been found to have a subtle increase in the figures of colonic cancer (SIR 3.4, 95% CI 1.4–7.0, Hemminki et al 1998), although later transfection studies utilizing mammalian cells indicated that the V317del mutant cells presented growth suppression properties similar to that observed in the wild-type cells (Chapman et al 2002). The same study proposed a role for the STAS domain in the control of growth (Chapman et al 2002), thus it remains to be clarified, whether a loss of the putative tumour suppressor activity of SLC26A3 with concomitant formation of malignancies could be rather associated with mutations disrupting the STAS domain. Hence, minimally increased figures of colonic cancer in the patients and carriers of the V317del-mutation could simply reflect consequences of altered intestinal physiology, such as a chronic inflammatory response due to low fecal pH. In that case, loss of the expression of SLC26A3 in malignancies could just reflect general dedifferentiation, rather than a primary role in the formation of cancer (Haila et al 2000).

References

Aichbichler BW, Zerr CH, Santa Ana CA, Porter JL, Fordtran JS 1997 Proton-pump inhibition of gastric chloride secretion in congenital chloridorrhea. N Engl J Med 336:106–109

Alvarez BV, Kieller DM, Quon AL, Markovich D, Casey JR 2004 Slc26a6: a cardiac chloride/ hydroxyl exchanger and predominant chloride/bicarbonate exchanger of the mouse heart. J Physiol 561:721–734

Canani RB, Terrin G, Cirillo P et al 2004 Butyrate as an effective treatment of congenital chloride diarrhea. Gastroenterology 127:630–634

Chapman JM, Knoepp SM, Byeon MK, Henderson KW, Schweinfest CW 2002 The colon anion transporter, down-regulated in adenoma, induces growth suppression that is abrogated by E1A. Cancer Res 62:5083–5088

Chernova MN, Jiang L, Shmukler BE et al 2003 Acute regulation of the SLC26A3 congenital chloride diarrhoea anion exchanger (DRA) expressed in Xenopus oocytes. J Physiol 549:3–19

Etani Y, Mushiake S, Tajiri H et al 1998 novel mutation of the down-regulated in adenoma gene in a Japanese case with congential chloride diarrhea. Mutations in brief no. 198. Hum Mutat 12:362

Greeley T, Shumaker H, Wang Z, Schweinfest CW, Soleimani M 2001 Downregulated in adenoma and putative anion transporter are regulated by CFTR in cultured pancreatic duct cells. Am J Physiol Gastrointest Liver Physiol 281:G1301–1308

Haila S, Saarialho-Kere U, Karjalainen-Lindsberg ML et al 2000 The congenital chloride diarrhea gene is expressed in seminal vesicle, sweat gland, inflammatory colon epithelium, and in some dysplastic colon cells. Histochem Cell Biol 113:279–286

Hemminki A, Höglund P, Pukkala E et al 1998 Intestinal cancer in patients with a germline mutation in the down-regulated in adenoma (DRA) gene. Oncogene 16:681–684

Hihnala S, Höglund P, Lammi L, Kokkonen J, Örmälä T, Holmberg C 2006 Long-term clinical outcome in patients with congenital chloride diarrhea. J Pediatr Gastroenterol Nutr, in press

Holmberg C 1986 Congenital chloride diarrhoea. Clin Gastroenterol 15:583–602

Holmberg C, Perheentupa J, Launiala K 1975 Colonic electrolyte transport in health and in congenital chloride diarrhea. J Clin Invest 56:302–310

Holmberg C, Perheentupa J, Launiala K, Hallman N 1977a Congenital chloride diarrhoea. clinical analysis of 21 Finnish patients. Arch Dis Child 52:255–267

Holmberg C, Perheentupa J, Pasternack A 1977b The renal lesion in congenital chloride diarrhea. J Pediatr 91:738–743

Höglund P, Haila S, Socha J et al 1996 Mutations of the down-regulated in adenoma (DRA) gene cause congenital chloride diarrhoea. Nat Genet 14:316–319

Höglund P, Auranen M, Socha J et al 1998a Genetic background of congenital chloride diarrhea in high-incidence populations: Finland, Poland, and Saudi Arabia and Kuwait. Am J Hum Genet 63:760–768

Höglund P, Haila S, Gustavson KH et al 1998b Clustering of private mutations in the congenital chloride diarrhea/down-regulated in adenoma gene. Hum Mutat 11:321–327

Höglund P, Holmberg C, Sherman P, Kere J 2001a Distinct outcomes of chloride diarrhoea in two siblings with identical genetic background of the disease: implications for early diagnosis and treatment. Gut 48:724–727

Höglund P, Sormaala M, Haila S et al 2001b Identification of seven novel mutations including the first two genomic rearrangements in SLC26A3 mutated in congenital chloride diarrhea. Hum Mutat 18:233–242

Höglund P, Hihnala S, Kujala M, Tiitinen A, Dunkel L, Holmberg C 2006 Disruption of the SLC26A3-mediated anion transport is associated with male subfertility. Fertil Steril 85:232–235

Jacob P, Rossmann R, Lamprecht G et al 2002 Down-regulated in adenoma mediates apical Cl-/HCO3- exchange in rabbit, rat, and human duodenum. Gastroenterology 122:709–724

Ko SB, Shcheynikov N, Choi JY et al 2002 A molecular mechanism for aberrant CFTR-dependent HCO(3)(-) transport in cystic fibrosis. EMBO J 21:5662–5672

Ko SB, Zeng W, Dorwart MR et al 2004 Gating of CFTR by the STAS domain of SLC26 transporters. Nat Cell Biol 6:343–350

Lamprecht G, Heil A, Baisch S et al 2002 The down regulated in adenoma (dra) gene product binds to the second PDZ domain of the NHE3 kinase A regulatory protein (E3KARP), potentially linking intestinal Cl-/HCO3- exchange to Na+/H+ exchange. Biochemistry 41:12336–12342

Lohi H, Makela S, Pulkkinen K et al 2002 Upregulation of CFTR expression but not SLC26A3 and SLC9A3 in ulcerative colitis. Am J Physiol Gastrointest Liver Physiol 283:G567–575

Melvin JE, Park K, Richardson L, Schultheis PJ, Shull GE 1999 Mouse down-regulated in adenoma (DRA) is an intestinal Cl(-)/HCO(3)(-) exchanger and is up-regulated in colon of mice lacking the NHE3 Na(+)/H(+) exchanger. J Biol Chem 274:22855–22861

Moseley RH, Höglund P, Wu GD et al 1999 Downregulated in adenoma gene encodes a chloride transporter defective in congenital chloride diarrhea. Am J Physiol 276:G185–192

Mäkelä S, Kere J, Holmberg C, Höglund P 2002 SLC26A3 mutations in congenital chloride diarrhea. Hum Mutat 20:425–438

Nuki G, Watson ML, Williams BC, Simmonds HA, Wallace R 1991 Congenital chloride losing enteropathy associated with tophaceous gouty arthritis. Adv Exp Med Biol 309A:203–208

Sato H 1993 Biology of the eccrine and apocrine sweat gland. In: Fitzpatrick T, Eisen AZ, Wolff K et al (eds) Dermatology in general medicine. McGraw-Hill, New York, p 154–164

Schweinfest CW, Henderson KW, Suster S, Kondoh N, Papas TS 1993 Identification of a colon mucosa gene that is down-regulated in colon adenomas and adenocarcinomas. Proc Natl Acad Sci USA 90:4166–4170

Silberg DG, Wang W, Moseley RH, Traber PG 1995 The Down Regulated in Adenoma (dra) gene encodes an intestine-specific membrane sulfate transport protein. J Biol Chem 270:11897–11902

Wheat VJ, Shumaker H, Burnham C, Shull GE, Yankaskas JR, Soleimani M 2000 CFTR induces the expression of DRA along with Cl(-)/HCO(3)(-) exchange activity in tracheal epithelial cells. Am J Physiol Cell Physiol 279:C62–71

DISCUSSION

Muallem: It doesn't surprise me that there isn't much effect on pancreatic function. We published a paper showing that when 95% of exocrine pancreas is absent the mice show no phenotype (Luo et al 2005). The pancreas has an enormous spare capacity.

Case: It is the same in humans. 85% loss of tissue has no obvious effects on digestion.

Muallem: I have a question relating to the cognitive functions of these patients. There is one patient in Dallas who has some cognitive problems. Have you seen this?

Höglund: Solitary cases with impaired cognitive function are known, but not in Finland. We believe that this is secondary and associated with severe electrolyte disturbances, especially hyponatremic crisis during the early phases of the disease.

Quinton: I was struck with the localization of this protein in the sweat gland. I don't know whether you can do this experiment in *cold* Finland, but is there an abnormality in sweating?

Höglund: Solitary patients show abnormally high concentration of chloride in the sweat test.

Quinton: But your localization is in the secretory coil while there is very little localization in the reabsorptive duct, which would not account for the higher Cl⁻ values in final sweat. I would have anticipated that there would have been a decreased secretory response in these patients.

Höglund: Usually the induced secretion appears totally normal.

Quinton: Did they have any problems with thermal regulation?

Höglund: It is not known whether they have troubles with thermal regulation, however some of the patients may lose salt during excessive sweating, although we don't have any measurements to prove this. The patients are telling us that during sport they feel better if they take extra doses of salt.

Kere: Finland is the best place to do these experiments because people go to saunas frequently! We would hear of complaints of non-sweating and dry skin if there was a problem.

Quinton: Is the protein expressed in the sweat glands of these patients?

Kere: We have not done biopsies. In chloride diarrhoea patients we see normal cDNA in the gut, and the protein should also be there, because it is an amino acid change.

Lee: Have you tried to look at the protein expression of the Finnish mutation? Since it is a single amino acid deletion, it reminds me of the case of the delta F508 in the cystic fibrosis transmembrane regulator protein (CFTR).

Höglund: In cell systems the mouse equivalent has been tested and it doesn't go to the membrane; it is retained in the cytoplasm.

Lee: Is it degraded by the proteasome?

Höglund: This isn't known.

Thomas: We have looked at this as well. Almost all these mutations are trafficking mutations. However, they don't seem to be turned over by the proteasome.

Mount: Is this the A3 mutation in congenital chloride diarrhoea (CLD)?

Thomas: All the mutations we have looked at are those within the STAT domains.

Seidler: A few years ago we looked at various gastrointestinal organs in the rabbit. We found very high expression of DRA in the gall bladder, which seemed to correlate with the very high capacity for resorbing fluid from the lumen. Do the patients exhibit large gall bladders when examined by ultrasonography?

Höglund: By ultrasound, we have found no evidence for enlarged gall bladders among children or adolescents with CLD.

Alper: In the past, I think it was Fordtran who proposed omeprazole as a treatment for cases that couldn't be supported by salt supplementation (Aichbichler et al 1997). Have you had any experience with omeprazole use? Or have you tried to use drugs that are designed to be antidiarrhoeals?

Höglund: The Aichbichler et al (1997) paper was a single report that showed that omeprazole reduced the amount of diarrhoea in an adult patient with CLD. We have heard several reports against its efficacy and believe that salt replacement therapy would be as effective as omeprazole, if the dose were adjusted optimally.

Alper: What about anti-diarrhoeals based on either K^+ channel blockers or anti-CFTR drugs?

Höglund: These haven't been tested.

Alper: Blocking secretion should cause less salt to be lost.

Soleimani: There are some claims in the literature that the symptoms of diarrhoea may diminish with age in CLD. An investigator is currently examining the expression of A6 in the small intestine of patients with CLD. Is there any evidence indicating the up-regulation of A6 in the small intestine of patients with A3 mutation?

Höglund: That's an interesting question but we haven't done any studies on this.

Soleimani: She is doing duodenal biopsy and we are giving her the antibody to do staining.

Seidler: Does the diarrhoea really diminish with age? I have read this also, but the basic defect does not change with age.

Höglund: It is perhaps because of the change in relative volume of the intestine: it's smaller in babies, and increases with age. Thus the total intestinal volume may increase more than general growth in proportion, so with age this loss is less disabling than it is in infancy. The amount of fluid passed doesn't diminish with age.

Mount: One common misconception is that the public single nucleotide polymorphism (SNP) databases are complete for every gene in the genome. For A2, 3 and 4 there should be enough collective experience and chromosome sequence that we would have a pretty good idea of the full spectrum of coding polymorphisms. It seems surprising that you only see four coding SNPs. Is that because you are looking mostly in restricted populations? How global is the racial coverage of the patients that you have screened over the years?

Höglund: We have screened patients primarily by SSCP, and then went along with sequencing. The sequence changes found have been tested in the control population. So no systematic sequencing among control populations has been performed.

Mount: But do you get patients from all over the world?

Höglund: Yes.

Mount: How many chromosomes have you sequenced of diseased patients in the last decade? For how many patients have you screened the entire gene for coding disease mutations?

Höglund: In the earlier phase of the study we screened using single-strand conformation polymorphism (SCCP). When we found the aberration we sequenced that exon only. Of course, there were cases where the SCCP results were normal. In those instances we went to direct sequencing. But I can't give figures of how many patients have been completely sequenced, however; every patient has been screened by SSCP at least.

Mount: How many different disease-associated mutations have been found in CLD?

Höglund: 30.

Chan: Your finding on male infertility from A3 knockout mice is interesting and puzzling. The finding in the seminal vesicle is consistent with sperm physiology since after going through the epididymis and vas deferens, sperm require bicarbonate to stimulate motility. However, I don't understand that you have the defect in A3 knockout epididymis because the epididymis is not supposed to have bicarbonate since the sperm need to be kept quiescent for their storage before ejaculation. This means that either the same transporter would act differently in the seminal vesicle and the epididymis, or its interaction with other proteins would be different at these two sites. It is possible that in the epididymis there is NHE working together with A3 to bring about NaCl and fluid reabsorption and in the meantime the NHE ejects protons that neutralize the luminal bicarbonate secreted by A3. However, in the seminal vesicle there may be no NHE and therefore the primary role of A3 is to secret bicarbonate required for stimulation of sperm motility. You said that in A3 knockout epididymis, there is large fluid accumulation in the lumen. Have you seen the same phenomenon in the seminal vesicle?

Höglund: By ultrasound, there has not been any evidence for swelling of the seminal vesicle, in contrast to the efferent duct of the testis.

Chan: This is consistent with a different role of A3 in the seminal vesicle from that in the epididymis.

Quinton: Does this involve a swelling of the vas deferens?

Höglund: As far as we know, A3 is not expressed in the vas deferens; there is no effect there.

Kere: That is where the CFTR is.

Seidler: I recently took apart the colon of an NHE3 knockout mouse. I was surprised at the enormous hypertrophy of the colonic crypts. Do you see anything similar in the CLD patients? Is there a change in morphology of the intestinal epithelium?

Höglund: Their colonic histology is completely normal.

Welsh: When the CFTR gene was discovered, people made the association with male infertility. Researchers then went to infertility clinics and screened people for mutations in the CFTR gene, and they discovered that CFTR gene variations are associated with infertility. Have you or anyone else done something similar with A3?

Höglund: We haven't, but we have been thinking about this.

Mount: There was a specific genetic interaction: there was a splice polymorphism and patients with this, along with other polymorphisms, would have infertility. The genetics is quite complicated (Cuppens et al 1998).

Welsh: What you are talking about is a polythymidine track in an intron just upstream of a splice acceptor site. The efficiency of splicing varies depending on the number of thymidines. There could be a disease-causing mutation on one chromosome; on the other chromosome you could have variations in the polythymidine track, along with a mild mutation in one haplotype downstream. What is occurring is a titration of the amount of CFTR. And because the vas deferens is quite sensitive to the loss of CFTR, you see disease there. It could be of some interest to do a similar study with A3. There is a lot of unexplained infertility.

Quinton: In cystic fibrosis patients the ejaculate is very acidic, with a pH of around 6. Normally it is about 8.

Höglund: The seminal vesicle has a central role in the neutralization of the acidic seminal plasma.

References

Aichbichler BW, Zerr CH, Santa Ana CA, Porter JL, Fordtran JS 1997 Proton-pump inhibition of gastric chloride secretion in congenital chloridorrhea. N Engl J Med 336:106–109

Cuppens H, Lin W, Jaspers M et al 1998 Polyvariant mutant cystic fibrosis transmembrane conductance regulator genes. The polymorphic (Tg)m locus explains the partial penetrance of the T5 polymorphism as a disease mutation. J Clin Invest 101:487–496

Luo X, Shin DM, Wang X, Konieczny SF, Muallem S 2005 Aberrant localization of intracellular organelles, Ca^{2+} signaling and exocytosis in Mist1 null mice. J Biol Chem 280:12668–12675

Expression, regulation and the role of SLC26 Cl⁻/HCO₃⁻ exchangers in kidney and gastrointestinal tract

Manoocher Soleimani

Department of Medicine, University of Cincinnati and Veterans Affair Medical Center, Cincinnati, OH, USA

Abstract. SLC26 isoforms are members of a large, conserved family of anion exchangers that display highly restricted and distinct tissue distribution. Cloning experiments have identified the existence of 10 SLC26 genes or isoforms (*SLC26A1–11*). The products of all, excepting SLC26A5 (prestin), function as anion exchangers with versatility with respect to transported anions. Modes of transport mediated by SLC26 members include the exchange of chloride for bicarbonate, hydroxyl, sulfate, formate, iodide, or oxalate with variable specificity. Several members of SLC26 family mediate chloride–bicarbonate exchange and display expression in a limited number of tissues including the gastrointestinal tract and/or kidney, with distinct subcellular (apical or basolateral) localization. These include SLC26A3 (DRA), SLC26A4 (pendrin), SLC26A6 (PAT1 or CFEX), SLC26A7 and SLC26A9. SLC26A3 and A9 are not expressed in the kidney but SLC26A4, A6 and A7 are. Genetically engineered null mice have highlighted the important role of two members of the SLC26 family, SLC26A4 and SLC26A6, in homeostatic function in kidney and/or intestine. In conjunction with expression studies, the evolving picture points to important roles for SLC26 family in chloride, bicarbonate, oxalate or sulfate transport and homeostasis in gastrointestinal tract, kidney and several other tissues. This review in particular focuses on the role and regulation of SLC26A6, A7 and A9 in the kidney and/or gastrointestinal tract.

2006 Epithelial anion transport in health and disease: the role of the SLC26 transporters family. Wiley, Chichester (Novartis Foundation Symposium 273) p 91–106

Cloning and characterization of SLC26 family members

Recent studies have identified a highly conserved family of anion exchangers, referred to as SLC26, which include at least 10 distinct genes (Bissig et al 1994, Hastabacka et al 1994, Everett et al 1997, Zheng et al 2000, Lohi et al 2000, 2002, Vincourt et al 2002, 2003). All excepting prestin (SLC26A5) show versatility in mediating multiple anion exchange modes, with several members showing capability to function as Cl⁻/HCO₃⁻ exchanger and displaying a unique expression pattern in kidney and/or gastrointestinal tract. DRA (SLC26A3) can function as a

Cl^-/HCO_3^-, Cl^-/OH^- and sulfate/OH^- exchanger (Melvin et al 1999, Wheat et al 2000, Ko et al 2002, Mount & Romero 2002, Chernova et al 2003). Pendrin can function in Cl^-/HCO_3^-, Cl^-/OH^- and Cl^-/formate exchange modes (Soleimani et al 2001, Soleimani 2001, Royaux et al 2001, Ko et al 2002, Scott & Karniski 2000). SLC26A7 and SLC26A9 mediate Cl^-/HCO_3^- exchange as well as oxalate and sulfate (Lohi et al 2002, Petrovic et al 2003a, 2004, Xu et al 2005b). Modes of transport mediated by SLC26A6 (PAT1 or CFEX) include Cl^-/formate exchange, Cl^-/HCO_3^- exchange and Cl^-/oxalate exchange (Knauf et al 2001, Wang et al 2002, Xie et al 2002, Ko et al 2004, Chernova et al 2005). The Cl^-/HCO_3^- exchange modes of DRA and PAT1 have been shown to be electro-neutral by some investigators but electrogenic by others (Melvin et al 1999, Ko et al 2004, Chernova et al 2003, 2005). The Cl^-/oxalate exchange mediated via PAT1 has been shown to be electrogenic by all investigators (Xie et al 2002, Jiang et al 2002, Chernova et al 2005).

SLC26A6 expression

SLC26A3 (DRA) and SLC26A6 (PAT1) are expressed in the intestine (Schweinfest et al 1993, Melvin et al 1999, Wang et al 2002, 2005, Jacob et al 2002). The expression pattern of PAT1 in the intestine, however, is completely opposite to that of DRA (Wang et al 2002). While DRA is predominantly expressed in the large intestine, the expression of PAT1 is very high in the small intestine and stomach but very low in the large intestine (Melvin et al 1999, Wang et al 2002). Both DRA and PAT1 are expressed on the apical membrane in the intestine and the pancreatic duct (Hoglund et al 1996, Melvin et al 1999, Greeley et al 2001, Wang et al 2002, Jacob et al 2002). In addition to the small intestine and pancreatic duct, PAT1 is also expressed on the apical membrane of kidney proximal tubule cells and in gastric parietal cells whereas DRA shows no expression in the kidney (Knauf et al 2001, Petrovic et al 2002, 2003b).

SLC26A6 null mice

To study its role in kidney and intestine physiology, gene targeting was used to prepare mice lacking *Slc26a6* (Wang et al 2005). Homozygous mutant *Slc26a6$^{-/-}$* mice appeared healthy and exhibited normal blood pressure, kidney function and plasma electrolyte profile (Wang et al 2005). In proximal tubules microperfused with a low bicarbonate-high chloride solution, the baseline rate of fluid absorption (J_v), an index of NaCl transport under these conditions, was the same in wild-type and null mice (Wang et al 2005). However, the stimulation of J_v by oxalate observed in wild-type mice was completely abolished in *Slc26a6* null mice (Wang et al 2005). Apical membrane chloride-base exchange activity, assayed with the pH-sensitive dye

BCPCF in microperfused proximal tubules, was decreased by ~60% in *Slc26a6*$^{-/-}$ animals (Wang et al 2005). In the duodenum, the baseline rate of bicarbonate secretion, measured in mucosal tissue mounted in Ussing chambers, was decreased by ~30%, whereas the forskolin-stimulated component of bicarbonate secretion was the same in wild-type and *Slc26a6*$^{-/-}$ mice (Wang et al 2005). It was concluded that SLC26A6 mediates oxalate-stimulated NaCl absorption and also contributes to apical membrane Cl⁻ base exchange in the kidney proximal tubule; it also plays an important role in bicarbonate secretion in the duodenum (Wang et al 2005). Recent studies demonstrate that PGE2-stimulated bicarbonate secretion in the duodenum is predominantly mediated via PAT1 (Seidler, Soleimani, personal observation). The schematic diagram in Fig. 1 depicts the localization of SLC26A6 (PAT1) in the villi of the duodenum and its role as the main apical Cl⁻/HCO₃⁻ exchanger and mediator of chloride absorption and bicarbonate secretion in the small intestine.

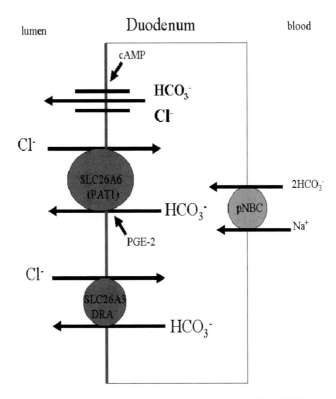

FIG. 1.   Schematic diagram depicting the localization of SLC26A6 (PAT1) in the villi of the duodenum. PAT1 is the major PGE2-activated apical anion exchanger, with DRA being a minor contributor to chloride absorption and bicarbonate secretion in the duodenum.

SLC26A7 expression

Cloning, functional expression and localization of SLC26A7 in the stomach

A recently cloned member of the SLC26A family is SLC26A7, which has been shown to be expressed in kidney and testis (Lohi et al 2002). More recent studies indicate that SLC26A7 is highly expressed in the stomach (Petrovic et al 2003a), with the stomach expression predominantly limited to parietal cells. Functional studies demonstrated that SLC26A7 mediates Cl^-/HCO_3^- exchange (Petrovic et al 2003a, 2004) and immunofluorescent studies localized A7 to the basolateral membrane of gastric parietal cells (Petrovic et al 2003a). Due to its unique expression in the stomach and its functional mode, it was postulated that SLC26A7 plays an important role in gastric acid secretion by exchanging the intracellular HCO_3^- for the extracellular Cl^- across the basolateral membrane of parietal cells, therefore providing H^+ for secretion via the apical gastric H-K-ATPase (Petrovic et al 2003a).

Basolateral Cl^-/HCO_3^- exchangers in acid secreting gastric parietal cells

The acid (HCl)-secreting parietal cells of the human, rat, rabbit and mouse stomach have been shown to express a basolateral Cl^-/HCO_3^- exchanger. This exchanger is essential for acid secretion by transporting the intracellular HCO_3^- into the peritubular space, therefore generating intracellular H^+ for secretion across the apical membrane via H-K-ATPase (Seidler et al 1989, Thomas & Machen 1991). Based on the localization of AE2 to the basolateral membrane of gastric parietal cells (Jons 1994 et al, Cox et al 1996) it was concluded that this anion exchanger is the main basolateral Cl^-/HCO_3^- exchanger in gastric parietal cells.

However, a more detailed analysis of the functional properties of the basolateral Cl^-/HCO_3^- exchanger in parietal cells shows significant differences vs. the known functional properties of AE2. For example, the basolateral Cl^-/HCO_3^- exchanger in gastric parietal cells is active at acidic pH_i and helps with intracellular pH recovery from acidosis, a property not demonstrated by AE2 (Seidler et al 1992, 1994, Paradiso et al 1989). Further, the basolateral Cl^-/HCO_3^- exchanger in parietal cells requires an inhibitory concentration of disulfonic stilbene DIDS that is more than ~10–100-fold higher than the concentration needed to inhibit AE2 (Seidler et al 1992, 1994). Taken together, these observations suggest that basolateral Cl^-/HCO_3^- exchanger(s) distinct from AE2 may play a major role in basolateral HCO_3^- extrusion in exchange for Cl^- entry into the gastric parietal cells. Given the localization of SLC26A7 on the basolateral membrane of the parietal cells and its function as a Cl^-/HCO_3^- exchanger we suggest that SLC26A7 plays an important role in bicarbonate transport across the basolateral membrane and acid secretion across the apical membrane (Petrovic et al 2003a). A recent study demonstrated that the deletion of AE2 resulted in achlorhydria (Gawenis et al 2004). One however should be cautious with linking the achlorhydria directly to AE2 deletion since AE2 null mice also showed more than 80% reduction in the expression of SLC26A7 (Gawenis et al 2004), the other basolateral Cl^-/HCO_3^- exchanger in parietal cells (Petrovic et al 2003a). The schematic diagram in Fig. 2 depicts the localization of AE2 and SLC26A7 on the basolateral membrane of gastric parietal cells and their contribution to basolateral bicarbonate exit and luminal acid secretion.

Expression and subcellular localization of SLC26A7 in the kidney

Northern hybridizations demonstrated the predominant expression of SLC26A7 in the outer medulla, with lower levels in the inner medulla and faint expression in the cortex (Petrovic et al 2004, Barone et al 2004). Western blotting confirmed the results of Northern hybridization (Petrovic et al 2004). Immunocytochemical staining demonstrated the localization of SLC26A7 on the basolateral membrane of a subset of outer medullary collecting duct cells (Petrovic et al 2004, Barone et al 2004). No labelling was detected in the cortex. Double immunofluorescence labelling with the AQP2 and SLC26A7 antibodies identified the SLC26A7-expressing cells as acid secreting α intercalated cells (Petrovic et al 2004, Barone et al 2004). Immunofluorescent labelling demonstrated the co-localization of AE1 and SLC26A7 on the basolateral membrane of α intercalated cells (Petrovic et al 2004, Barone et al 2004). Functional studies in oocytes demonstrated that increasing the osmolarity of the media (to simulate the physiologic *milieu* in the medulla) increased the Cl^-/HCO_3^- exchanger activity mediated via SLC26A7 (Petrovic et al 2004). It was postulated that SLC26A7 is a basolateral Cl^-/HCO_3^- exchanger in

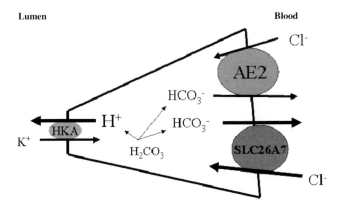

FIG. 2. Schematic diagram depicting the localization of AE2 and SLC26A7 on the basolateral membrane in gastric parietal cells and their contribution to bicarbonate exit and luminal acid secretion.

intercalated cells of the outer medullary collecting duct (OMCD) and may play an important role in bicarbonate reabsorption in the medullary collecting duct (Petrovic 2004 et al, Barone et al 2004).

Basolateral Cl^-/HCO_3^- exchangers in acid secreting α intercalated cells in the collecting duct

Functional studies in microperfused kidney collecting duct have identified the presence of a basolateral Cl^-/HCO_3^- exchanger in α intercalated cells of the cortical and outer medullary collecting tubules (Alpern et al 1991, Schuster 1991). Using antibodies specific for the membrane-spanning domain of the human erythrocyte band 3, Wagner et al (1987) have demonstrated that a highly homologous exchanger is expressed on the basolateral membranes of the collecting duct of human kidney. Immunohistochemical studies have localized AE-1 to the basolateral membrane of the α intercalated cells in cortical collecting duct (CCD) and OMCD (Alper et al 1989). In support of a major role for AE1 as the main basolateral Cl^-/HCO_3^- exchanger in α intercalated cells in CCD, studies in human indicate that mutation in AE1 results in classical distal renal tubular acidosis, suggesting a major role for AE1 in bicarbonate reabsorption in the CCD (Karet et al 1998, Tanphaichitr et al 1998). However, deletion of AE1 in mouse does not completely abrogate the basolateral Cl^-/HCO_3^- exchanger activity in α intercalated cells in OMCD (Wagner et al 2004), strongly supporting the existence of another Cl^-/HCO_3^- exchanger on the basolateral membrane of these cells. Given its basolateral localization in α intercalated cells and its function as a Cl^-/HCO_3^- exchanger we suggest that SLC26A7 is a major basolateral Cl^-/HCO_3^- exchanger in α intercalated cells and plays an important role in bicarbonate absorption in OMCD (Petrovic et al 2004, Barone et al 2004). The schematic diagram in Fig. 3 depicts the localization of AE1 and SLC26A7 on the basolateral membrane of α intercalated cells and their role in bicarbonate absorption in OMCD.

Differential regulation of SLC26A7 and AE1 in the kidney

Two recent studies have examined the regulation of SLC26A7 and AE1 in the kidney in pathophysiological states. In the first series of studies, rats were subjected to water deprivation, which is known to increase medullary interstitial osmolarity, and then examined. The results demonstrated that the expression of SLC26A7 increased whereas the expression of AE1 decreased in water deprivation, indicating differential regulation of these two anion exchangers in water deprivation (Barone et al 2004). In the second series of studies, Brattleboro rats which lack endogenous vasopressin (AVP) were treated with the synthetic AVP analogue (dDAVP) for 8 days and examined. The results demonstrated that AVP treatment, which increases

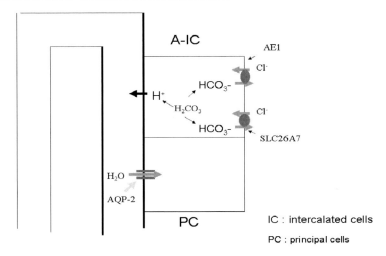

FIG. 3.   Schematic diagram depicting the localization of AE1 and SLC26A7 on the basolateral membrane of α intercalated cells and their role in bicarbonate absorption in OMCD.

the interstitial osmolarity and normalizes the urine out, osmolarity and AQP2 expression, heavily up-regulated the expression of SLC26A7 with no significant effect on AE1 expression (Petrovic et al 2005). Taken together, these results demonstrate differential regulation of SLC26A7 and AE1 in OMCD in pathophysiological states.

SLC26A7 trafficking in cultured kidney cells

Immunocytochemical staining demonstrated that in addition to the membrane, SLC26A7 shows significant intracellular, sub-membrane localization in both kidney and stomach cells (Petrovic et al 2003a, 2004, Barone et al 2004), raising the possibility of additional functional regulation by altering its abundance at the cell surface by changes in traffic between cell membrane and intracellular compartments. To analyse this possibility, we cloned full length human SLC26A7 cDNA, which was then fused in frame with N-terminally-tagged GFP, transiently expressed in Madin-Darby canine kidney (MDCK) epithelial cells and visualized by confocal microscopy (Xu et al 2006). In monolayer cells grown to polarity in isotonic media and at pH 7.4, SLC26A7 showed punctate distribution through the cell cytoplasm, with no expression on the membrane. However, in media made hypertonic (50 mM NaCl added to the isotonic media, pH 7.4) for 16 h, GFP-SLC26A7 was detected exclusively in the plasma membrane (Xu et al 2005b). The presence of MAP kinase inhibitors completely blocked the trafficking of SLC26A7 to the plasma membrane. Double labelling with transferrin receptor in isotonic and hypertonic media demonstrated that SLC26A7 is targeted to the transferrin receptor positive recycling endosomes (Xu et al 2005b). Deletion of the last 16 amino acids on the C-terminal end prevented the targeting of SLC26A7 to the recycling endosomes in isotonic media and to the membrane in hypertonic media. Further, while SLC26A1 (SAT1) is targeted to the membrane in isotonic and hypertonic media, SLC26A7/SLC26A1 chimera encoding the N-terminal fragment of A7 and the C-terminal tail of A1 failed to localize to the recycling endosomes in isotonic media and to the membrane in hypertonic media, confirming that the carboxyl terminal end of SLC26A7 is responsible for its targeting to the endosomes and trafficking to the membrane.

In separate studies, SLC26A7 showed predominant localization in plasma membrane in potassium depleted isotonic media (0.5 mM/L KCl in isotonic media, pH 7.4) vs. cytoplasmic distribution in normal potassium isotonic media (4 mEq/L in isotonic media, pH 7.4). Taken together, these results demonstrate that SLC26A7 is present in recycling endosomes and its targeting to the basolateral membrane is increased in hypertonicity and potassium depletion. The trafficking to the cell surface demonstrates novel functional regulation of SLC26A7 in states associated with increased medullary tonicity or hypokalemia, and serves to maintain or enhance bicarbonate absorption across the basolateral membrane of OMCD cells, thereby regulating or dysregulating, respectively, the acid base balance.

SLC26A9 expression

Cloning, functional expression and localization of SLC26A9 in the stomach

Recently, SLC26A9 was cloned based on homology to other SLC26 transporters (Lohi et al 2002). SLC26A9 was detected in the lung and facilitated the uptake of chloride, sulfate and oxalate in oocytes (Lohi et al 2002). A very recent study demonstrated that A9 is abundantly expressed in the stomach (Xu et al 2005b). *In situ* hybridization and immunofluorescence labelling studies demonstrated that SLC26A9 is expressed on the apical membrane of gastric surface epithelial cells and cells of the gastric gland which seemed to be parietal cells (Xu et al 2005b). Functional studies in cultured cells demonstrated that SLC26A9 mediates Cl^-/HCO_3^- exchange and chloride-independent bicarbonate transport (Xu et al 2005b).

Apical Cl^-/HCO_3^- exchange, Cl^--independent HCO_3^- transport and HCO_3^- secretion in gastric surface epithelial cells

The acidic solution that is secreted from the gastric parietal cells can achieve a pH as low as 1.5 (Feldman 1998). To protect itself against the corrosive effects of luminal acid, the gastric surface epithelium secretes a HCO_3^--rich fluid into the mucus gel layer (Phillipson et al 2002, Engel et al 1995). HCO_3^- secretion by surface epithelial cells was reported to be dependent on luminal Cl^-, suggesting that Cl^-/HCO_3^- exchange was involved (Feldman 1998). These results have been disputed (Takeuchi et al 1982), however, and there is evidence for a HCO_3^- conductance pathway (Curci et al 1994). Both the Cl^-/HCO_3^- exchange and HCO_3^- conductance pathway remain attractive and relevant possibilities as major bicarbonate secretory pathways in neutralizing the H^+ diffusion into the mucus layer. In particular, Cl^-/HCO_3^- exchange remains an attractive possibility because the diffusion of H^+ and Cl^- into the mucus layer would both neutralize outwardly transported HCO_3^- and provide Cl^- for inward transport, thereby maintaining a strong driving force for Cl^-/HCO_3^- exchange. This hypothesis has received support by the recent finding of a HCO_3^- secretory mechanism in rat gastric mucosa that is sensitive to DIDS (Phillipson et al 2002), an inhibitor of some of the electroneutral Cl^-/HCO_3^- exchangers and related transporters. Recent studies from our laboratories identified AE4 as a Cl^-/HCO_3^- exchanger expressed on the apical membrane of gastric surface epithelial cells and villus cells in the duodenum (Xu et al 2003). AE4, however, is significantly resistant to inhibition by DIDS (Tsuganezawa et al 2001), whereas acid-stimulated bicarbonate secretion in gastric surface epithelial cells is significantly inhibited by DIDS (Phillipson et al 2002). This suggests that bicarbonate transporter(s) other than AE4 play a major role in bicarbonate secretion in the stomach. Given the apical localization and functional studies, as well as inhibition by ammonium (Xu et al 2005b) we suggest that SLC26A9 plays an important role in bicarbonate secretion by gastric surface epithelial cells.

In conclusion, SLC26 Cl^-/HCO_3^- exchangers play important roles in the normal physiology and homeostatic functions of kidney and gastrointestinal tract. Coupled with expression studies, the evolving picture points to important roles for the SLC26 family in chloride, bicarbonate, oxalate or sulfate transport and homeostasis in gastrointestinal tract, kidney and several other tissues.

Acknowledgements

These studies were supported by the National Institute of Health Grant DK 62809, a Merit Review Grant, a Cystic Fibrosis Foundation grant, and grants from Dialysis Clinic Incorporated (to MS).

References

Alper SL, Natale J, Gluck S, Lodish HF, Brown D 1989 Subtypes of intercalated cells in rat kidney collecting duct defined by antibodies against erythroid band 3 and renal vacuolar H^+-ATPase. Proc Natl Acad Sci USA 86:5429–5433

Alpern RJ, Stone DK, Rector FC 1991 Jr Renal acidification mechanisms. In: Brenner BM, Rector FC, Saunders WB Jr (eds) The Kidney. WB Saunders Co, Philadelphia p 318–378

Barone S, Amlal H, Xu J et al 2004 Differential regulation of basolateral Cl−/HCO3− exchangers SLC26A7 and AE1 in kidney outer medullary collecting duct. J Am Soc Nephrol 15:2002–2011

Bissig M, Hagenbuch B, Stieger B, Koller T, Meier PJ 1994 Functional expression cloning of the canalicular sulfate transport system of rat hepatocytes. J Biol Chem 269:3017–3021

Chernova MN, Jiang L, Shmukler BE et al 2003 Acute regulation of the SLC26A3 congenital chloride diarrhoea anion exchanger (DRA) expressed in Xenopus oocytes. J Physiol 549:3–19

Chernova MN, Jiang L, Friedman DJ 2005 Functional comparison of mouse slc26a6 anion exchanger with human SLC26A6 polypeptide variants: differences in anion selectivity, regulation, and electrogenicity. J Biol Chem 280:8564–8580

Cox KH, Adair-Kirk TL, Cox JV 1996 Variant AE2 anion exchanger transcripts accumulate in multiple cell types in the chicken gastric epithelium. J Biol Chem 271:8895–8902

Coyle B, Reardon W, Herbrick J-A et al 1998 Molecular analysis of the PDS gene in Pendred syndrome (sensorineural hearing loss and goitre). Human Mol Gen 7:1105–1112

Curci S, Debellis L, Caroppo R, Fromter E 1994 Model of bicarbonate secretion by resting frog stomach fundus mucosa. I. Transepithelial measurements. Pfluegers Arch 428:648–654

Engel E, Guth PH, Nishizaki Y, Kaunitz JD 1995 Barrier function of the gastric mucus gel. Am J Physiol 269:G994–999

Everett LA, Glaser B, Beck JC et al 1997 Pendred syndrome is caused by mutations in a putative sulphate transporter gene (PDS). Nat Genet 17:411–422

Feldman M 1998 Gastric secretion: normal and abnormal. In: Feldman MD, Scharschmidt BF, Sleisenger MH, Klein SWB (eds) Gastrointestinal and liver disease: pathophysiology/diagnosis/management (2nd ed), Saunders, Philadelphia p 587–603

Gawenis LR, Ledoussal C, Judd LM et al 2004 Mice with a targeted disruption of the AE2 Cl⁻/HCO₃⁻ exchanger are achlorhydric. J Biol Chem 279:30531–30539

Greeley T, Shumaker H, Wang Z, Schweinfest CW, Soleimani M 2001 Downregulated in adenoma and putative anion transporter are regulated by CFTR in cultured pancreatic duct cell. Am J Physiol 281:G1301–1308

Hastabacka J, de la Chapelle A, Mahtani MM et al 1994 The diastrophic dysplasia gene encodes a novel sulfate transporter: positional cloning by fine-structure linkage disequilibrium mapping. Cell 78:1073–1087

Höglund P, Haila S, Socha J et al 1996 Mutations of the down-regulated in adenoma (DRA) gene cause congenital chloride diarrhea. Nat Genet 14:316–319

Jacob P, Rossmann H, Lamprecht G et al 2002 Down-regulated in adenoma mediates apical Cl−/HCO3− exchange in rabbit, rat, and human duodenum. Gastroenterology 122:709–724

Jiang Z, Grichtchenko II, Boron WF, Aronson PS 2002 Specificity of anion exchange mediated by mouse Slc26a6. J Biol Chem 277:33963–33967

Jons T, Warrings B, Jons A, Drenckhahn D 1994 Basolateral localization of anion exchanger 2 (AE2) and actin in acid-secreting (parietal) cells of the human stomach. Histochemistry 102:255–263

Karet FE, Gainza FJ, Gyory AZ et al 1998 Mutations in the chloride-bicarbonate exchanger gene AE1 cause autosomal dominant but not autosomal recessive distal renal tubular acidosis. Proc Natl Acad Sci USA 95:6337–6342

Knauf F, Yang CL, Thomson RB, Mentone SA, Giebisch G, Aronson PS 2001 Identification of a chloride-formate exchanger expressed on the brush border membrane of renal proximal tubule cells. Proc Natl Acad Sci USA 98:9425–9430

Ko SB, Shcheynikov N, Choi JY et al 2002 A molecular mechanism for aberrant CFTR-dependent HCO(3)(−) transport in cystic fibrosis. EMBO J 21:5662–5672

Ko SB, Zeng W, Dorwart MR et al 2004 Gating of CFTR by the STAS domain of SLC26 transporters. Nat Cell Biol 6:343–350

Lohi H, Kujala M, Kerkela E, Saarialho-Kere U, Kestila M, Kere J 2000 Mapping of five new putative anion transporter genes in human and characterization of SLC26A6, a candidate gene for pancreatic anion exchanger. Genomics 70:102–112

Lohi H, Kujala M, Makela S et al 2002 Functional characterization of three novel tissue-specific anion exchangers SLC26A7, -A8, and -A9. J Biol Chem 277:14246–14254

Melvin JE, Park K, Richardson L, Schultheis PJ, Shull GE 1999 Mouse down-regulated in adenoma (DRA) is an intestinal Cl^-/HCO_3^- exchanger and is upregulated in colon of mice lacking the NHE-3 Na^+/H^+ exchanger. J Biol Chem 274:22855–22861

Mount DB, Romero MF 2004 The SLC26 gene family of multifunctional anion exchangers. Pflugers Arch 447:710–721

Paradiso AM, Townsley MC, Wenzl E, Machen TE 1989 Regulation of intracellular pH in resting and in stimulated parietal cells. Am J Physiol 257:C554–561

Petrovic S, Wang Z, Ma L et al 2002 Colocalization of the apical Cl^-/HCO_3^- exchanger PAT1 and gastric H-K-ATPase in stomach parietal cells. Am J Physiol Gastrointest Liver Physiol 283:G1207–1216

Petrovic S, Ju X, Barone S et al 2003a Identification of a basolateral Cl^-/HCO_3^- exchanger specific to gastric parietal cells. Am J Physiol Gastrointest Liver Physiol 284:G1093–1103

Petrovic S, Ma L, Wang Z, Soleimani M 2003b Identification of an apical $Cl-/HCO3-$ exchanger in rat kidney proximal tubule. Am J Physiol Cell Physiol 285:C608–617

Petrovic S, Barone S, Xu J et al 2004 SLC26A7: a basolateral Cl^-/HCO_3^- exchanger specific to intercalated cells of the outer medullary collecting duct. Am J Physiol Renal Physiol 286:F161–169

Petrovic S, Amlal H, Ma L et al 2005 Vasopressin induces the expression of the Cl^-/HCO_3^- exchanger SLC26A7 in the kidney medullary collecting duct of Brattleboro rats. Am J Physiol Renal Physiol Dec 13 [Epub ahead of print]

Phillipson M, Atuma C, Henriksnas J, Holm L 2002 The importance of mucus layers and bicarbonate transport in preservation of gastric juxtamucosal pH. Am J Physiol Gastrointest Liver Physiol 282:G211–219

Royaux IE, Wall SM, Karniski LP et al 2001 Pendrin, encoded by the Pendred syndrome gene, resides in the apical region of renal intercalated cells and mediates bicarbonate secretion. Proc Natl Acad Sci USA 98:4221–4226

Schuster VL 1991 Cortical collecting duct bicarbonate secretion. Kid Int (suppl) 33:S47–50

Schweinfest CW, Henderson KW, Suster S, Kondoh N, Papas TS 1993 Identification of a colon mucosa gene that is down-regulated in colon adenomas and adenocarcinomas. Proc Natl Acad Sci USA 90:4166–4170

Scott DA, Karniski LP 2000 Human pendrin expressed in Xenopus laevis oocytes mediates chloride/formate exchange. Am J Physiol Cell Physiol 278:C207–211

Seidler U, Carter K, Ito S, Silen W 1989 Effect of CO2 on pHi in rabbit parietal, chief, and surface cells. Am J Physiol 256(3 Pt 1):G466–475

Seidler U, Roithmaier S, Classen M, Silen W 1992 Influence of acid secretory state on Cl(−)-base and Na(+)-H+ exchange and pHi in isolated rabbit parietal cells. Am J Physiol 262(1 Pt 1):G81–91

Seidler U, Hubner M, Roithmaier S, Classen M 1994 pHi and HCO3− dependence of proton extrusion and Cl(−)-base exchange rates in isolated rabbit parietal cells. Am J Physiol 266(5 Pt 1):G759–776

Soleimani M 2001 Molecular physiology of the renal chloride-formate exchanger. Curr Opin Nephrol Hypertens 10:677–683

Soleimani M, Greeley T, Petrovic S et al 2001 Pendrin: an apical Cl$^-$/OH$^-$/HCO$_3^-$ exchanger in the kidney cortex. Am J Physiol 280:F356–364

Takeuchi K, Merhav A, Silen W 1982 Mechanism of luminal alkalinization by bullfrog fundic mucosa. Am J Physiol 243:G377–388

Tanphaichitr VS, Sumboonnanonda A, Ideguchi H et al 1998 Novel AE1 mutations in recessive distal renal tubular acidosis. Loss-of-function is rescued by glycophorin A. J Clin Invest 102:2173–2179

Thomas HA, Machen TE 1991 Regulation of Cl/HCO3 exchange in gastric parietal cells. Cell Regul 2:727–737

Tsuganezawa H, Kobayashi K, Iyori M et al 2001 A new member of the HCO$_3$-transporter superfamily is an apical anion exchanger of beta-intercalated cells in the kidney. J Biol Chem 276:8180–8189

Vincourt JB, Jullien D, Kossida S, Amalric F, Girard JP 2002 Molecular cloning of SLC26A7, a novel member of the SLC26 sulfate/anion transporter family, from high endothelial venules and kidney. Genomics 79:249–256

Vincourt JB, Jullien D, Amalric F, Girard JP 2003 Molecular and functional characterization of SLC26A11, a sodium-independent sulfate transporter from high endothelial venules. FASEB J 17:890–892

Wagner S, Vogel R, Lietzke R, Koob R, Drenckhahn D 1987 Immunochemical characterization of a band 3-like anion exchanger in collecting duct of human kidney. Am J Physiol 253: F213–221

Wagner S, Stehberger P, Stuart-Tilley AK, Shumkler BS, Alper SL, Wagner CA 2004 Impaired distal renal acidification in a mouse model for distal renal tubular acidosis lacking the AE1 Cl/HCO3− exchanger (Slc4a1). J Am Soc Nephrol 15:70A

Wang Z, Petrovic S, Mann E, Soleimani M 2002 Identification of an apical Cl(−)/HCO3(−) exchanger in the small intestine. Am J Physiol Gastrointest Liver Physiol 282:G573–579

Wang Z, Wang T, Petrovic S et al 2005 Kidney and intestine transport defects in Slc26a6 null mice. Am J Physiol Cell Physiol 288:C957–965

Wheat VJ, Shumaker H, Burnham C, Shull GE, Yankaskas JR, Soleimani M 2000 Cystic fibrosis transmembrane conductance regulator induces the expression of DRA (Down-Regulated in Adenoma) along with Cl$^-$/HCO$_3^-$ exchange activity in tracheal epithelial cells. Am J Physiol: Cell 279:62–70

Xie Q, Welch R, Mercado A, Romero MF, Mount DB 2002 Molecular characterization of the murine Slc26a6 anion exchanger: functional comparison with Slc26a1. Am J Physiol Renal Physiol 283:F826–838

Xu J, Barone S, Petrovic S et al 2003 Identification of an apical Cl−/HCO3− exchanger in gastric surface mucous and duodenal villus cells. Am J Physiol 285:G1225–1234

Xu J, Worrell RT, Li HC, Petrovic S, Barone SL, Soleimani M 2006 The Cl$^-$/HCO$_3^-$ exchanger SLC26A7 is localized in recycling endosomes in kidney medullary collecting cells and is targeted to the basolateral membrane in hypertonicity and potassium depletion. J Am Soc Neph, in press

Xu J, Henriksnas J, Barone S et al 2005b SLC26A9 is expressed in gastric surface epithelial cells, mediates Cl$^-$/HCO$_3^-$ exchange and is inhibited by NH$_4^+$ Am J Physiol: Cell Physiol 289: C493–505

Zheng J, Shen W, He DZ, Long KB, Madison LD, Dallos P 2000 Prestin is the motor protein of cochlear outer hair cells. Nature 405:149–155

DISCUSSION

Muallem: We published recently a paper on the properties of A7 (Kim et al 2005). It is a pure Cl⁻ channel as far as we can see. The reason I mention this is that it raises the question of what will a Cl⁻ channel be doing in the basolateral membrane of the parietal cells? I have a question for Ursula Seidler who did this work: isn't prostaglandin E2 (PGE2) supposed to work through cAMP?

Seidler: It is a complicated story in the duodenum. The acid-activated bicarbonate secretory response *in vivo* is mediated by an axonal reflex, so it can only be studied *in vivo*. This acid-activation of bicarbonate secretion is gone in the PGE2/EP3 receptor knockout mouse. It is the EP3 receptor that would be linked to inhibition of cAMP generation. In the *in vitro* situation, in which we have studied these epithelia, we put TTX on and indomethacin, and the response of PGE2 on bicarbonate secretion is probably mediated by the EP4 receptor subtype, which is located in the surface epithelium. The EP4 receptor has a lower affinity to PGE2 than the EP3 receptor, and it is linked to the generation of cAMP in the cell. This may not be its only second messenger system. When we look at all the data there is some debate about whether PGE2 has any additional signal transduction pathways. We would think from the accumulated data in the literature that this is the case. The problem in the duodenum is that there are so many different PGE2 receptors, and there is a difference in the endothelial layer, so when you accumulate the data you can't be sure that you are seeing a pure cAMP response. Interestingly, the PGE2-mediated short circuit current response is completely gone in the CFTR knockout mouse. This suggests that PGE2 opens the CFTR conductance to chloride. PGE2 can stimulate a bicarbonate secretory response, which looks electroneutral, but this is in intact tissue so we can't be sure about the second messenger systems that are involved. There are two very strange observations here. First, why can PGE2 open the chloride conductance of CFTR, but not the bicarbonate conductance? This could be related to a very rapid rise and decline in cAMP, which has been shown to happen with PGE2 and not forskolin. Second, why can forskolin not stimulate this chloride-dependent electroneutral PAT1-mediated, bicarbonate secretory response? These data were unexpected and they are still puzzling.

Muallem: You are using two modalities to increase cAMP, one of which (stimulation by PGE2) is completely abolished by A6 (as one would have expected) and the other which is forskolin. I would tend to believe the results with PGE2 because this is a hormone, and forskolin is a drug. I am aware that PGE2 does induce $[Ca^{2+}]_i$ increase but not in the duodenum and not the type of receptors that work in the intestine. Did you measure short-circuit current stimulated by PGE2 in those A6 knockout mice?

Seidler: The stimulation of short-circuit current by PGE2 was the same in the knockout; only bicarbonate secretion was inhibited.

Soleimani: This would be explainable if A6 were working as an electroneutral chloride–bicarbonate exchanger.

Muallem: It is absolutely not.

Soleimani: There are two conclusions from these studies. One is that activation of A6 is independent of CFTR. The other is that it is most likely electroneutral.

Alper: In terms of independence, we have data that A6 can be activated by cAMP in the absence of exogenous or heterologous CFTR.

Welsh: How is this thing controlled? I would like to understand better the regulatory processes.

Alper: The important thing to remember is that although AE2 is inhibited by acid pH inside or outside the cell, in the presence of the ambient ammonia concentrations found in gut and renal medulla we might expect not to see acid inhibition. Also, in the context of the hypotonicity of the medulla, that portion of AE2 which is presumably active in principle cells will still be active. It is another issue in the context of AE1. There the regulation is lacking. The other regulation you are showing with A7 is likely to be very interesting and important. The other thing about AE2 in the intestine and stomach is that there are isoforms and variants that we are characterizing that have shifted pH dependences. Expressed in ensemble, these will very much extend the range over which pH regulates chloride–bicarbonate exchange.

Soleimani: In the parietal cells, based on our experiments, A7 and AE2 are present; the question is, what do they do? If you look at the developmental regulation of AE2, from day 1 after birth to the adulthood it is abundantly present. A7, on the other hand, shows no detectable expression at birth and on day 3 after birth, but it shows detectable expression on day 17 after birth, which is around weaning time, which corresponds to the initiation of secretion of the gastric acid. An AE2 knockout mouse has been made and A7 knockout mouse is in the process of being generated. The question for us is whether AE2 is more important than A7 or vice versa when it comes to gastric acid secretion. I'd like to believe that A7 is more important for acid secretion. AE2 knockouts are very small—they display developmental arrest—and it seems that the same thing happens with their parietal cells, which are immature. Perhaps AE2 plays more of a housekeeping role, and is important developmentally. A7, on the other hand, may be more important with regard to acid secretion in response to stomach stimuli.

Alper: The components of acid secretion are present in those parietal cells. The number of parietal cells is just about normal. The cytological immaturity is a soft call. The achlorhydria is not a soft call.

Soleimani: If you look at the stomach of AE2 knockout, you will see that A7 is significantly down-regulated, by 60–70%. This raises the question of the role of AE2 and A7 in decreased acid secretion by gastric parietal cells. What is the cause of decreased gastric acid secretion in the AE2 knockout? How much of that is due to AE2 deficiency and how much is due to the down-regulation of A7? Why is A7 down-regulated?

Alper: With respect to acid activation, as you define it by recovery from the CO_2 pulse, it turns out that AE1, EAE1, KAE1 and AE2 do exactly the same things. Your recoveries were done with elevated, out-of-equilibrium bicarbonate. When we do this in equilibrated CO_2 bicarbonate, A7, A1 and A2 have fairly comparable rates of recovery. When we use your conditions of 33–35 mM bicarbonate, indeed we see accelerated rates of recovery. It is not a unique property of SLC26.

Muallem: Have you ever measured Cl^-/HCO_3^- exchange with A7 using microelectrode as we did and recorded exactly how the oocytes respond to it?

Alper: In my paper I will show data on chloride–bicarbonate exchange measured without microelectrodes, and I'll show data on currents and fluxes.

Welsh: Is all regulation controlled solely by insertion and retrieval? Is the important variable simply the amount of transporter in the membrane? Or is there phosphorylation of this transporter? Is it modified in some other way while it is in the membrane?

Soleimani: We haven't looked at the regulation of the A7 transporter in detail. These ideas all need to be tested. We have been trying to find the membrane targeting domain of A7, and we have made A7/A3 chimeras as well as A7/A1 chimeras. We are also trying to find out the domain or motif that senses the osmolarity. Our data suggest that it is likely to be located in the C-terminal end.

Romero: I think the regulation by osmolarity is intriguing. When you stimulate the duodenal cells and look at A7, in the short-circuit chronic measurements, in your set-up can you dissect out the apical or basolateral membrane conductances? The model that you are proposing (assuming that under some conditions it is a chloride–bicarbonate exchanger and under other conditions it acts perhaps as an anion channel) predicts that in addition to bicarbonate secretion you would also see a specific change in the basolateral membrane conductance, if in fact this is being trafficked to the basolateral membrane. This could account for some of the effects. With a number of these members, I am wrestling with whether these functions coexist at the same time, or is there some modality of regulation? For example, with A6, sometimes it does chloride–formate exchange but it can also exchange chloride for oxalate or bicarbonate. Many of the others have similar switching of characteristics. Is the regulation through second messengers or is it cell type specific? There are two questions: is there a change in the membrane conductance of these basolateral membranes that A7 has been trafficked to? And do you have any handle on whether there are switches in modalities that happen physiologically, either in transfected cells or in an intact system?

Seidler: The presented functional data on A7 were not performed in intact tissues, like the stomach. For the stomach, it would be very difficult to look at basolateral membrane conductance. What we do in intact gastric mucosa in Ussing-chambers is look at transepithelial potential measurements. In the stomach this is unbelievably complex and no one has yet figured out the changes in potential difference and what this relates to. To look at trafficking of A7, it would be possible to isolate parietal cells and take them into culture or use them in isolated gland situation, and then isolate various membrane fractions. You can actually look at trafficking that way, to see translocation of K^+ channels into either the apical or basolateral membrane, for example. This hasn't been done for A7. Functional data for this transporter from the stomach are lacking.

Wall: What happens to A7 expression with changes in acid–base balance?

Soleimani: That is a good question. In collaboration with Dr Petrovic, we are examining these questions. It seems that in acidosis there may be an enhancement of trafficking of A7 to the membrane. If we look at the AE1 knockout there is a residual chloride–bicarbonate exchange activity on the basolateral membrane of α intercalated cells in outer medullary collecting duct (OMCD), which we believe is mediated via A7 since AE1 and A7 colocalize to this membrane domain. Moreover, in the presence of increased osmolarity in microperfused tubules, we observe that the chloride–bicarbonate exchanger activity across the basolateral membrane of OMCD cells becomes more DIDS resistant than in isotonic environment. We interpret this to indicate that increasing the osmolarity activates A7 (and maybe down regulates AE1). Studies in OMCD of A7 knockout mice should clarify the role of A7 in kidney tubules.

Gray: Can you comment on the pancreatic duct work you have done on the A6 knock out animals?

Soleimani: We have collaborated with Dr Min Goo Lee on this issue. He sees that in perfused pancreatic duct, apical chloride–bicarbonate exchanger activity is significantly reduced in A6 knockout mice. cAMP-stimulated bicarbonate secretion, however, is not reduced. We are planning to do the same experiment with the double A3/A6 knockout to make sure that A3 is not compensating for A6 deletion.

Reference

Kim KH, Shcheynikov N, Wang Y, Muallem S 2005 SLC26A7 is a Cl⁻ channel regulated by intracellular pH. J Biol Chem 280:6463–6470

Anion exchangers in flux: functional differences between human and mouse SLC26A6 polypeptides

Seth L. Alper, Andrew K. Stewart, Marina N. Chernova, Alexander S. Zolotarev, Jeffrey S. Clark and David H. Vandorpe

Molecular and Vascular Medicine and Renal Units, Beth Israel Deaconess Medical Center, Department of Medicine, Harvard Medical School, Boston, MA 02215, USA

Abstract. The SLC26 anion transporter polypeptides exhibit considerably greater sequence diversity among near-species orthologues than is found among the SLC4 bicarbonate transporters, and among SLC26 transporters is most marked among SLC26A6 orthologues. This observation prompted systematic functional comparison in *Xenopus* oocytes of mouse Slc26a6 with several human SLC26A6 polypeptide variants. Mouse and human polypeptides exhibited similar rates of bidirectional [^{14}C]oxalate flux, Cl^-/HCO_3^- exchange, and Cl^-/OH^- exchange, and similar cAMP-stimulation and enhancement of that stimulation by wild-type but not ΔF508 CFTR. However, high rates of $^{36}Cl^-$ and ^{35}S-sulfate transport by mouse Slc26a6 contrasted with low transport rates of the human proteins. The high $^{36}Cl^-$ transport phenotype cosegregated with the transmembrane domain of mouse Slc26a6 in chimera studies. Mouse Slc26a6 and human SLC26A6 each mediated electroneutral Cl^-/HCO_3^- and Cl^-/OH^- exchange. But, whereas $Cl^-/$oxalate exchange by mouse Slc26a6 was electrogenic, that mediated by human SLC26A6 appeared electroneutral. Oocyte expression of either mouse or human orthologue elicited currents that were pharmacologically distinct from the monovalent anion exchange activities measured in the same lots of oocytes. The human SLC26A6 polypeptide variants SLC26A6c and SLC26A6d were inactive in isotopic flux assays. Understanding of SLC26 transport mechanisms and pathophysiology will benefit from recognition of substantial differences in transport properties among orthologues.

2006 Epithelial anion transport in health and disease: the role of the SLC26 transporters family. Wiley, Chichester (Novartis Foundation Symposium 273) p 107–125

Electroneutral Cl^-/HCO_3^- exchange fulfils several major functions in eukaryotic cells (Alper 2002, Romero et al 2004). Cl^-/HCO_3^- exchange regulates intracellular pH in the presence of usual prevailing transmembrane gradients of Cl^- and HCO_3^- by mediating homeostatic pH_i recovery from an alkaline load. In polarized epithelial cells, basolateral Cl^-/HCO_3^- exchange can extrude HCO_3^- from cell to interstitial fluid to support apical H^+ secretion, and/or can load Cl^- into the cell

107

for transepithelial Cl^- secretion. Apical Cl^-/HCO_3^- exchange can also mediate transepithelial HCO_3^- secretion in the presence of basolateral HCO_3^- loading of the cell by other transporters, or can load luminal Cl^- into the cell to accomplish transepithelial Cl^- reabsorption. Since cellular and organismic demands for acid/base balance and for volume homeostasis can conflict, anion exchangers must be regulated coordinately with additional cellular transport systems to allow at least independent control of intracellular volume and pH.

Cellular Cl^-/HCO_3^- exchangers are encoded by two major gene superfamilies, SLC4 and SLC26 (Alper 2002, Romero et al 2004, Mount & Romero 2004). The first SLC26 genes identified encoded fungal and plant sulfate transporters. The polypeptide products of the 10 human SLC26 genes transport a wide variety of anions, including Cl^-, HCO_3^-, OH^-, sulfate, formate, iodide and oxalate. Mutations in four of the SLC26 genes, A2, A3, A4 and A5, cause familial disease, with phenotypes replicated or approximated by mouse knockouts (Dawson & Markovich 2005, Mount & Romero 2004). Mouse knockouts of the SLC26 genes not yet implicated in human genetic disease are beginning to reveal *in vivo* roles for these anion transporters (Wang et al 2005).

Several SLC26 polypeptides are localized to the apical membrane of polarized epithelial cells, where they colocalize with the cystic fibrosis transmembrane regulator Cl^- channel, CFTR. SLC26 polypeptide-mediated Cl^-/HCO_3^- exchange activity is up-regulated by cAMP-stimulated CFTR (Ko et al 2002, 2004, Chernova et al 2005), likely by direct interaction with the R domain (Ko et al 2004). In contrast, SLC4 polypeptide-mediated Cl^-/HCO_3^- exchange is insensitive to CFTR activity. Moreover, the C-terminal cytoplasmic STAS domain of SLC26A3 can directly activate CFTR channel function (Ko et al 2004). SLC26A6 and SLC26A3, in particular, have been proposed to serve as major pathways of HCO_3^- secretion by pancreatic and duodenal epithelia. CFTR mutations associated with 'pancreatic sufficiency' have also been speculated to represent mutations which allow continued stimulation of SLC26 mediated Cl^-/HCO_3^- exchange by CFTR (Choi et al 2001, Ko et al 2002, Chernova et al 2003, 2005) and/or continued STAS domain stimulation of CFTR (Ko et al 2004).

Human SLC4 polypeptide sequences share >90% amino acid identity with their orthologous rodent sequences, and no functional differences between human and rodent SLC4 polypeptides have been described. In contrast, with the notable exception of the highly conserved putative mechanotransducer of the cochlear outer hair cell, SLC26A5/prestin, human SLC26 polypeptide sequences exhibit comparatively low amino acid identity with orthologous rodent sequences. Most remarkably, human SLC26A6/PAT1 and mouse Slc26a6/CFEX share only 78% amino acid identity, raising questions about extrapolation from mouse physiology to the human. This concern was intensified by earlier reports that human SLC26A6 was functionally inactive (Waldegger et al 2001) or minimally active (Lohi et al 2003), and by restriction of detailed functional studies of mammalian Slc26a6 polypeptides to the rodent (Knauf et al 2001, Xie et al 2002, Ko et al 2002, 2004, Wang et al 2002).

These concerns along with the important role proposed for SLC26A6 in the pathophysiology of cystic fibrosis, prompted our hypothesis that the interspecies divergences in amino acid sequence might reveal differences in substrate selectivity, mechanism, or regulation of anion transport. If true, these differences might then reveal regions of the SLC26 protein contributing to regulation and to anion selectivity. We found that four human SLC26A6 polypeptide variants are active anion transporters when expressed in *Xenopus* oocytes. Human SLC26A6 transports oxalate in exchange for Cl^- at rates 40–80% those of mouse Slc26a6. Human SLC26A6 mediates nominal Cl^-/HCO_3^- exchange at a rate 80% that of mouse Slc26a6. Curiously, however, human SLC26A6 transports $^{36}Cl^-$ at barely detectable levels, and $[^{35}S]$sulfate at rates of 5–10% those of mouse Slc26a6. The high $^{36}Cl^-$ transport phenotype segregates with the Slc26a6 transmembrane domain in studies of chimeras. The orthologous mouse and human polypeptides also differ in their acute regulation.

Human and mouse SLC26A6 differ in their transport of $^{36}Cl^-$ and of $^{35}SO_4^{2-}$

We compared expression in *Xenopus* oocytes of four polypeptide variants of human SLC26A6 (L+Q, L−Q, S+Q, S−Q; Chernova et al 2005) with mouse Slc26a6(S−Q)/CFEX (Knauf et al 2001). All polypeptides accumulated and reached the oocyte surface at roughly equivalent levels, but without equivalence of function (Fig. 1). In contrast to robust Cl^-/Cl^- exchange mediated by mouse Slc26a6 and measured as bidirectional $^{36}Cl^-$ fluxes (Fig. 1a–c), all human SLC26A6 polypeptides exhibited $^{36}Cl^-$ influx at rates only 5% those of mouse Slc26a6 (Fig. 1a). Rates of $^{36}Cl^-$ efflux by human SLC26A6 polypeptides into Cl^- bath were less that 5% those of mouse, and (as evident in Fig. 1b and 1c) often undetectable. Cl^-/sulfate exchange by human SLC26A6 measured as $[^{35}S]$sulfate uptake was only 10% the rate exhibited by mouse Slc26a6 (Fig. 1d). Cl^-/formate exchange by human SLC26A6 was undetectable as measured by $[^{14}C]$formate efflux in contrast to substantial efflux by mouse Slc26a6 at pH 7.4 and 9.0 (Clark & Alper, unpublished). The substantial Cl^-/nitrate and Cl^-/Br^- exchange activities of mouse Slc26a6 were similarly minimal or undetectable in oocytes expressing human SLC26A6 (Chernova et al 2005). In contrast, Cl^-/oxalate exchange by human SLC26A6 measured as $[^{14}C]$oxalate uptake from 0.1 mM oxalate bath was 40% that of mouse Slc26a6 (Fig. 1e), and $[^{14}C]$oxalate efflux into Cl^- bath was 80% that of mouse Slc26a6 (Fig. 1f). Thus, in contrast to the 'multi-anion exchanger' properties of mouse Slc26a6, human SLC26A6 behaves in isotopic flux experiments as though it is a specific Cl^-/oxalate exchanger, with little or no capacity for Cl^-/Cl^- homo-exchange or for hetero-exchange of Cl^- with sulfate, formate, nitrate or bromide.

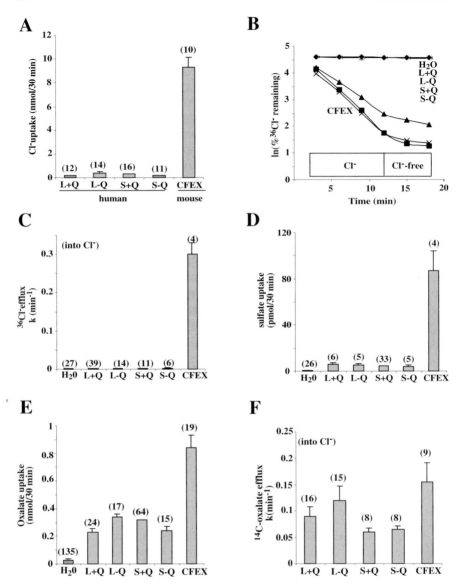

FIG. 1. Mouse Slc26a6 (CFEX) and human SLC26A6 variant polypeptides exhibit different anion selectivities. (A) Unidirectional uptake of $^{36}Cl^-$. Values are corrected for uptake by water-injected oocytes. (B) Time course of unidirectional $^{36}Cl^-$ efflux by representative oocytes expressing mouse Slc26a6 (CFEX) or the indicated human SLC26A6 variant polypeptides, and by one water-injected oocyte. (C) Summary of $^{36}Cl^-$ efflux rate constants. (D) Unidirectional [^{35}S]sulfate uptake by oocytes expressing mouse Slc26a6(CFEX) or human SLC26A6 variants. (E) Unidirectional [^{14}C]oxalate uptake by oocytes expressing mouse Slc26a6(CFEX) or human SLC26A6 variants. (F) [^{14}C]oxalate efflux rate constants of oocytes expressing mouse Slc26a6(CFEX) or human SLC26A6 variants. (Modified from Chernova et al 2005.)

Human and mouse SLC26A6 differ little in their Cl^-/HCO_3^- exchange activities

The oxalate/Cl^- exchange activity of human SLC26A6 shows that earlier reports of lack of function were not definitive. However, the central role suggested for human SLC26A6 in pancreatic and gastrointestinal bicarbonate secretion and in the defective bicarbonate secretion of cystic fibrosis has been based upon studies with rodent Slc26a6. Extrapolation of this role to human SLC26A6 requires activity as a Cl^-/HCO_3^- exchanger. Figure 2 demonstrates that human SLC26A6 expressed in *Xenopus* oocytes exhibits Cl^-/HCO_3^- exchange activity measured by BCECF excitation ratio fluorimetry, in marked contrast to its minimal Cl^-/Cl^- exchange activity. The rate of both Cl^-/HCO_3^- exchange and of Cl^-/OH^- exchange exhibited by human SLC26A6 were ~80% those exhibited by mouse Slc26a6 (Fig. 2). The rate of Cl^-/HCO_3^- exchange measured by BCECF fluorescence ratio (6.2 mM/s) and that from $^{36}Cl^-$ efflux into 5% CO_2/24 mM HCO_3^-/72 mM gluconate (>8.6 mM/s) are in reasonable agreement. However, ^{36}Cl efflux from oocytes expressing human SLC26A6 could not be reliably measured for a similar comparison. Extensive control experiments excluded DNA sequence, expression vector, influx conditions, source, lot, or toxic contamination of isotope as sources of the very different Cl^-/Cl^- exchange rates of mouse Slc26a6 and human SLC26A6 (Chernova et al 2005). Exchange of intracellular HCO_3^- for extracellular oxalate rapidly slowed within 5 min (Fig. 2), perhaps reflecting combined replacement of extracellular Cl^- and competition by the accumulation of intracellular oxalate.

Cl^-/HCO_3^- exchange by human SLC26A6 and by mouse Slc26a6 each exhibited modest cAMP-stimulation in *Xenopus* oocytes, and that stimulation was for each enhanced by coexpression with CFTR. However, coexpression with mutant CFTR ΔF508 abolished stimulation by cAMP (Chernova et al 2005). The enhancement by CFTR of co-expressed Slc26a6-mediated Cl^-/HCO_3^- exchange activity in oocytes is consistent with observations in HEK-293 cells (Ko et al 2002). [14C]oxalate transport by human SLC26A6 was strongly inhibited by butyrate-induced intracellular acidification and activated by NH_4^+. In contrast, the murine protein was minimally inhibited by butyrate and unaffected by NH_4^+.

The ability to mediate Cl^-/HCO_3^- exchange in the absence of Cl^-/Cl^- exchange is not shared with human SLC26A3, but is not unique

The dichotomy between Cl^-/Cl^- exchange and Cl^-/HCO_3^- exchange shown above for human SLC26A6 is unprecedented in our previous studies of SLC4-mediated Cl^-/Cl^- and Cl^-/HCO_3^- exchange. However, Fig. 3 shows that not all human SLC26 anion exchangers exhibit this dichotomy. Human SLC26A3/DRA mediates Cl^-/Cl^- exchange as evidenced by unidirectional $^{36}Cl^-$ influx and efflux, as well as Cl^-/HCO_3^- exchange as measured by BCECF ratio fluorimetry and by

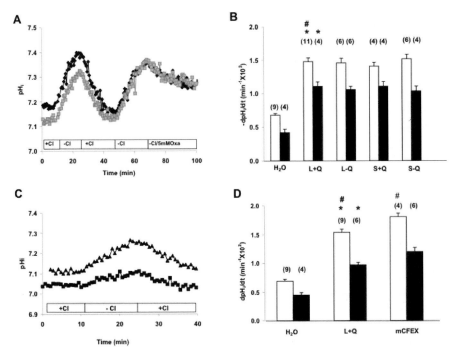

FIG. 2.   Transport of HCO_3^- and (nominal) OH^- by human SLC26 variants. (A) HCO_3^- exchange with Cl^- (first two bath change cycles) and with 5 mM oxalate (final bath change) in representative oocytes expressing mouse Slc26a6 (CFEX, black diamonds) or SLC26A6(L+Q) (grey squares). (B) Initial rates of Cl^-/HCO_3^- exchange (white bars, measured upon Cl^- readdition) and of oxalate/HCO_3^- exchange (black bars, measured upon oxalate addition) by oocytes injected with water or expressing the indicated human SLC26A6 variant. *, $P < 0.001$ vs. same assay in water-injected oocytes; #, $P < 0.005$ vs. oxalate/HCO_3^- exchange by oocytes expressing same SLC26A6 variant. Mean initial pH_i values for all groups were statistically indistinguishable. (C) Nominal Cl^-/OH^- exchange by representative oocytes expressing mouse Slc26a6/CFEX (triangles) or human SLC26A6(L+Q) (squares). (D) Rates of nominal Cl^-/HCO_3^- exchange (white bars) and Cl^-/OH^- exchange (black bars) in oocytes injected with water or expressing SLC26A6(L+Q) or mouse Slc26a6/CFEX. Values in bar graphs are mean + SEM for (n) oocytes. *, $P < 0.001$ vs. same assay in water-injected oocytes; #, $P < 0.005$ vs. Cl^-/OH^- exchange by same polypeptide. Mean initial resting pH_i values for all groups were indistinguishable in the presence of CO_2/HCO_3^-. Mean initial resting pH_i values in room air were similarly indistinguishable among groups. (Reproduced from Chernova et al 2005.)

pH-sensitive microelectrode recording (Chernova et al 2003, see also Melvin et al 1999, Lamprecht et al 2005). However, preliminary studies of human SLC26A7, A8, A9 and A11 indicate a pattern similar to that of human SLC26A6: oocytes injected with 10 ng cRNA express easily detectable Cl^-/HCO_3^- exchange accompanied by minimal or undetectable Cl^-/Cl^- exchange (bidirectional $^{36}Cl^-$ flux).

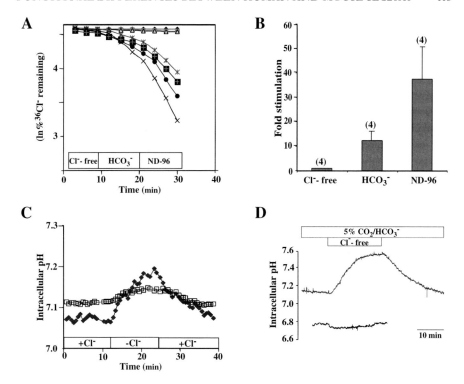

FIG. 3.   Human SLC26A3/DRA mediates Cl⁻/Cl⁻ and Cl⁻/HCO₃⁻ exchange. (A) [³⁶Cl]efflux traces from water-injected oocytes (upper three traces) and from SLC26A3-expressing oocytes (lower four traces) during sequential incubations in baths without Cl⁻ containing 24 mM HCO₃⁻, and containing 96 mM Cl⁻ (ND-96). (B) Fold-stimulation of ^{36}Cl⁻ efflux upon changing from Cl⁻-free baths to baths containing the indicated anion ($n = 4$). (C) pH_i of representative oocytes previously injected with water (open squares) or with hDRA cRNA (filled diamonds) during bath Cl⁻ removal and restoration. pH_i was measured by BCECF fluorescence ratio imaging. (D) pH_i of a representative oocyte previously injected with water (lower trace) or with human SLC26A3 cRNA (upper trace) during bath Cl⁻ removal and restoration. pH_i was measured by pH-sensitive microelectrode. (Modified from Chernova et al 2003.)

The marked difference in rates of Cl⁻/Cl⁻ and Cl⁻/sulfate exchange rates between human and mouse SLC26A6 segregates with the polypeptides' polytopic transmembrane domains

We constructed chimeric proteins to test whether species-specific anion selectivity could be attributed to the transmembrane domain or to the C-terminal cytoplasmic region containing the STAS domain and the far C-terminal PDZ domain recognition motif (Fig. 4). All constructs were functional as oxalate transporters, but the presence of either transmembrane or cytoplasmic domain of the human conferred

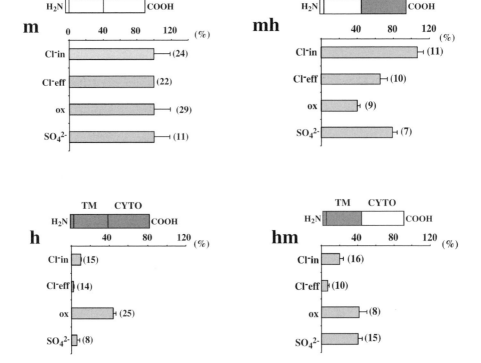

FIG. 4. Magnitude of [36]Cl[−] flux by SLC26A6 polypeptides cosegregates with the transmembrane domain. [36]Cl[−] influx (Cl[−] in) and efflux (Cl[−] eff), [^{14}C]oxalate influx (ox) and [35]SO$_4^{2-}$ influx were measured in oocytes expressing wild-type mouse Slc26a6 (CFEX) (m), wild-type human SLC26A6(S–Q) (h), the chimera of the mouse transmembrane domain attached to the human C-terminal cytoplasmic domain (mh), and the chimera of the human transmembrane domain attached to the mouse C-terminal cytoplasmic domain (hm). Mean values + SEM for (n) oocytes from two or more frogs, normalized to values of wild-type mouse Slc26a6/CFEX. (Reproduced from Chernova et al 2005.)

human oxalate transport rates. However, high rates of [36]Cl[−] efflux segregated strictly with the transmembrane domain of the mouse protein but not the human protein. The sulfate transport phenotype was less straightforward. The mouse transmembrane domain attached to either cytoplasmic domain mediated high sulfate transport rates, but the human transmembrane domain/mouse cytoplasmic domain chimera exhibited an intermediate phenotype. Cl[−]/HCO$_3^-$ exchange activity of all chimeras was equivalent to the similar exchange rates of the wild-type mouse and human proteins (not shown).

Thus, high Slc26a6-mediated Cl[−]/Cl[−] and Cl[−]/sulfate exchange rates can be attributed to residues in the mouse transmembrane domain that differ in the human transmembrane domain. The transmembrane domain sequences of both orthologous polypeptides are permissive for robust bidirectional transport of Cl^- in exchange for HCO_3^- or for oxalate.

Electroneutral monovalent anion exchange and species-specific electrogenic oxalate/Cl^- exchange by Slc26a6

Two groups have observed that Slc26a6-expressing oocytes reversibly hyperpolarize in response to bath Cl^- removal (Ko et al 2002, Xie et al 2002), whereas Slc26a3-expressing oocytes reversibly depolarize (Ko et al 2002). Ko et al (2002) also reported that oocytes expressing either Slc26a3 or Slc26a6 exhibit two electrode voltage clamp currents attributable to Cl^-/OH^- and to Cl^-/HCO_3^- exchange. These data were interpreted as reflecting electrogenic anion exchange. The impact of inclusion of apical electrogenic anion exchange of appropriate stoichiometry into models of stimulated pancreatic bicarbonate secretion has remained uncertain (Steward et al 2005).

In contrast, we have observed much smaller currents in voltage-clamped oocytes expressing either human SLC26A6 or mouse Slc26a6 in Cl^- media equilibrated with either room air or CO_2/HCO_3^- (Fig. 5). Oocytes from the same lots exhibited robust $[^{14}C]$oxalate flux and Cl^-/HCO_3^- exchange (not shown). Several observations suggest that these currents do not reflect electrogenic anion exchange by either Slc26a6 orthologue. First, the currents represented charge movement that could account for <5% of measured $^{36}Cl^-$ influx or $^{36}Cl^-$ efflux. Second, the reversal potential of the measured current was insensitive to replacement of bath Cl^- with gluconate, whether in the absence or presence of CO_2/HCO_3^-. Third, the small currents did not exhibit the pharmacological profile of the $^{36}Cl^-$ and oxalate fluxes (DIDS, NS3623, tenidap). Fourth, high K^+ depolarization did not alter rates of Cl^-/HCO_3^- exchange or of efflux of $^{36}Cl^-$ efflux mediated by mouse Slc26a6 orthologue (Chernova et al 2005). These results suggested electroneutrality of Cl^-/OH^- and Cl^-/HCO_3^- exchanges mediated by mouse Slc26a6 and human SLC26A6. Similar results supporting electroneutral Cl^-/OH^- and Cl^-/HCO_3^- exchange were recorded in voltage-clamped oocytes expressing mouse Slc26a3, human SLC26A3 and rat Slc26a4/pendrin (not shown) and in BCECF-loaded HEK-293 cells expressing mouse Slc26a3 (Melvin et al 1999) or human SLC26A3 (Lamprecht 2005).

Our two electrode voltage clamp current measurements are validated by two additional observations. First, addition of 1 mM oxalate to Slc26a6-expressing oocytes in gluconate bath greatly increased outward current and shifted the reversal potential by −24 mV (Fig. 5) and in a second series of oocytes by −35 mV (Chernova et al 2005). This current reflects inward negative charge movement predicted by exchange of extracellular divalent oxalate with intracellular monovalent Cl^- in agreement with Jiang et al (2002). Second, this oxalate-induced current was DIDS-sensitive (Chernova et al 2005, Jiang et al 2002).

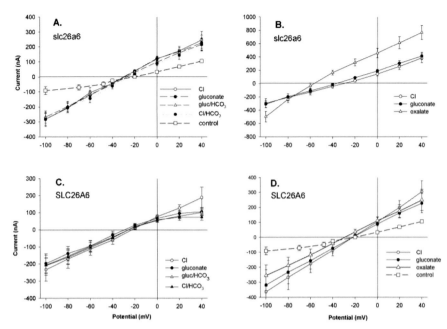

FIG. 5.   Oocyte currents associated with expression of human SLC26A6 and of mouse Slc26a6 do not represent electrogenic exchange of monovalent anions. (A) I–V curves of oocytes expressing mouse Slc26a6/CFEX ($n = 8$) recorded during sequential bath changes from Cl^- to gluconate to 24 mM HCO_3^-/5% CO_2 (plus 72 mM gluconate) to 24 mM HCO_3^-/5% CO_2 (plus 72 mM Cl^-). I–V curve of water-injected oocytes (control, $n = 4$) was recorded in Cl^- bath. (B) I–V curves of oocytes expressing mouse Slc26a6/CFEX ($n = 12$) recorded during bath change from Cl^- to gluconate, followed by addition of 1 mM oxalate. Reversal potential hyperpolarizes −24 mV upon addition of oxalate. (C) I–V curves of oocytes expressing human SLC26A6(L+Q) ($n = 6$) recorded during sequential bath changes from Cl^- to gluconate to 24 mM HCO_3^-/5% CO_2 (plus 72 mM gluconate) to 24 mM HCO_3^-/5% CO_2 (plus 72 mM Cl^-). (D) I–V curves of oocytes expressing human SLC26A6(L+Q) ($n = 10$) recorded during bath change from Cl^- to gluconate, followed by addition of 1 mM oxalate. The I-V trace of water-injected oocytes (control, $n = 4$) is reproduced from panel A. (Reproduced from Chernova et al 2005.)

Addition of 1 mM oxalate to human SLC26A6-expressing oocytes in gluconate bath did not increase current or shift its reversal potential. This suggests that human SLC26A6-mediated oxalate/Cl^- exchange is electroneutral, unlike that of mouse Slc26a6. Since human SLC26A6-mediated [^{14}C]oxalate fluxes are 40–80% as rapid as that of mouse Slc26a6, associated currents should not be below the level of detection in an assay such as that shown in Fig. 5. Consistent with this species difference in electrogenicity, [^{14}C]oxalate efflux by human SLC26A6 into Cl^- was unaffected by the change to high K^+ bath. In contrast, mouse Slc26a6-mediated [^{14}C]oxalate efflux into Cl^- was inhibited 50% by high K^+ depolarization, as expected for electrogenic transport of one negative charge per transport cycle (not shown). These findings suggest that the oocyte depolarization produced by high K^+ bath change is of sufficient degree to alter the rates of electrogenic anion exchange exposed to steady-state anion gradients. The insensitivity of rates of $^{36}Cl^-/Cl^-$ exchange by Slc26a6 or by Slc26a3 thus supports an electroneutral mechanism of monovalent anion exchange.

Initial characterization of SLC26-related polypeptides in lower organisms

The *Caenorhabditis elegans* genome includes four SLC4-related *abts* genes encoding at least five transcripts and eight SLC26-related *sulp* genes encoding at least 14 transcripts. Sherman et al (2005) have localized GFP promoter fusion products from each of these genes, and have localized translational GFP polypeptide fusions of these anion transporter genes for half of them. The SULP4 polypeptide expressed in *Xenopus* oocytes increased uptake of $^{36}Cl^-$ by 2.5-fold, and uptake of [^{35}S]sulfate by eightfold (Sherman et al 2005).

The SLC26-related *Sulp* sulfate permease gene family is found widely among sequenced unicellular eukaryotic, bacterial and archaebacterial genomes. In marine cyanobacteria, a *Sulp* gene with a carbonic anhydrase domain in place of a STAS domain mediates Na-dependent HCO_3^- uptake with a postulated role in carbon fixation (Price et al 2004). Similar SulP–carbonic anhydrase open reading frame fusions are present in several other bacterial genomes (Felce & Saier 2005).

We have cloned and inducibly overexpressed in *Escherichia coli* a 6-His-tagged version of the Sulp gene product Rv1739c from *Mycobacterium tuberculosis*. Rv1739c overexpression increased uptake of [^{35}S]sulfate, but not $^{36}Cl^-$ or [^{14}C]oxalate. Induced [^{35}S]sulfate uptake is activated by acid pH, exhibits a K_m (app) of 4.1 μM at pH 6, and is inhibited by N-ethylmaleimide and CCCP. Uptake is also inhibited by cold sulfate, thiosulfate, selenate, poorly inhibited by molybdate, and is not affected by Cl^- or HCO_3^-. Rv1739c associated sulfate uptake is maintained in a protein in which all three Cys residues are changed to Ser. The Rv1739c polypeptide has solubilized and affinity purified in dodecylmaltoside, with a yield of 1 mg/l of induced culture (not shown).

Conclusion

The SLC26 family gene products include the apical membrane anion exchanger polypeptides which appear to mediate much of the long sought CFTR-stimulated luminal Cl^-/HCO_3^- activity of epithelia. This is confirmed in the case of the human congenital disease chloride-losing diarrhoea, secondary to loss-of-function mutations in SLC26A3/DRA. In the absence of a human SLC26A6 deficiency disease, the role of apical SLC26A6 in epithelial bicarbonate secretion remains less clear.

Further study of the physiology of the recently described *Slc26a6*[−/−] mouse will be central to this understanding. However, as summarized in this report, the amino acid sequence divergence between mouse Slc26a6 and human SLC26A6 polypeptides reflects significant differences in anion selectivity, acute regulation, and possibly also mechanism of divalent/monovalent anion exchange. Human SLC26A6, in particular, exhibits an unusual dichotomy between very low rates of Cl^-/Cl^- exchange and robust Cl^-/HCO_3^- exchange activity. Although both human and mouse SLC26 polypeptides mediate Cl^-/HCO_3^- exchange sensitive to activation of coexpressed CFTR, the consequences to our extrapolation of findings in the genetically modified mouse to pathophysiology in the human remain to be seen. The reported electrogenicity of Cl^-/HCO_3^- exchange by Slc26a3 and Slc26a6 remains controversial. Indeed, the importance of electrogenic vs. electroneutral apical Cl^-/HCO_3^- exchange (or any Cl^-/HCO_3^- exchange) during maximal HCO_3^- secretion by the human pancreatic duct is still being debated (Sohma et al 2000, Ko et al 2002, Whitcomb & Ermentrout 2004, Steward et al 2005). Whatever the electrogenicity of Cl^-/HCO_3^- exchange, the species specificity of SLC26-mediated anion transport properties may help explain the difference between the <70 mM peak pancreatic [HCO_3^-] secretion of rat (and likely mouse), and up to 145 mM or more in human, dog, cat and guinea pig.

Acknowledgement

This work was supported by NIH grant DK43495.

References

Alper SL 2002 Genetic diseases of acid-base transporters. Annu Rev Physiol 64:899–823

Chernova MN, Jiang L, Shmukler BE et al 2003 Acute regulation of the SLC26A3 congenital chloride diarrhea anion exchanger (DRA) expressed in Xenopus oocytes. J Physiol 549:3–19

Chernova MN, Jiang L, Friedman DJ et al 2005 Functional comparison of mouse slc26a6 anion exchanger with human SLC26A6 polypeptide variants. J Biol Chem 280:8564–8580

Choi JY, Muallem D, Kiselyov K, Lee MG, Thomas PJ, Muallem S 2001 Aberrant CFTR–dependent HCO3– transport in mutations associated with cystic fibrosis. Nature 410:94–97

Dawson PA, Markovich D 2005 Pathogenetics of the human SLC26 transporters. Curr Med Chem 12:385–396

Felce J, Saier MH 2005 Carbonic anhydrases fused to anion transporters of the SulP family: evidence for a novel type of bicarbonate transporter. J Mol Microbiol Biotechnol 8:169–176

Jiang Z, Grichtchenko II, Boron WF, Aronson PS 2002 Specificity of anion exchange mediated by mouse Slc26a6. J Biol Chem 277:33963–33967

Knauf F, Yang CL, Thomson RB, Mentone SA, Giebisch G, Aronson PS 2001 Identification of a chloride-formate exchanger expressed on the brush border membrane of renal proximal tubule cells. Proc Natl Acad Sci USA 98:9425–9430

Ko SB, Shcheynikov N, Choi JY et al 2002 A molecular mechanism for aberrant CFTR-dependent HCO3− transport in cystic fibrosis. EMBO J 21:5662–5672

Ko SB, Zeng W, Dorwart MR et al 2004 Gating of CFTR by the STAS domain of SLC26 transporters. Nat Cell Biol 6:343–350

Lamprecht G, Baisch S, Schoenleber E, Gregor M 2005 Transport properties of the human intestinal anion exchanger DRA (down-regulated in adenoma) in transfected HEK293 cells. Pflugers Arch 449:479–490

Lohi H, Lamprecht G, Markovich D et al 2003 Isoforms of SLC26A6 mediate anion transport and have functional PDZ interaction domains. Am J Physiol Cell Physiol 284:C769–779

Melvin JE, Park K, Richardson L, Schultheis PJ, Shull GE 1999 Mouse down-regulated in adenoma (DRA) is an intestinal Cl−/HCO3− exchanger and is upregulated in colon of mice lacking the NHE3 Na+/H+ exchanger. J Biol Chem 274:22855–22861

Mount DB, Romero MF 2004 The SLC26 gene family of multifunctional anion exchangers. Pflugers Arch 447:710–721

Price GD, Woodger FJ, Badger MR, Howitt SM, Tucker L 2004 Identification of a SulP-type bicarbonate transporter in marine cyanobacteria. Proc Natl Acad Sci USA 101:18228–18233

Romero MF, Fulton CM, Boron WF 2004 The SLC4 family of HCO3− transporters. Pflugers Arch 447:495–509

Sherman T, Chernova MN, Clark JS, Jiang L, Alper SL, Nehrke KE 2005 The abts and sulp families of anion transporters from Caenorhabditis elegans. Am J Physiol Cell Physiol 289:C341–351

Sohma Y, Gray MA, Imai Y et al 2000 HCO3 transport in a mathematical model of the pancreatic ductal equilibrium. J Membr Biol 176:77–100

Steward MC, Ishiguro H, Case RM 2005 Mechanisms of bicarbonate secretion in the pancreatic duct. Annu Rev Physiol 67:377–409

Waldegger S, Moschen I, Ramirez A et al 2001 Cloning and characterization of SLC26A6, a novel member of the solute carrier 26 gene family. Genomics 72:102–112

Wang Z, Petrovic S, Mann E, Soleimani M 2002 Identification of an apical Cl/HCO3 exchanger in the small intestine. Am J Physiol Gastroenterol Physiol 282:G573–579

Wang Z, Wang T, Petrovic S et al 2005 Renal and intestinal transport defects in slc26a6-null mice. Am J Physiol Cell Physiol 288:C957–965

Whitcomb DC, Ermentrout GB 2004 A mathematical model of the pancreatic duct cell generating high bicarbonate concentrations in pancreatic juice. Pancreas 29:e30–40

Xie Q, Welch R, Mercado A, Romero MF, Mount DB 2002 Molecular characterization of the murine Slc26a6 anion exchanger: functional comparison with Slc26a1. Am J Physiol Renal Physiol 283:F826–838

DISCUSSION

Welsh: Has anyone purified any member of this family and studied it in a reconstitution system, such as a vesicle or lipid membrane? This might be one approach to resolving some of the differences.

Alper: We have purified one of the *Mycobacterium tuberculosis* proteins and we are now trying to reconstitute it.

Muallem: It is mind boggling to me that you can see such an incredible difference in $^{36}Cl^-$ fluxes between the human and mouse isoforms, and yet still see the same Cl^-/HCO_3^- exchange activity in the various clones that you are using. We have turned the human exchangers upside down and we can see absolutely no activity. This is also showing very well with the oxalate transporters: we see no change in reversal potential. We see no oxalate transport by the human transporter and we have no idea why or how you can see all this Cl^-/HCO_3^- exchange activity and it does not bother you.

Alper: What you are saying is that you see no oxalate transport electrophysiologically.

Muallem: Yes, but we do everything electrophysiologically. We measure the pH and we take into account the enormous buffer capacity of the entire oocyte volume. We don't measure just the pH under the plasma membrane of the oocytes.

Alper: Nor do we.

Muallem: But the currents you are measuring can only be described as puny compared with ours.

Alper: Yes, I agree, they are small. But we get CFTR currents that are the same magnitude that everyone else gets.

Muallem: I would suggest that something is wrong when you get huge Cl^- fluxes with one isoform and almost none with the other. Yet you get the same Cl^-/HCO_3^- exchange activity with both isoforms. This should have raised a huge red flag.

Alper: It did, and is why we have waited a long time to publish.

Seidler: When you express CFTR in an oocyte and it enhances the rate of chloride–bicarbonate exchange for all the transporters, what is the CFTR doing? Is it a chloride shunt pathway through the membrane?

Alper: I glossed over the fact that in our hands, expression of CFTR alone actually increases some nominal chloride–bicarbonate exchange activity in oocytes. These are oocytes with injection of only CFTR cRNA. This enhanced activity is inhibited by the CFTR inhibitor. When it is co-expressed with A7 the elevated activity is inhibited by the CFTR inhibitor only to the extent that it would inhibit the activity contributed by CFTR expression in the absence of A7. This is an indication that this enhancement of activity may not just be direct stimulation. It is conceivable that it only represents addition of two activities which are not pharmacologically identical, and may not even be the same mechanistically. So when we do chloride removal and restoration assays, what are the possible interpretations of the result? It may not be only coupled chloride–bicarbonate exchange in the context in which we have been thinking of it in the past.

Mount: If you look at KCl co-transport, there is endogenous oocyte activity and oocytes express a *Xenopus* KCC (Mercado et al 2001). In terms of anion fluxes, if you telescope down we do see changes in sulfate transport in response to *cis*-inhibition, for example, that are consistent with endogenous SLC26 transporters in the oocyte. We have the *Xenopus* A6 orthologue that is highly functional. As a brief comment: we tend to think of the oocyte as a bland cell, but these cells do have their own low repertoire of other transporters.

Alper: One way to look at this is that CFTR is activating an endogenous SLC26, although the sensitivity to the CFTR inhibitor 172 raises a potential question mark about this.

Aronson: With respect to the discordance between the ^{36}Cl$^-$ flux and chloride–bicarbonate exchange, when you look at ^{36}Cl$^-$ uptake you don't really know what it is exchanging for. At least when outside chloride is removed, you are looking at apparent chloride–bicarbonate exchange, and chloride is effluxing. In this case you would think that there should be some correspondence between the chloride flux and the acid base equivalence. You didn't show that calculation.

Alper: For mouse A6, the rate constant for ^{36}Cl$^-$ efflux into bicarbonate/gluco-nate is 10% of that into ND96. J_{H+} measured from dpH$_i$/dt and buffer capacity comes within 1.5-fold of the efflux rate constant of ^{36}Cl$^-$ into bicarbonate. This efflux rate constant is only 1/10 of what is seen effluxing into chloride. This is a modestly satisfying calculation, but this is only for mouse. In human A6, isotopic chloride efflux is too low to account for measured nominal chloride/bicarbonate exchange.

Aronson: So there is far too little chloride efflux to account for the bicarbonate effect. What is the explanation? Is it possible that outside chloride is stimulating bicarbonate exit, but there is a chloride-dependent bicarbonate transport that is not coupled to the movement of chloride?

Alper: I doubt that is true when we are looking at chloride restoration, but it could be true. When you are looking at the nominal (MQAE) chloride measurement itself, it could represent another anion. There may be some important dependence for activity on bicarbonate, and chloride–chloride exchange is one function but chloride–bicarbonate exchange is another. The latter is conserved among all of these homologues.

Aronson: How can you have chloride–bicarbonate exchange bringing bicarbonate in without the chloride leaving?

Alper: It is in the presence of bicarbonate. It is in the nature of the isotopic efflux assay.

Aronson: A different internal anion might be effluxing.

Alper: Yes, or it might be that bicarbonate is there, and for reasons we don't understand we can't get that ^{36}Cl$^-$ assay to work to validate this hypothesis. But if so, only for human A6. The various measured modalities of transport are consistent for mouse a6.

Quinton: All these studies were taking place in the oocyte. What kind of carbonic anhydrase is present?

Alper: We can immuno-detect carbonic anhydrase 2 (CA2). We know that acetazolamide treatment of the oocyte produces the same 50% inhibition of recombinant AE1-mediated chloride–bicarbonate exchange as it produces in 293 cells expressing human AE1. We know that *Xenopus laevis* expresses CA4 mRNA, but we don't know if CA4 activity is present in the oocyte.

Quinton: Since the carbonic anhydrases appear to have such a significant effect on the anion exchangers, I can't help but wonder whether there must be something unique about each of the 14 or more carbonic anhydrase isoforms that exist relative to different HCO_3^- transporters.

Romero: When I was in Walter Boron's lab, we did a lot of work with carbonic anhydrases when we were studying aquaporins. The endogenous oocyte has a very little bit of acetozolamide and ethoxolamide inhibitable pH decay or recovery when it is treated. The oocyte by itself doesn't seem to have a lot of carbonic anhydrase activity as detected by the more soluble of the carbonic anhydrase inhibitors at pretty industrial concentrations.

Case: I'd like to refer back to fluxes and stoichiometry. We are interested in the pancreatic duct which secretes large quantities of bicarbonate. We were excited by the fact that the luminal anion exchanger might have a 2:1 stoichiometry, because we are always faced with the problem of how we get this high bicarbonate concentration. We postulated that if it was 2:1 this helps us get a bit further towards the 140 mM bicarbonate seen in pancreatic secretion; but what would really help is if there were two bicarbonates going out and one bicarbonate going in. Then it would effectively be a sort of bicarbonate channel (Steward et al 2005). In your oocyte expression systems you said you incubated in nitrate overnight, to get a low intracellular chloride concentration. So when you add bicarbonate, what happens to intracellular pH?

Alper: We haven't yet measured that. It's a good question.

Markovich: I am not surprised by the different transport rates among different species. We see the same.

Alper: It is not the difference between species that is surprising, but rather the selective modality differences.

Markovich: It is, because your chimera data showed that there is the ability to change the transport capability. The big change is due to the sequence of the transporter, and also the functional significance of that protein in that species. You see different tissues express A6 between human and mouse.

Alper: I am not sure this is ultimately going to be functionally important. If it is chloride–bicarbonate exchange, the human and the mouse are quite similar. All of those chimeras have now been tested for chloride–bicarbonate exchange in BCEF-loaded oocytes, and they are all pretty comparable. All that is changed is unidirectional isotopic flux rate constants.

Welsh: When you measure pH with the fluorophores are you measuring pH just beneath the membrane?

Alper: We are measuring a crescent-like area of interest that extends in 10–20% of the cell diameter, and we are doing this in two areas, taking the average of these.

When we have watched regions of interest in towards the centre of the cell we haven't seen great changes. We don't have definitive kinetics on the spatial gradient of pH versus time after solution changes. When we have done preliminary micro-electrode experiments we have been able to show that distance in affects the pH measured.

Romero: Joachim Deitmer's lab (Kaiserslautern, Germany) has shown that with multiple pH electrodes (Bröer et al 1998). The further in the cell you are, the slower the pH changes for monocarboxylate transporters (MCT1).

Alper: The region of interest we select is a strip under the membrane.

Welsh: Are you doing this with conventional epifluorescence?

Alper: Yes, we have not yet used laser confocal microscopy for submembrane pH_i measurements, but we hope to validate this in the future.

Seidler: For the mouse and human SLC26 analogue you showed different K_m values for chloride, but similar ones for bicarbonate. Couldn't this explain the differences in $^{36}Cl^-$ efflux rates versus your chloride–bicarbonate exchange where you have relatively high concentrations of both ions? If you do a $^{36}Cl^-$ flux study you probably use low concentrations.

Alper: No, our influx studies are done with 104 mM extracellular chloride. Among the things we varied to make sure that we weren't crazy was the concentrations: we replicated our assays with every published concentration condition.

Muallem: I say it again: if one transporter is not active, that's what you get.

Alper: It is active, because it is active as a bicarbonate mover. With A6 it is active as a bicarbonate mover: do you see this?

Muallem: With the human clone we see no Cl^-/HCO_3^- exchange, no Cl^- transport, no oxalate transport and no change in current.

Alper: Oxalate fluxes by mouse a6 are electrogenic, but with human A6 appear to be electroneutral.

Soleimani: I wanted to raise two questions. First, when you express a transporter in oocytes, if you don't observe any function it does not mean that that protein is inactive. Some transporters show better expression or display activity in mammalian expression systems such as cultured cells. We observe this with regard to A9 that failed to give a robust chloride–bicarbonate exchange in oocytes, but showed a pronounced function in cultured HEK cells. The same is true with Pendrin, at least in our hands. Contrary to these examples, both A6 and A7 function very well in oocytes. There is something about the structure of the oocyte membrane and the accessory proteins related to this that may be at work here.

Alper: The temperature dependence of the transport allows us to accelerate or slow the rate, but it doesn't change the ratio of mouse to human.

Soleimani: Yes, and the osmolarity of the oocyte is around 200 mM, whereas cultured cells have an osmolarity of ~300 mM. There are a good number of differences, and one has to use an array of expression systems to find the answers to these questions.

Welsh: The oxalate seems to be a good test to try to figure out what is different here.

Muallem: In the mouse with oxalate we see a transient but clear alkalinization, which disappears as oxalate accumulates in the oocytes. We see a hyperpolarizing signal, from about −30 mV to −70 mV.

Welsh: What is oxalate doing?

Muallem: That's a good question: we aren't sure how it is working. To hyperpolarize the membrane there needs to be more than just Cl^-/oxalate exchange. It does do this, but the first hyperpolarization seems the result of a channel-like activity because the Cl^- almost doesn't change in the cell.

Welsh: What is different about what you are doing?

Alper: There seems to be nothing different. The only difference is that one of us has performed one assay that the other hasn't, which is [^{14}C]oxalate flux. By this assay, human SLC26A6 is 40–80% as robust as the mouse. Shmuel wouldn't see this electrophysiologically for apparently electroneutral human A6, but he would for electrogenic mouse a6.

Muallem: This doesn't make sense to me.

Mount: Look at analogous experiments with analogous orthologues. One example is comparing our paper on mouse A6 with Peter Aronson's, where in chloride–bicarbonate exchange, you saw a modest hyperpolarization but nowhere near the amount that we reported. This was with very different amounts of extracellular CO_2 and bicarbonate. When you do mouse A6 and look at bicarbonate influx in the presence of extracellular CO_2 and bicarbonate, do you see the hyperpolarization that Michael sees?

Alper: We have not done pH and voltage recording simultaneously.

Mount: What are comparable experiments?

Alper: We have only done voltage clamp. We have not done the voltage-sensitive microelectrode versus time with solution changes.

Muallem: In one of your figures you showed zero Cl^- and no change in membrane potential.

Alper: That's right. This is under voltage clamp.

Aronson: To get back to the issue of pH and bicarbonate, I'd point out that SAT1, which most people agree is mainly a divalent anion exchanger, exchanges sulfate for oxalate and allegedly sulfate for two bicarbonate. It is possible that it is not really changing sulfate for two bicarbonates, but for carbonate. Suppose that carbonate can actually exchange, and assume that A6 can exchange monovalents for each other and certain divalents for monovalents. Then under more alkaline pH conditions favouring the formation of carbonate, you can get carbonate exchange for chloride which would be electrogenic, whereas under less alkaline pH conditions where the carbonate is very low, chloride would exchange for bicarbonate which would be electroneutral. So whether apparent chloride-bicarbonate exchange mediated by A6 is electrogenic would depend on the pH.

Mount: You are getting ahead of my question, which is a completely naïve technical question: why, with the same clone and vectors in oocytes, are other people not seeing the same hyperpolarizations with mouse A6?

Soleimani: You can go a step further. We see A7 as a chloride–bicarbonate exchanger in the same oocytes, and Dr Muallem doesn't see this. It is the same set-up.

Mount: That's the heart of my question: it is not the same set-up. If we just look at the CO_2 and bicarbonate concentrations, they are not the same.

Alper: Peter's results are coincident with ours, despite the fact that he used 1.5% CO_2 and 10 mM bicarbonate, and we used 5% CO_2 and 25 mM bicarbonate. Michael uses 33 mM bicarbonate and 5% CO_2.

Mount: Part of the problem is that no one is using exactly the same conditions.

Alper: Why do you choose to use a non-physiological solution, which might evoke a more robust response?

Romero: Because that's what gives a pH that is 7.4–7.5 at room temperature with an oocyte. That is, for pH 7.5 and 5% CO_2, the $[HCO_3^-]$ is 33 mM.

References

Broer S, Schneider HP, Broer A, Rahman B, Hamprecht B, Deitmer JW 1998 Characterization of the monocarboxylate transporter 1 expressed in Xenopus laevis oocytes by changes in cytosolic pH. Biochem J 333:167–174

Mercado A, de Los Heros P, Vazquez N, Meade P, Mount DB, Gamba G 2001 Functional and molecular characterization of the K–Cl cotransporter of Xenopus laevis oocytes. Am J Physiol Cell Physiol 281:C670–680

Steward MC, Ishiguro H, Case RM 2005 Mechanisms of bicarbonate secretion in the pancreatic duct. Annu Rev Physiol 67:377–409

Physiology of electrogenic SLC26 paralogues

Michael F. Romero, Min-Hwang Chang, Consuelo Plata, Kambiz Zandi-Nejad*, Adriana Mercado*, Vadjista Broumand*, Caroline R. Sussman and David B Mount*

*Physiology & Biophysics, Case Western Reserve University, School of Medicine, Cleveland, OH 44106-4970 and *Renal Division, Brigham and Women Hospital and Division of General Internal Medicine, VA Boston Healthcare System, Harvard Medical School, Boston, MA 02115, USA*

Abstract. SLC26 anion exchangers transport monovalent and divalent anions, with a diversity of anion specificity and stoichiometry. Our microelectrode studies indicate that several SLC26 members are electrogenic. We reported that Slc26a6 functions as a Cl^-/formate, Cl^-/oxalate, Cl^-/OH^- and electrogenic Cl^-/$nHCO_3^-$ exchanger. Recently, we have also confirmed that Slc26a7 does not behave as a Cl^-/HCO_3^- exchanger but does function as an electrogenic anion conductance, perhaps a channel. We have also cloned murine Slc26a9, which is strongly expressed in the respiratory tract and stomach. Radioisotope uptakes in *Xenopus* oocytes indicate that Slc26a9 is a highly selective anion exchanger, transporting Cl^- but neither formate, oxalate, nor SO_4^{2-}. We also utilized electrophysiology to voltage clamp (VC) and/or measure intracellular pH (pH_i), Cl^- ($[Cl^-]_i$) and Na^+ ($[Na^+]_i$), in response to various ion replacements. Cl^- removal in HCO_3^- depolarizes oocytes (to $> +60mV$), alkalinizes oocytes, and decreases aCl_i^-. Slc26a9 thus functions as an electrogenic nCl^-/HCO_3^- exchanger, suggesting a role in pulmonary and gastric HCO_3^- secretion and/or CO_2 transport. VC experiments revealed channel-like currents ($>10\mu A$ at $-60mV$ and $>80\mu A$ at $+60mV$) mediated by Slc26a9 in the presence and absence of HCO_3^-. Our experiments and those of others continue to reveal additional characteristics and unique roles for this new class of electrogenic anion transporters.

2006 Epithelial anion transport in health and disease: the role of the SLC26 transporters family. Wiley, Chichester (Novartis Foundation Symposium 273) p 126–147

SLC26 transporters: multifunctional anion transporters

The rapidly growing SLC26 gene family encodes moderately homologous anion exchangers with a diversity of physiological roles. The first family member, SLC26A1 (Sat1), encodes an SO_4^{2-}/oxalate exchanger identified by expression cloning (Bissig et al 1994). Other family members were subsequently identified through positional cloning of the disease genes for diastrophic dysplasia (SLC26A2 or DTDST) (Hastbacka et al 1992), congenital chloride diarrhoea (SLC26A3/DRA) (Hoglund et al 1996), and Pendred syndrome (SLC26A4/pendrin) (Everett et al 1997). Subtractive cloning of the 'molecular motor' of cochlear outer hair

126

cells subsequently yielded SLC26A5 (prestin) (Zheng et al 2000), and SLC26A6 was identified as a candidate gene for both the apical bicarbonate transporter of the exocrine pancreas (Lohi et al 2000) and the Cl^--formate exchanger of the proximal tubule (Knauf et al 2001). Further work has completed the identification of the entire gene family, which now numbers 10; an eleventh family member (SLC26A10) appears to be a transcribed pseudogene.

The anions transported by the SLC26 gene family include SO_4^{2-}, Cl^-, I^-, formate, oxalate, OH^- and HCO_3^- (Scott & Karniski 2000, Bissig et al 1994, Soleimani et al 2001, Satoh et al 1998, Karniski et al 1998, Melvin et al 1999). The evolving functional characterizations of this gene family have revealed distinctive patterns of anion specificity and *cis*-inhibition. SLC26A3 and SLC26A6 are thus capable of transporting most if not all of the substrates above (Melvin et al 1999, Xie et al 2002). In contrast, SLC26A1 does not transport Cl^- or formate, but does transport SO_4^{2-} and oxalate (Bissig et al 1994, Satoh et al 1998, Karniski et al 1998, Xie et al 2002). SLC26A4 in turn transports monovalent anions such as Cl^-, I^- and formate, but not divalent anions such as SO_4^{2-} and oxalate (Scott et al 1999, Scott & Karniski 2000). Several paralogues function in Cl^-/OH^- and Cl^-/HCO_3^- exchange modes, with increasingly important roles in transepithelial Na^+/Cl^- and HCO_3^- cotransport. Where direct comparisons have been made, the SLC26 proteins appear to be more potent exchangers of Cl^- with HCO_3^- than with OH^- (Xie et al 2002, Ko et al 2002, Wang et al 2002). The transport characteristics and anion specificities of the individual SLC26 paralogues and how these paralogues might contribute to physiology has recently been reviewed (Mount & Romero 2004).

The physiological role(s) of individual SLC26 anion exchangers is presumably a function of both substrate specificity and expression pattern, and the diverse physiology of this gene family is illustrated by the phenotypes associated with loss-of-function. The SLC26A3 protein is expressed at the brush border membrane of differentiated colonic epithelial cells (Haila et al 2000), and Cl^--base exchange via SLC26A3 is thought to function in transepithelial salt transport by the colon. Accordingly, recessive loss-of-function mutations in the SLC26A3 gene result in severe congenital diarrhoea (Hoglund et al 1996). In contrast, mutations in SLC26A2 are associated with disordered skeletal development in diastrophic dysplasia and related disorders. In these patients, loss-of-function mutations in SLC26A2 impair SO_4^{2-} uptake by chondrocytes, and the resulting decrease in the sulfation of extracellular matrix decreases the response to matrix-dependent developmental cues such as fibroblast growth factor (Hastbacka et al 1992, Satoh et al 1998). Loss-of-function in SLC26A4 (pendrin) is in turn characterized by deafness with or without congenital goitre. SLC26A4 appears to function as a Cl^-/I^- exchanger at the apical membrane of thyroid follicles (Scott et al 1999, Royaux et al 2000) and to thus participate in the organification of thyroglobulin. Although defects in SLC26A4 frequently result in goitre in patients with the complete

Pendred syndrome, mice deficient in Slc26a4 do not have abnormalities in thyroid structure or function (Everett et al 2001). Slc26a4 null mice do however have deafness and variable vestibular dysfunction, with endolymphatic dilatation in the inner ear (Everett et al 2001). Within the kidney, Slc26a4 is expressed at the apical membrane of type B intercalated cells, where it functions in renal bicarbonate secretion (Royaux et al 2001). Slc26a6 is at the apical membrane of gastric glands, colocalizing with the H^+/K^+ pump (Petrovic et al 2002). Recently Slc26a7 has been found to act as a Cl^-/HCO_3^- exchanger by one group (Petrovic et al 2003) and a 'Cl$^-$ channel' by another (Kim et al 2005). Regardless, Slc26a7 is found in basolateral membranes of parietal cells (Petrovic et al 2003) and intercalated cells of the renal outer medullary collecting duct (Petrovic et al 2004).

Respiratory acid–base compensation SLC4 transporters

Systemic acid-base homeostasis (blood pH) is a balance between renal function (control of HCO_3^- and H^+ absorption and secretion) and the lungs (control of P_{CO_2} via ventilation). NBCe1 mutations cause a severe pRTA: blood pH $\geq$ 7.1, blood [HCO_3^-] 3–11 mM. Yet, P_{CO2} levels in these NBCe1 patients reveal that minimal respiratory compensation occurs ($P_{CO_2} \geq 32$ mmHg rather than <25 mmHg calculated by Henderson-Haselbalch). The patient with the Q29X-NBCe1 mutation has this uncompensated phenotype, implicating the kNBCe1 as a 'controller protein' in both renal and respiratory control of systemic acid–base homeostasis. Simplistically, this compensation could be in (a) the central chemoreceptors (pre-Bötzinger complex), (b) the peripheral chemoreceptors (the glomus cells), (c) the lung epithelia and/or (d) nerves/muscles controlling breathing (e.g. diaphragm and phrenic nerve). We know that NBCe1 mRNA is found in lung (Romero et al 1998), yet thus far NBCe1 has only been localized to basolateral membranes of Calu-3 cells (Devor et al 1999). Likewise an electroneutral NBC (NBCn1) has been cloned from pulmonary arteries (Choi et al 2000). A report using ducks as a model indicates that pulmonary chemoreceptor discharge is stimulated by the stilbene inhibitor DIDS (Shoemaker & Hempleman 2001), which can block Na^+ independent and Na^+ dependent SLC4 HCO_3^- transporters.

The AE2 anion exchanger (Slc4a2) is found in the lung (Kemp & Boyd 1993) and appears important for mucous cells of the gut and lung (Rossmann et al 2001). AE2 is thought to mediate the apical exit of HCO_3^- from epithelia of larger airways. Bronchi treated with anion inhibitors with ACh stimulation are blocked with mucin, presumably from lack of proper fluid clearance. Transport of anionic oxidants (superoxides) can cause changes in pulmonary vascular tone which is attributed to AE2 (Nozik-Grayck et al 2003). Anion exchange activity is also found in alveolar type II cells (Kemp & Boyd 1993), though the molecular entity is unclear. Recently, AE1–3 (Vince & Reithmeier 2000), NBCe1 and NBCn1 have been shown to interact with carbonic anhydrases (CA)(see for review Romero et al 2004). It is also known that mice or human null for CA-II have respiratory acidosis. Thus, interaction of HCO_3^- transporters with carbonic anhydrases may also be involved in altering gas exchange at the level of the small airways.

SLC26 transporters in the lung

Slc26a6 (Xie et al 2002) and *Slc26a9* mRNAs are found in lung tissue. Clearly, anion exchange as indicated above is also important, but Slc26 paralogues have not previously been implicated in this process. As with the Slc4 transporters, a recent report indicates that CA-II can have a stimulatory interaction with DRA (SLC26A3) (Sterling et al 2002). We identified and cloned *Slc26a9* from mouse and human lung.

Results and discussion

Slc26a9 cloning and Mrna

Given the existence of nine Slc26 anion exchangers in the *Drosophila* genome (not shown), the likelihood was high that the mammalian gene family included more than the published six members. Indeed, the existence of mammalian expressed sequence tags (ESTs) derived from novel family members was noted some time ago (Everett & Green 1999). To clone these novel family members we used genomic database mining, gene annotation, RT-PCR and EST characterization. Human cDNAs are noted with all capitals, while other species are lower case.

Human *SLC26A9* exons were initially identified in the draft sequences of the BAC clones RP11-196022 and RP11-370I5. The first coding exon was identified in these contigs using GENSCAN, neither gene annotation programs nor homology to other SLC26 proteins were useful for the identification of the last coding exon. Therefore, 3′ sequences from the *SLC26A9* genomic contigs were used to search the human EST database for potential 3′-UTR ESTs. One such EST (#2915384) was sequenced in its entirety; this insert extends from bp 2502 to 4526 of the final cDNA sequence. The open reading frame was then cloned from human lung using Takara LAPCR polymerase and a sense primer in exon 1 an antisense primer derived from the sequence of the 3′-EST. A 500 000 bp contig containing the entire mouse *Slc26a9* gene was identified using a *blastn* search of the Celera mouse genomic database with the human *SLC26A9* cDNA. A 3049 bp cDNA encompassing the entire open reading frame was then amplified by RT-PCR from mouse lung, using Takara LA polymerase and the sense primer with the antisense primer. Northern analysis revealed that robust *SLC26A9* and *Slc26a9* expression in lung.

Cl⁻ transport by Slc26a9

The hallmark of Slc26 transporters is anion transport, particularly Cl⁻. We have measured Cl⁻ uptake by several Slc26 paralogues (Fig. 1A). Using isotopes, we

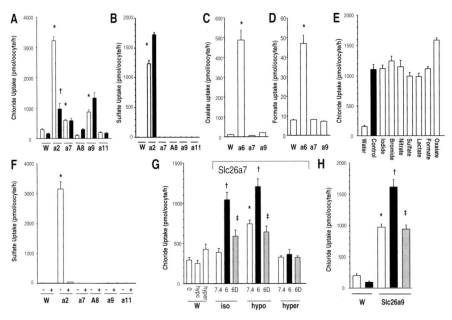

FIG. 1.   Functional characterization of SLC26/Slc26 paralogues: chloride, sulfate, oxalate and formate uptakes. (A) $^{36}Cl^-$ uptake (pmol/oocyte/h) in injected *Xenopus* oocytes, with extracellular pH of both 7.4 (open bars) and 6.0 (black fill). *Xenopus* oocytes injected as noted with water (W), Slc26a2 (a2), Slc26a7 (a7), human SLC26A8 (A8), Slc26a9 (a9) and Slc26a11 (a11). Uptakes were performed over one hour, in the absence of extracellular Cl^-. Slc26a2-, Slc26a7- and Slc26a9-injected oocytes demonstrate uptake of $^{36}Cl^-$ that is significantly higher than that of water-injected controls. (B) [^{35}S]sulfate uptake (pmol/oocyte/h) in *Xenopus* oocytes injected as noted with W, a2, a7, A8, a9 and a11. Uptakes were performed over one hour, in the absence of extracellular Cl^-, with extracellular pH of both 7.4 (open bars) and 6.0 (black fill). (C) Oxalate uptake in oocytes injected with water, a6, a7 or a9. (D) Formate uptake in oocytes injected with water, a6, a7 or a9. (E) Lack of *cis*-inhibition of $^{36}Cl^-$ uptake in Slc26a9-injected oocytes by various anionic substrates. Oocytes were incubated in $^{36}Cl^-$ uptake medium for one hour, in the absence (control) or presence of 25 mM concentrations of the indicated anions. (F) Effect of extracellular Cl^- on $^{35}SO_4^{2-}$ uptake. Oocytes injected as noted were incubated in $^{35}SO_4^{2-}$ uptake medium for one hour, in the absence (−) or presence (+) of 25 mM Cl^-. Sulfate uptake by a2-injected oocytes is *cis*-inhibited by extracellular Cl^-; unlike Slc26a1 (Xie et al 2002), latent $^{35}SO_4^{2-}$ uptake in Slc26a7/a8/a9 was not elicited by extracellular Cl^-. (G) $^{36}Cl^-$ uptake (pmol/oocyte/h) in water (W) and mouse a7-injected oocytes, as a function of extracellular osmolality. Oocytes were incubated during the pre-uptake period and uptake period at the indicated osmolality—120 mOsm/kg (hypotonic, '−'), 210 mOsm/kg (isotonic, '0') and 300 mOsm/kg (hypertonic, '+')—at either pH 7.4 (open bars), pH 6.0 (black bars), or in the presence of 1 mM DIDS at pH 6.0 (grey bars). (H) DIDS inhibition of $^{36}Cl^-$ uptake (pmol/oocyte/h) in Slc26a9-injected oocytes. Oocytes were incubated during the preuptake period and uptake period at either pH 7.4 (open bars), pH 6.0 (black bars), or in the presence of 1 mM DIDS at pH 6.0 (grey bars). The uptake experiments included 12–18 oocytes in each experimental group, and results were reported as means ± SEM. The * symbol denotes a statistically significant difference ($p < 0.05$) from water-injected control oocytes; '†' denotes $P < 0.05$ from pH 7.4; '‡' denotes $p < 0.05$ from pH 6.0 (Xie et al 2002, used with permission from The American Physiological Society).

showed that Slc26a9 had larger Cl^- uptake at pH 6 than 7.4, but uptake was higher for Slc26a3 and Slc26a6 (not shown). With Slc26a9 there was a 40% dependence on extracellular Na^+ (not shown). Cl^- uptake by Slc26a9 cannot be *cis*-inhibited by other halides or monovalent anions (Fig. 1E), suggesting that it is highly specific for Cl^-. Slc26a2 transports sulfate whereas this activity is reproducibly absent in Slc26a7, Slc26a8, Slc26a9 or Slc26a11 (Fig. 1B). We previously found that Slc26a1-mediated sulfate transport is activated considerably in the presence of extracellular Cl^- (Xie et al 2002); however, 25 mM Cl^- in the extracellular medium did not bring out a latent sulfate transport activity by Slc26a7/a8/a9, and *cis*-inhibited Slc26a2 (Fig. 1F). Likewise, though Slc26a6-expressing oocytes transport oxalate (Fig. 1C) and formate (Fig. 1D), these anions are not substrates for Slc26a7 or Slc26a9. Cl^- uptake by Slc26a7 is moderately regulated by osmolarity and extracellular pH (Fig. 1G). Slc26a7-mediated Cl^- transport (pH 6) can be inhibited by ~50% by DIDS (Fig. 1G). While we have not been able to show that Slc26a9 is regulated by osmolarity, Cl^- uptake (pH 6) is partially blocked by DIDS (Fig. 1H). In summary, Slc26a9 functions as a highly selective Cl^- transporter. Unlike Slc26a6 (Xie et al 2002), Cl^- uptake in Slc26a9-expressing cells is reproducibly increased by an acid-outside uptake solution.

To better understand the ion transport mediated by Slc26a9, we used micro-electrodes to monitor intracellular Cl^- ($[Cl^-]_i$) and membrane potential (V_m). These experiments revealed that resting $[Cl^-]_i$ for H_2O-oocytes was 43 ± 6 mM, while Slc26a9 $[Cl^-]_i$ was lower (27 ± 4 mM). Addition of 5% $CO_2/33$ mM HCO_3^- increases $[Cl^-]_i$ about the same amount for both groups (H_2O, 7.3 ± 3.1 vs. Slc26a9, 7.4 ± 1.4 mM). Removal of bath Cl^- ('$0Cl^-$' gluconate replacement), results in an immediate, rapid fall of $[Cl^-]_i$ for Slc26a9 (~20 mM) vs. H_2O (2.0 mM). $0Cl^-$ also elicits a robust depolarization (>100 mV for Slc26a9 vs. ± 2 mV for H_2O). This was our first indication that Slc26a9 was electrogenic. Removal of Na^+ ($0Na^+$, choline or *N*-methyl-D-glucamine, NMDG), did not cause measurable differences in $[Cl^-]_i$ (−0.5 mM for Slc26a9 vs. −0.7 mM for H_2O).

Slc26a9 is a novel, electrogenic Cl^-/HCO_3^- exchanger

To determine whether Slc26a9 functions as a Cl^-/HCO_3^- exchanger, we measured intracellular pH (pH_i) in response to the manipulation of both HCO_3^- and Cl^- (Fig. 2C) as previously described (Romero et al 1998, Xie et al 2002). The addition of CO_2/HCO_3^- to the bath solution causes oocyte pH_i to decrease due to CO_2 plasma membrane diffusion followed by intracellular hydration and dissociation forming intracellular H^+ and HCO_3^-. Water-injected oocytes exposed to 5% $CO_2/33$ mM HCO_3^- (pH 7.5) acidify by ~0.44 pH units at 460×10^{-5} pH units/s similar to oocytes expressing AE2 (Fig. 2A), Slc26a6 (Fig. 2B) (Xie et al 2002), Slc26a9 (Fig. 2C) and SLC26A7 (Fig. 2D). Unlike AE2-oocytes, HCO_3^- caused a slight but

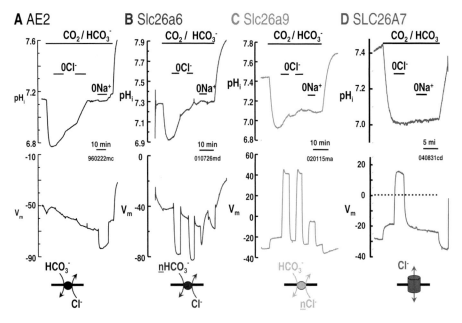

FIG. 2. Functional expression electrogenic Slc26 paralogues. Using ion-selective micro-electrodes, we measured intracellular pH (pH$_i$) and membrane potential (V$_m$) of oocytes injected with human AE2/SLC4A2 (A), mouse Slc26a6 (B), mouse Slc26a9 (C) or human SLC26A7 (D). Addition of 5% CO$_2$/33 mM HCO$_3^-$ (pH 7.5) to the bath elicits a slight depolarization and results in cell acidification in oocytes injected with water (controls, not shown). That is, CO$_2$ diffuses across the membrane, forms H$_2$CO$_3$, and then subsequently generates H$^+$ and HCO$_3^-$ from fast dissociation. AE2 (A) is not electrogenic yet alkalinizes during Cl$^-$ removal (Cl$^-$/HCO$_3^-$ exchange) but not during Na$^+$ removal. While both Slc26a6 and Slc26a9 alkalinize with Cl$^-$ removal, Slc26a6 hyperpolarizes (B) and Slc26a9 depolarizes (C). SLC26A7 depolarizes (D) with Cl$^-$ removal, yet fails to alkalinize. Panel A was modified from Romero (2001) and B (Slc26a6) (Xie et al 2002, used with permission from The American Physiological Society).

abrupt depolarization in Slc26a6, Slc26a9 and SLC26A7-injected oocytes, reminiscent of the anion conductance observed in oocytes expressing NDAE1 (Romero et al 2000), not observed in AE2-expressing cells. Replacement of Cl$^-$ (gluconate) did not change the pH$_i$ or V$_m$ observed in the water control. However, Figs 2A–D show that Cl$^-$ removal increased pH$_i$ in AE2, Slc26a6 and Slc26a9 oocytes but not SLC26A7 oocytes. Surprisingly, while gluconate replacement evoked no V$_m$ change for AE2, Slc26a6 hyperpolarized (−22.7 ± 2.9 mV, n = 9), and Slc26a9 (+60–100 mV) and SLC26A7 (+35–40 mV) depolarized. Cl$^-$ removal in solutions bubbled with 100% O$_2$, reveal a similar, yet less robust, alkalinization (+23–30 × 10^{-5} pH units/s) and depolarization (>100 mV). Thus, Slc26a9 appears to have an electrogenic Cl$^-$/HCO$_3^-$ exchange, i.e. nCl$^-$:1HCO$_3^-$ (see Fig. 2C, Fig. 5).

We also replaced Na^+ with choline to test the cation dependence of Slc26a9; Na^+ removal reduced pH_i (-15–30×10^{-5} pH units/s) and depolarized the oocytes (>25 mV). This activity is similar to that observed with an electrogenic Na^+/HCO_3^- cotransporter like NBCe1 (Dinour et al 2004, Romero et al 1998) and is distinct from known Na^+ driven Cl^--HCO_3^- exchangers like NDAE1 (Romero et al 2000). Prior to CO_2 removal, pH_i rises to 7.2 (almost the non-HCO_3^- level). Removal of CO_2/HCO_3^- elicits a robust alkalinization and pH_i overshoot to 7.9 (7.79 ± 0.11, $n = 7$), indicative of cellular HCO_3^- loading (+0.37 ± 0.06 pH units above starting pH_i) (Romero et al 2000, 1998). This overshoot was not observed in controls (not shown). The Slc26a9 oocytes also showed an abrupt hyperpolarization ($\sim$5–7 mV) with removal of CO_2/HCO_3^-.

Na^+ transport by Slc26a9

To determine whether Na^+ was a cofactor or actually transported, we made Na^+ electrodes to measure intracellular $[Na^+]$ ($[Na^+]_i$) in *Xenopus* oocytes. Resting $[Na^+]_i$ was similar for water (3.4 ± 0.2 mM) and Slc26a9 (3.8 ± 0.6 mM) as previously reported. Removal of Cl^- or Na^+ did not alter $[Na^+]_i$ for water controls. However, Na^+ removal for Slc26a9 oocytes decreased $[Na^+]_i$ ($\sim$1 mM) while also depolarizing the cell, similar to NBCe1 (Romero et al 1998). Likewise, readdition of Na^+ in the absence of Cl^- raised $[Na^+]_i$ which could be further increased by readdition of Cl^-. V_m changes were similar to those seen in Fig. 2. Xu and co-workers found that Cl^-/HCO_3^- exchange activity of Slc26a9 was inhibited by the NH_4^+ cation (Xu et al 2005), implying more general cation transport or regulation of Slc26a9. Our $[Na^+]_i$ and V_m data indicate that Slc26a9 transports Na^+ and has an activity consistent with electrogenic Na^+/nCl^- or $Na^+/nHCO_3^-$ cotransport.

Voltage clamp studies with Slc26a9

We performed two-electrode voltage clamp (VC) experiments to directly examine the currents mediated by Slc26a9 (I_{a9}; Fig. 3A,C) and compare them to those of the SLC26A7 'channel/transporter' (I_{A7}; Fig. 3B,D) (Kim et al 2005). Initial clamping in ND96 (NaCl ringer) resulted in a huge negative current (-8 to $-20 \mu A$) which decayed over $\sim$5 min (not shown). Interestingly, though SLC26A7 has been characterized as an anion channel (Kim et al 2005), the Slc26a9 currents are much larger than SLC26A7 (Fig. 3B). Figure 3C shows the current sweeps resulting from clamping at -60 mV and using a pulse protocol. For these VC experiments, we cannot inject more than 0.5 ng Slc26a9 cRNA/oocyte or we often exceed the capacity of our clamping amplifier. Current at +60 mV is often $> +60 \mu A$ and there appear to be tail-currents (not shown). Removal of Cl^- (Fig. 3A,C; inverted

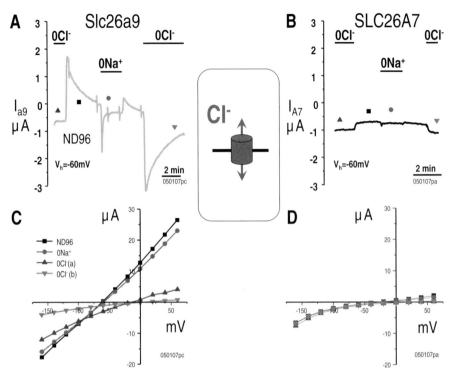

FIG. 3. Whole oocyte currents evoked by Slc26a9 vs. SLC26A7. Two electrode voltage clamp experiment on an oocyte injected with 0.5 ng/oocyte Slc26a9 and 10 ng/oocyte of human SLC26A7. SLC26A7 has been reported by Muallem's group to be an anion channel without Cl^-/HCO_3^- exchanger activity (Kim et al 2005). For these experiments, oocytes were pre-incubated in $0Cl^-$ ND96 (triangle) for 3–5 min prior to clamping to reduce the steady-state currents. (A) The Slc26a9 currents at −60 mV with the indicated solution changes. (B) The same experiment for SLC26A7. (C) Current-voltage (IV) relationship for Slc26a9. (D) SLC26A7-IV. Oocytes were held at −60 mV and stepped from −160 to +60 mV in 20 mV steps. The currents for ND96 (squares), $0Na^+$ ND96 (circle), and $0Cl^-$ ND96 (inverted triangle) are plotted.

triangle) dramatically reduced I_{a9} but currents were ± 5 μAs whereas control oocytes had currents ±300 nA (not shown). Na^+ removal elicited obvious currents for Slc26a9 (Fig. 3A,C) but not SLC26A7 (Fig. 3B,D). Removal of Na^+ and Cl^- in a CO_2/HCO_3^- solution has similar though reduced currents (not shown).

Immunolocalization of Slc26a9 in stomach

After cloning SLC26A9 and Slc26a9, we generated rabbit Slc26a9 peptide antibodies. Antibodies were immunopurified and then tested on Western blots from oocytes expressing Slc26a1, Slc26a3, Slc26a6, Slc26a8 and Slc26a9. All of the antibodies were specific for their target transporter (not shown). One Slc26a9 antibody specifically labels on mouse and rat tissues. Figure 4 shows phase contrast (A) and epifluorescent images of rat stomach (B,C).

Colocalization of Slc26a9 with the β-HK pump in mouse stomach showed only slight overlap between Slc26a9 and the HK pump (Fig. 4) as previously reported (Xu et al 2005). That is, cells secreting acid do not appear to contain much if any Slc26a9. Along the gastric gland axis, Slc26a9 expression is high at the base (not shown) and surface cells but is conspicuously absent from the middle part of the gland. Given that Slc26a9 can have several different functional modes, this distribution may reflect Na^+ transport but not Cl^-/HCO_3^- exchange or anion conductance at the gastric gland base, and anion conductance or Cl^-/HCO_3^- exchange activity by the surface cells.

Potential cellular physiology of electrogenic SLC26 transporters

Our understanding of the SLC26 family has increased greatly over the past several years (see (Mount & Romero 2004) for review). Our group has focused primarily on newer Slc26 members (Slc26a6–a11). We have found that several of these members (Slc26a6, SLC26A7 and Slc26a9) are electrogenic. Our recent studies have also shown that some members (Slc26a6 and Slc26a9) can exchange Cl^- for HCO_3^- (Fig. 5). Other members have anion conductances (SLC26A7 and Slc26a9). We also have some indications that there may be yet other activities that these

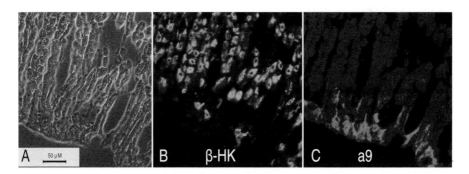

FIG. 4.   Immunolocalization of Slc26a9 in stomach. Adult mouse stomach was fixed in PLP, cryosectioned, and double-labelled (A) with a mouse monoclonal antibody against the β-HK pump (CalBiochem, used 1:1000) visualized with secondary antibody conjugated to Cy2 (B) and a rabbit antibody against Slc26a9 (used 1:2000) visualized with secondary antibody conjugated to Cy3 (C). Nuclei were stained with DAPI (not shown). A phase contrast image of the same field of view is shown (A). Note predominant labelling of Slc26a9 at the tips of microvilli and of the HK pump deeper in stomach crypts. No labelling was observed in sections exposed to secondary antibodies conjugated to Cy3 and Cy2 alone or when the Slc26a9 primary antibody was competed with peptide immunogen (not shown).

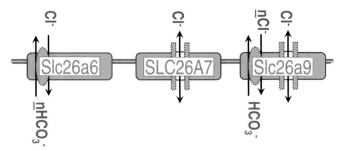

FIG. 5. Model of electrogenic Slc26 transporters. Schematic representation of the transport
activities of the electrogenic Slc26 transporters: Cl^-/HCO_3^- exchanger and anion channel.

transporters/channels possess. Future investigations will undoubtedly include dis-
secting in what tissues and membranes these transporters are found, if the varied
activities are independently regulated and if tissues expressing these proteins
display all the known activities. At the mechanistic level, a major challenge will be
to characterize the molecular determinants of the variable anion specificity, pharmacology, and electrophysiology of these fascinating transporters.

Acknowledgements

We thank Nathan Angle, Gerald Babcock and Montelle Sanders for excellent technical support.
We thank Dr Stephan W. Jones for helpful discussions and Dr Aleksandra Sindić for helpful
comments on the manuscript. We thank Dr Shmuel Muallem for providing his human SLC26A7
plasmid and Dr Seth Alper for the AE2 plasmid. We also acknowledge the many scientists in
the field whose work could not be cited due to space limitations. This work was supported by
NIH grants DK-56218 and DK-60845 (MFR); DK038226 and DK57708 (DBM), an Advanced
Career Development Award from the Veterans Administration (DBM); a new investigator grant
from the Wadsworth Foundation (CRS); and postdoctoral fellowships from the American
Heart Association (M-HC and KZ-N).

References

Bissig M, Hagenbuch B, Stieger B, Koller T, Meier PJ 1994 Functional expression cloning of
    the canalicular sulfate transport system of rat hepatocytes. J Biol Chem 269:3017–3021
Choi I, Aalkjaer C, Boulpaep EL, Boron WF 2000 An electroneutral sodium/bicarbonate
    cotransporter NBCn1 and associated sodium channel. Nature 405:571–575
Devor DC, Singh AK, Lambert LC et al 1999 Bicarbonate and chloride secretion in Calu-3
    human airway epithelial cells. J Gen Physiol 113:743–760
Dinour D, Chang M-H, Satoh J-I et al 2004 A novel missense mutation in the sodium bicar-
    bonate cotransporter (NBCe1/SLC4A4) causes proximal tubular acidosis and glaucoma
    through ion transport defects. J Biol Chem 279:52238–52246
Everett LA, Green ED 1999 A family of mammalian anion transporters and their involvement
    in human genetic diseases. Hum Mol Genet 8:1883–1891

Everett LA, Glaser B, Beck JC et al 1997 Pendred syndrome is caused by mutations in a putative sulphate transporter gene (PDS). Nat Genet 17:411–422

Everett LA, Belyantseva IA, Noben-Trauth K et al 2001 Targeted disruption of mouse Pds provides insight about the inner-ear defects encountered in Pendred syndrome. Hum Mol Genet 10:153–161

Haila S, Saarialho-Kere U, Karjalainen-Lindsberg ML et al 2000 The congenital chloride diarrhea gene is expressed in seminal vesicle, sweat gland, inflammatory colon epithelium, and in some dysplastic colon cells. Histochem Cell Biol 113:279–286

Hastbacka J, de la Chapelle A, Kaitila I et al 1992 Linkage disequilibrium mapping in isolated founder populations: diastrophic dysplasia in Finland. Nat Genet 2:204–211

Hoglund P, Haila S, Socha J et al 1996 Mutations of the down-regulated in adenoma (DRA) gene cause congenital chloride diarrhoea. Nat Genet 14:316–319

Karniski LP, Lotscher M, Fucentese M et al 1998 Immunolocalization of sat-1 sulfate/oxalate/bicarbonate anion exchanger in the rat kidney. Am J Physiol 275:F79–87

Kemp PJ, Boyd CA 1993 Anion exchange in type II pneumocytes freshly isolated from adult guinea-pig lung. Pflugers Arch 425:28–33

Kim KH, Shcheynikov N, Wang Y, Muallem S 2005 SLC26A7 Is a Cl$^-$ Channel regulated by intracellular pH. J Biol Chem 280:6463–6470

Knauf F, Yang CL, Thomson RB et al 2001 Identification of a chloride-formate exchanger expressed on the brush border membrane of renal proximal tubule cells. Proc Natl Acad Sci USA 98:9425–9430

Ko SB, Shcheynikov N, Choi JY et al 2002 A molecular mechanism for aberrant CFTR-dependent HCO$_3^-$ transport in cystic fibrosis. EMBO J 21:5662–5672

Lohi H, Kujala M, Kerkela E et al 2000 Mapping of five new putative anion transporter genes in human and characterization of SLC26A6, a candidate gene for pancreatic anion exchanger. Genomics 70:102–112

Melvin JE, Park K, Richardson L, Schultheis PJ, Shull GE 1999 Mouse down-regulated in adenoma (DRA) is an intestinal Cl$^-$/HCO$_3^-$ exchanger and is up-regulated in colon of mice lacking the NHE3 Na$^+$/H$^+$ exchanger. J Biol Chem 274:22855–22861

Mount DB, Romero MF 2004 The SLC26 gene family of multifunctional anion exchangers. Pflügers Arch 447:710–721

Nozik-Grayck E, Huang YC, Carraway MS, Piantadosi CA 2003 Bicarbonate-dependent superoxide release and pulmonary artery tone. Am J Physiol Heart Circ Physiol 285:H2327–2335

Petrovic S, Wang Z, Ma L et al 2002 Colocalization of the apical Cl$^-$/HCO$_3^-$ exchanger PAT1 and gastric H-K-ATPase in stomach parietal cells. Am J Physiol Gastrointest Liver Physiol 283:G1207–1216

Petrovic S, Ju X, Barone S et al 2003 Identification of a basolateral Cl$^-$/HCO$_3^-$ exchanger specific to gastric parietal cells. Am J Physiol Gastrointest Liver Physiol 284:G1093–1103

Petrovic S, Barone S, Xu J et al 2004 SLC26A7: a basolateral Cl$^-$/HCO$_3^-$ exchanger specific to intercalated cells of the outer medullary collecting duct. Am J Physiol Renal Physiol 286:F161–169

Romero MF 2001 The electrogenic Na$^+$/HCO$_3^-$ cotransporter, NBC. J Pancreas 2:182–191

Romero MF, Fong P, Berger UV, Hediger MA, Boron WF 1998 Cloning and functional expression of rNBC, an electrogenic Na$^+$-HCO$_3^-$ cotransporter from rat kidney. Am J Physiol 274:F425–432

Romero MF, Henry D, Nelson S et al 2000 Cloning and characterization of a Na$^+$ driven anion exchanger (NDAE1): a new bicarbonate transporter. J Biol Chem 275:24552–24559

Romero MF, Fulton CM, Boron WF 2004 The SLC4 gene family of HCO$_3^-$ transporters. Pflügers Arch 447:495–509

Rossmann H, Bachmann O, Wang Z et al 2001 Differential expression and regulation of AE2 anion exchanger subtypes in rabbit parietal and mucous cells. J Physiol 534:837–848

Royaux IE, Suzuki K, Mori A et al 2000 Pendrin, the protein encoded by the Pendred syndrome gene (PDS), is an apical porter of iodide in the thyroid and is regulated by thyroglobulin in FRTL-5 cells. Endocrinology 141:839–845

Royaux IE, Wall SM, Karniski LP et al 2001 Pendrin, encoded by the Pendred syndrome gene, resides in the apical region of renal intercalated cells and mediates bicarbonate secretion. Proc Natl Acad Sci USA 98:4221–4226

Satoh H, Susaki M, Shukunami C et al 1998 Functional analysis of diastrophic dysplasia sulfate transporter. Its involvement in growth regulation of chondrocytes mediated by sulfated proteoglycans. J Biol Chem 273:12307–12315

Scott DA, Karniski LP 2000 Human pendrin expressed in Xenopus laevis oocytes mediates chloride/formate exchange. Am J Physiol Cell Physiol 278:C207–211

Scott DA, Wang R, Kreman TM, Sheffield VC, Karniski LP 1999 The Pendred syndrome gene encodes a chloride-iodide transport protein. Nat Genet 21:440–443

Shoemaker JM, Hempleman SC 2001 Avian intrapulmonary chemoreceptor discharge rate is increased by anion exchange blocker 'DIDS'. Respir Physiol 128:195–204

Soleimani M, Greeley T, Petrovic S et al 2001 Pendrin: an apical $Cl^-/OH^-/HCO_3^-$ exchanger in the kidney cortex. Am J Physiol Renal Physiol 280:F356–364

Sterling D, Brown NJ, Supuran CT, Casey JR 2002 The functional and physical relationship between the DRA bicarbonate transporter and carbonic anhydrase II. Am J Physiol Cell Physiol 283:C1522–1529

Vince JW, Reithmeier RA 2000 Identification of the carbonic anhydrase II binding site in the Cl^-/HCO_3^- anion exchanger AE1. Biochemistry 39:5527–5533

Wang Z, Petrovic S, Mann E, Soleimani M 2002 Identification of an apical Cl(−)/HCO3(−) exchanger in the small intestine. Am J Physiol Gastrointest Liver Physiol 282:G573–579

Xie Q, Welch R, Mercado A, Romero MF, Mount DB 2002 Molecular and functional characterization of the Slc26A6 anion exchanger, functional comparison to Slc26a1. Am J Physiol Renal Physiol 283:F826–838

Xu J, Henriksnas J, Barone S et al 2005 SLC26A9 is expressed in gastric surface epithelial cells, mediates Cl^-/HCO_3^- exchange and is inhibited by NH_4^+. Am J Physiol Cell Physiol 289: C493–505

Zheng J, Shen W, He DZ et al 2000 Prestin is the motor protein of cochlear outer hair cells. Nature 405:149–155

DISCUSSION

Ashmore: The huge chloride leak seen with A9 is reminiscent of something that besets the hair cell people. This is the leak current that appears in outer hair cells and whose origin is unknown. Is there any evidence that A9 is stretch sensitive?

Romero: That's an interesting point; I haven't really thought of this. The one caveat is that we haven't looked in the ear to see whether there is an A9 message or protein. If Slc26a9 were something that were stretch sensitive, clearly with properties like this you wouldn't need a lot of Slc26a9 to cause a significant change.

Ashmore: The suggestion seems to be that A9 and A5 may both act as a chloride channels.

Welsh: Did you only see those currents after you removed extracellular chloride?

Romero: No. Readdition of Cl⁻ and removal or readdition of Na⁺ also elicits currents in oocytes expressing Slc26a9.

Oliver: Was the reversal potential of this huge current consistent with it being a pure chloride conductance?

Romero: If you remove all the chloride, yes. If you go from full chloride to zero chloride it looks like a pure anion conductance. I can also tell you that bromide will substitute for the chloride current and may even increase the current a little bit.

Case: What about bicarbonate, in the absence of chloride?

Romero: The currents are a bit diminished in the presence of chloride with bicarbonate.

Case: It is quite important for us in the pancreas. There is very little chloride in the cells and we still have to get the bicarbonate out.

Romero: This is not in the pancreas!

Muallem: In our hands it is not an HCO_3^- conductance.

Romero: There is a bit of effect on the current, but we did do the experiment where we reduced chloride levels initially to the level of our bicarbonate replacement. When we do this, there is not much or any of a residual voltage change or current going from 71 mM chloride plus 33 mM gluconate to 71 mM chloride, 33 mM bicarbonate. This issue of stoichiometry is one that I don't want to tackle because it is all over the place.

Welsh: You showed that the current is not very rectified. When you take away chloride from the outside, intracellular chloride falls. Yet when you replace the extracellular chloride, it doesn't come back.

Romero: The current comes back.

Welsh: How can that be?

Romero: That's a good question.

Alper: What happens to Shmuel Muallem's bicarbonate switch in the absence of extracellular chloride?

Romero: There doesn't seem to be an obvious bicarbonate conductance.

Muallem: We don't see it in the presence or absence of Cl⁻. This is one of the things I don't understand. We remove Cl⁻, lose the current, add Cl⁻ and see the current come back completely, but unfortunately we never thought of measuring Cl⁻ in oocytes expressing A9.

Welsh: You would have to have an enormous shift in pH for those currents.

Muallem: In our hands, the current is pH dependent. I'm referring to both A7 and A9.

Welsh: I was talking about A9.

Romero: Part of the reason for showing the chloride uptake data is that at lower extracellular pH, chloride uptake is increased with A9.

Mount: When we started doing chloride uptake experiments, we did simple experiments with pH 7.4 outside versus pH 6. Like Peter Aronson showed with

A6, there was no change. So we thought, simplistically, that increasing the gradient for hydroxyl exit would increase it. This had been shown with vesicle data. For A9 over an embarrassingly long time, there is a reproducible 40% increase in uptake. We see the same level with adding Na^+ for chloride uptake. All our other chloride uptakes are done in the nominal absence of extracellular Na^+.

Muallem: But you do acidify the oocytes when you reduce the external pH to 6.

Welsh: You remove extracellular chloride. Intracellular chloride has gone way down, yet you see large inward and outward currents. What is carrying the current? Does your reversal potential change? If you remove chloride and then wait a while, what is the reversal potential after intracellular chloride has depleted?

Romero: This is a good point, and I am not sure we have looked at this. I have been perplexed by this feature as well. It is clearly an anion conductance. For example, when we are looking to see whether other anions would replace chloride, we replace chloride with gluconate and go back to normal chloride, and then replace chloride with bromate, and then go back to normal chloride and so on. We haven't overlaid the chloride removals that were spread apart far in time.

Muallem: I can give you an answer. The membrane depolarized to about +45 mV, and then adding chloride takes this back to about −30 mV. Add it back and it comes back to exactly what it is supposed to. Which cells in the lung express A9?

Soleimani: The tracheal epithelial cells. We have looked at this.

Muallem: Is it the same as CFTR?

Soleimani: Yes, exactly the same. This is one reason A9 knockout is important. If you do antibody staining in nasal epithelia A9 is localized exactly where CFTR is. We have not done double staining with CFTR and A9 but based on published reports on CFTR, it seems that A9 and CFTR may co-localize. Further, there is significant amount of A9 labelling from nasal epithelium to the tracheal epithelium, both in human as well as mouse.

Muallem: CFTR activates A9.

Alper: One further factor is that of book-keeping. I'm glad that you mentioned that the chloride electrode measurements were done under open circuit conditions. This would compute to about 5 nM chloride per minute, which is 300 nM per hour. How does this compare with the magnitude of your influx measurements? Have you ever done an efflux measurement with isotope?

Mount: We have found with the KCCs that even under conditions that are linear we get dramatically different uptake of chloride versus rubidium. I am not comfortable making those kinds of calculations. For a transporter we accept as electroneutral, which transports 1 K^+ for 1 Cl^-, we never get concordance of chloride uptake for rubidium uptake. Partly because of this I have stayed away from making these kinds of calculations.

Alper: Those are pretty low fluxes, against a background of chloride.

Mount: For us, it is a lot.

Alper: What is the magnitude of the chloride flux you have shown here?

Mount: This is one thing that we have been puzzling about. With human A3, we get 1000–2000 picomoles per oocyte per hour compared with 1000 picomoles per oocyte per hour for a9, which has such gargantuan chloride currents.

Alper: So that is 15 picomoles per minute versus 5 nmoles per minute in open circuit recording.

Romero: We would really like to look at chloride efflux.

Alper: There could be bidirectional currents.

Romero: The chloride electrode data show that chloride falls dramatically when you are forcing it to go out, but when you give chloride back there is modest if not almost non-existent recovery of intracellular chloride.

Alper: If you start in zero chloride conditions, recording under voltage clamp, by addition of chloride you do see some significant changes.

Romero: We see a current. We haven't done the chloride electrode and clamping at the same time.

Welsh: It would be interesting to do the clamping.

Seidler: I am trying to reconcile your data on A9 with what is known in stomach physiology, particularly in relation to surface cells. People have looked unsuccessfully for chloride–bicarbonate exchange in these cells. One thing that is known is that if you put a stomach into an Ussing chamber there is a gradual build up of a dramatic negative potential difference. This is attributed to an as-yet uncharacterized cation conductance. It could be anything, though, even A9. Has anyone looked at amiloride sensitivity of A9?

Romero: No, we haven't. I have been thinking in anion mode. From our work with the Na^+ bicarbonate transporters in the SLC4 family we know that they are not particularly sensitive to industrial doses of amiloride and its variant compounds.

Seidler: Your transporter would be able to mediate chloride and Na^+ uptake or secretion. Have you looked at the change in membrane potential to see the relative movements of anion and cations?

Romero: The stoichiometry would indicate that it is more anions than Na^+.

Soleimani: We have looked at A9 activity in cultured HEK cells. We haven't looked in detail at the membrane potential, but one of the things that goes along with what you said is that when we lower the extracellular pH in the presence of bicarbonate there is more rapid intracellular acidification. This happens even when we do not have lumenal chloride on the outside. We have interpreted this to suggest that this transporter can transport bicarbonate in a chloride-independent fashion. We have referred to it as chloride-independent bicarbonate transport. We were not able to replicate these experiments in oocytes, because when we did similar studies and lowered extracellular pH in oocytes in the presence of bicarbonate we saw a huge intracellular acidification in water-injected oocytes. This suggests that there is an endogenous process in oocytes which is responsible for acidification under these circumstances. In HEK cells, the magnitude of acidification is faster in A9-expressing cells vs. non-transfected cells. Even though this pathway (chloride-independent bicarbonate transport) is very similar to what has been observed in apical membrane of tracheal epithelium and gastric surface epithelial cells one can not link A9 to these processes *in vivo* yet. Interestingly, in the presence of 4 mM ammonium the activity of this transporter is significantly decreased.

Romero: Is this in HEK cells?

Soleimani: Yes, HEK cells.

Romero: So this is alkalinization.

Soleimani: Yes, when we remove chloride the alkalinization is significantly blunted in A9.

Romero: For the benefit of others, the reason I am making this clarification is because when we give ammonium chloride to an oocyte, we don't see an alkalinization, we only see acidification. In contrast, in HEK cells we see an alkalinization in the presence of ammonium.

Soleimani: Lena Holm, our collaborator at Uppsala University, Sweden, has applied this information to gastric bicarbonate secretion *in vivo*. When she adds ammonium at the same concentration (4 mM) on the lumenal side of the stomach and stimulates acid secretion, bicarbonate secretion by surface cells is significantly inhibited. This is far fetched, but we like to suggest that this finding is relevant to the pathophysiology of peptic ulcer by *Helicobacter pylori. H. pylori* generates ammonium under its enzyme urease, which splits the urea to ammonium. If you calculate the concentration of ammonium in the stomach of people infected with *H. pylori*, it comes up to more than 7–10 mM, well above the concentration needed to inhibit gastric bicarbonate secretion.

Romero: This would fit with the extracellular pH dependency of the chloride uptakes, as well. Even though we haven't done the experiments, you would predict that if you raise extracellular pH the activity of Slc26a9 would go down significantly.

Quinton: Have you estimated what the single channel conductance of this thing might be?

Romero: No, but we have done excised patch experiments. With these sorts of currents we should be able to see something. In fact, in a normal patch electrode we can see currents at −60 mV ranging from 20–50 picoamps coming through a patch of cells that is expressing this transporter. We are setting up to look to see whether there is real channel activity here. With currents like these we should be able to see channel activity. Certainly the wave forms, and what seem to be gating currents, are very reminiscent of what many people see with chloride channels.

Quinton: Why do you want to hook this up to CFTR?

Romero: We haven't done the colocalization experiments. If there is a regulation, as with other members of the family, between CFTR and Slc26a9, this alters the picture dramatically of what might be going on in the tissue. In the absence of having done the experiments myself I don't know exactly what the distribution is. I was interested to see that A9 is seen down at the base of the glands and in the surface cells, with very little in the middle. The gradient of CFTR goes down in the stomach as we get higher up. There may be a titration effect in different types of cells along the gastric gland, assuming that there is interaction.

Welsh: Are any of these SLC26 members intracellular nucleotide dependent?

Mount: There is an assumption that they are because of the STAS domain (Aravind & Koonin 2000), which in bacteria binds GTP. They don't have a Walker motif, but they have something that could behave like that.

Muallem: The Na^+/H^+ exchanger NHE1 is dead when the cells are depleted of ATP and it is not necessary to completely deplete the cells of ATP to completely shut off NHE1. I wouldn't be surprised if the SLC26 transporters are also regulated by phosphorylation and by ATP in a mechanism that is more than simply phosphorylation.

Welsh: Phosophorylation requires only very low levels of ATP.

Alper: Niflumate inhibits $^{36}Cl^-$ flux by several SLC26 transporters. Does it also inhibit the current you observe?

Romero: We haven't tried this. Niflumic acid does not do anything to chloride uptake.

Alper: Have you noticed, while you are dumping the oocyte chloride within a matter of minutes, what happens to the volume of the oocyte?

Romero: I haven't looked at the oocytes while these experiments have been done. That's a good point.

Alper: It is such a rapid shrinkage you could imagine the thing crinkling.

Ashmore: That has a bearing on your chloride measurements, doesn't it? There is a microenvironment around your chloride electrode.

Mount: You can't necessarily make the calculation to the whole volume of the oocyte. That is a good point.

Gray: I wanted to return to the stoichiometry of anion transport by SLC26A6. You showed that chloride/hydroxyl exchange was occurring. I thought from the membrane potential measurements that you showed, the degree of hyperpolarization in the low chloride solution was different (less) from what was observed during low chloride substitution in a bicarbonate containing solution?

Romero: It is less. The pH changes are also less. It fits.

Gray: So you think the stoichiometry is the same?

Romero: We see alkalinization with removal of chloride, with and without bicarbonate, and we see hyperpolarization in both cases. In a non-bicarbonate condition the alkalinization is decreased and the hyperpolarization is decreased. They are moving in the same direction.

Gray: I wondered if there was perhaps some chance that carbonate was being transported.

Romero: We haven't looked at that.

Muallem: It seems that when you look at the Na^+ effect, the only thing that Na^+ blocks is the time-dependent current.

Romero: No, it is also the magnitude.

Muallem: Is this another mode of transport? What is going on here?

Romero: You are right; it does block a time-dependent current.

Gray: Presumably you could have two independent conductances here. If you take the Na^+ away, this just leaves the instantaneous activated current, which is very similar to what Shmuel has published for SLC26A7.

Muallem: Yes. Just look at the instantaneous current. It is about $20\,\mu A$.

Romero: What I am referring to is the steady-state current. There is still a little bit of increase of current here, but the timescale is low: we only hold these test pulses for 30 ms. If we do it longer there are other issues.

Muallem: The only thing that seems to disappear is the time-dependent activation of the Na^+ current.

Romero: You would argue that the time-dependent current is Na^+ dependent; that's an interesting idea.

Muallem: Do you see the tail currents when you hold membrane potential for shorter periods?

Romero: We need to start thinking of channel protocols. Let's hold at different voltages and look for tail currents and things like this, to get at potential differences. This is part of the direction that we are going to go with this.

Muallem: The tail current disappears in the absence of Na^+. This leads me to think that you are changing the transporter from one mode to another mode in a voltage-dependent manner.

Gray: You are not getting rectification of that current. If you are only taking external chloride away acutely, and you have a chloride conductance, you would predict some inward current rectification. You are actually shutting down a conductance when chloride is removed.

Romero: If we look at the IV curves, we see large currents when chloride is present and smaller currents, yet non-zero currents, when there is no chloride.

Gray: The reversal potential is different as well.

Muallem: The instantaneous current is still the same. The time-dependent current has disappeared.

Romero: That is a good point.

Argent: Can you enlarge on the ammonium effect? How many of these transporters demonstrate this?

Romero: We haven't looked at this. The reason I mentioned it is that the effect that Manoocher brought up was consistent with the pH effect that David has on chloride uptake. If you lower extracellular pH, chloride uptake goes up. In Manoocher's experiment, when they add ammonium they see a decrease in activity. It may be a general cation effect on the activity of Slc26a9.

Alper: I'll speak to that question with respect to A3 and A6. We haven't yet done this with A7–11. With A3 and A6 intracellular acidification by butyrate will suppress activity, while intracellular acidification by ammonium activates. We believe it is not the acidification but ammonium that is responsible. The same phenomenon applies to AE2 but not to AE1, and it applies to a certain degree to AE3.

Soleimani: There are two different processes here. We are looking at the chloride–bicarbonate exchange mode and you are looking at chloride–chloride exchange mode.

Alper: We have also shown that these things apply to chloride–bicarbonate exchange ($^{36}Cl^-$ efflux into HCO_3^-/gluconate bath). Activity seems to increase. It is happening independent of acidification.

Romero: If we haven't learned anything else, we should know by now that it is dangerous to make generalizations about how all the members of the family work. We need to look one-by-one at ion dependencies, effects of pH versus ammonia and so on.

Case: You started off by saying how different they were among different species, and then you go and do all your experiments in a frog egg! The question is not only how different are they all, but what are they doing where they are?

Welsh: I was thinking along the same lines. Can studies in animals help us, or moving back, studies in disease? We have had a lot of discussion about the mechanisms by which these transporters work. Are there any insights to be gained from going back to the physiology?

Soleimani: Some of the work, such as that done by Dr Seidler with PGE2 stimulation in animals is very relevant to human disease. The idea is that you use an alternative to stimulate bicarbonate in cystic fibrosis patients. This would be a reasonable approach. When CFTR is absent there are a number of defects, not only in the intestine or trachea, but also in the stomach. This is the reason why there has been interest in Ca^{2+}-activated chloride channels in the cystic fibrosis field.

Seidler: We know quite a bit about the gastric surface and the environment around it. Assume that A9 sits in the apical membrane of the gastric surface cells: it won't have much contact with the lumenal acid because it is protected, and it will have an apical membrane voltage of around −60 mV. What would A9 do in this situation? Would it resorb Na^+ and chloride, or would it secrete bicarbonate.

Romero: At −60 mV the anions would be moving in.

Seidler: Would the sodium drive some amount of chloride uptake?

Case: There won't be any sodium there, will there?

Seidler: Yes, there will be. Around the surface cell, intricate studies using the surface electrode have measured the pH during stimulation of acid secretion. There seems to be a rather neutral pH around 7, and the Na^+ concentration will be isotonic. The only thing really known about these surface cells is that they have a huge cation conductance. While the tissue is doing this it secretes a low level of bicarbonate. I wonder whether Na^+ and chloride uptake, a phenomenon described 20 years ago when people commonly used frog stomach, could be mediated, at least in part, via A9.

Soleimani: There is a good amount of lumenal chloride in the stomach. Independent of membrane potential, several groups have shown that A9 is a chloride–bicarbonate exchanger.

Alper: There is chloride-dependent bicarbonate movement.

Soleimani: If there is abundant chloride in the lumen, then it is possible that A9 can operate in chloride–bicarbonate exchange mode.

Aronson: The problem is that you have a stoichiometry that is more than one chloride per bicarbonate. Intracellular chloride is above equilibrium and intracellular bicarbonate is above equilibrium. If the stoichiometry of chloride/bicarbonate exchange is high enough, then the chloride gradient will have a larger effect than the bicarbonate gradient as a driving force and chloride will go out and bring bicarbonate in.

Mount: Whatever cation effect there may be, it may be buried by this.

Muallem: The HCO_3^- movement seen is very small. We see it only when there is zero Cl^- outside. If you have $100\,mM$ Cl^- on the surface of epithelial cells and a membrane potential of $-60\,mV$, I can't imagine that you will move much HCO_3^- through these transporters.

Case: But it must go.

Muallem: I don't have an obvious explanation for this. I have the same problem with CFTR.

Romero: With A9, it is only when you take away chloride that you see alkalinization.

Muallem: When you measure the rate of HCO_3^- transport at zero Cl^-, we see very slow transport compared with any of the classical Cl^-/HCO_3^- exchangers. Even CFTR transports more HCO_3^- under zero Cl^- conditions.

Seidler: That is exactly what is found in the surface cell. There is very sluggish chloride–bicarbonate exchange. Bicarbonate secretion of the isolated tissue is very low.

Muallem: Many of these anion channels will allow anions through them. As far as I know, HCO_3^- is an anion that will go through a channel. The question is, is this a functional HCO_3^- channel under physiological conditions? From the fluxes and currents that we see it is almost inconceivable to see much HCO_3^- moving through it under physiological conditions. It seems to me to operate as a very nice Cl⁻ channel. If it is a stretch activated channel I can give you a good reason why it should be in the surface epithelial cells. As acid is secreted and the gland lumen acidifies you can activate an acid activated Cl⁻ channel and recirculate Cl⁻ across the luminal membrane to support Cl⁻/HCO₃⁻ exchange and protect the surface cells in the gland from the acid secreted by parietal cells.

Soleimani: Is that parietal cells or surface epithelial cells?

Muallem: The surface cells. In the lung, if the glands start expanding and apply stretch you can activate A9 to circulate Cl⁻.

Soleimani: We haven't yet proven that A9 is the main apical bicarbonate transporter in surface epithelial cells. If our studies prove that to be the case then one way to activate A9 is when the lumen becomes acidic in response to acid secretion by parietal cells, which means that bicarbonate concentration drops to negligible levels. Coupled to high luminal chloride, that sets the chemical gradients for exchange of luminal chloride for intracellular bicarbonate.

Reference

Aravind L, Koonin EV 2000 The STAS domain—a link between anion transporters and antisigma-factor antagonists. Curr Biol 10:R53–55

Role of SLC26-mediated Cl⁻/base exchange in proximal tubule NaCl transport

Peter S. Aronson

Departments of Medicine and of Cellular and Molecular Physiology, Yale University School of Medicine, New Haven, CT 06520-8029, USA

Abstract. The majority of the Na⁺ and Cl⁻ filtered by the kidney is reabsorbed in the proximal tubule. In this nephron segment, a significant fraction of Cl⁻ is transported via apical membrane Cl⁻/base exchange: Cl⁻/formate exchange in parallel with Na⁺/H⁺ exchange and H⁺/formate cotransport, and Cl⁻/oxalate exchange in parallel with oxalate/sulfate exchange and Na⁺/sulfate cotransport. Apical membrane Cl⁻–OH⁻ or Cl⁻/HCO₃⁻ exchange has also been observed. NHE3 mediates most if not all apical membrane Na⁺/H⁺ exchange in the proximal tubule. We evaluated SLC26 family members as candidates to mediate proximal tubule Cl⁻/base exchange. We could not detect pendrin (SLC26A4) expression in the proximal tubule, and found no change in transtubular NaCl absorption in pendrin null mice. We did find expression of SLC26A6 (CFEX, PAT1) on the apical membrane of proximal tubule cells, and demonstrated that SLC26A6 is capable of mediating the Cl⁻/base exchange activities described to take place across the brush border membrane. Microperfusion studies on SLC26A6 null mice demonstrated that SLC26A6 is essential for oxalate-dependent NaCl absorption but does not contribute to baseline transport, suggesting it primarily mediates Cl⁻/oxalate exchange rather than Cl⁻–OH⁻ or Cl⁻/HCO₃⁻ exchange in the proximal tubule. Expression of SLC26A7 was also detected on the brush border membrane of proximal tubule cells. Finally, we demonstrated an essential role for the scaffolding protein PDZK1 in apical membrane expression of SLC26A6.

2006 Epithelial anion transport in health and disease: the role of the SLC26 transporters family. Wiley, Chichester (Novartis Foundation Symposium 273) p 148–163

The majority of the Na⁺ and Cl⁻ filtered by the kidney is reabsorbed in the proximal tubule. Our initial studies to elucidate the underlying molecular mechanisms began over 20 years ago and used isolated brush border membrane vesicles as an experimental system to define transport activities physiologically. In recent years, we have sought the molecular identities of the apical membrane transporters responsible for these transport activities, and we have attempted to assess the contributions of these molecularly defined pathways to mediating net NaCl transport in the proximal tubule under physiological conditions. These recent studies have identified an essential role for SLC26A6 (CFEX, PAT1) in mediating modes of apical membrane Cl^-/base exchange that contribute importantly to NaCl absorption in the proximal tubule.

Apical membrane ion exchange activities in the proximal tubule

The presence of Na^+/H^+ exchange activity in brush border vesicles isolated from rat renal cortex was first demonstrated by Murer and colleagues over 25 years ago (Murer et al 1976). We used a similar preparation of microvillus membrane vesicles isolated from rabbit renal cortex to confirm and extend these findings (Kinsella & Aronson 1980). We demonstrated that this transporter is sensitive to inhibition by amiloride and that it can function in additional modes such as to mediate Na^+/NH_4^+ exchange (Kinsella & Aronson 1981a, 1981b).

By immunofluorescence microscopy and immunoblotting of membrane fractions, we identified expression of at least three NHE isoforms in proximal tubule cells. Expression of NHE1 was detected along the basolateral membrane in the proximal tubule and in multiple other nephron segments (Biemesderfer et al 1992). In contrast, NHE3 was observed on the apical surface in the proximal tubule as well as the loop of Henle (Biemesderfer et al 1993, 1997). Recently, we demonstrated that NHE8 is also expressed on the brush border membrane of proximal tubule cells (Goyal et al 2003, 2005).

We used several different approaches to assess the contribution of NHE3 to mediating apical membrane Na^+/H^+ exchange and transtubular HCO_3^- absorption in this segment of the nephron. The profile of observed inhibitor sensitivity suggested that NHE3 accounts for virtually all of the measured Na^+/H^+ exchange activity in isolated brush border membrane vesicles (Wu et al 1996). Similarly, the profile of sensitivity to inhibitors suggested a major role for NHE3 in mediating volume and HCO_3^- absorption in microperfused proximal tubules (Wang et al 2001b). In addition, we found that the rates of fluid and HCO_3^- absorption are reduced by about 60% in microperfused tubules of NHE3 null compared to wild-type mice (Schultheis et al 1998), confirming a major role for NHE3 in mediating $NaHCO_3$ reabsorption in the proximal tubule.

Although a major component of Cl^- reabsorption in the proximal tubule is passive and paracellular, several lines of evidence indicated an important contribution of transcellular mechanisms (Aronson & Giebisch 1997). We therefore used renal brush border membrane vesicles to test for the presence of mechanisms for apical Cl^- entry that had previously been found in other nephron segments and in other epithelia. We could not detect significant Na^+/Cl^- or $Na^+/K^+/2Cl^-$ cotransport, or Cl^-/OH^- or Cl^-/HCO_3^- exchange (Seifter et al 1984). But we did find appreciable Cl^-/formate and Cl^-/oxalate exchange activities in renal brush border

membrane vesicles (Karniski & Aronson 1985, 1987). We demonstrated that Cl⁻/
oxalate exchange is electrogenic, whereas Cl⁻/formate exchange is electroneutral
(Karniski & Aronson 1987).

Given the submillimolar concentrations of formate and oxalate in biological
fluids, we reasoned that Cl⁻/formate exchange and Cl⁻/oxalate exchange could
only mediate substantial quantities of Cl⁻ absorption across the apical membrane
of proximal tubule cells if there were mechanisms available to recycle these organic
anions from the lumen back into the cell and thereby to continuously replenish
their intracellular concentrations.

When we screened for possible mechanisms of lumen to cell formate transport,
we could not demonstrate direct Na⁺/formate cotransport, but instead found that
imposing an inward H⁺ gradient markedly stimulated formate uptake and caused
its transient uphill accumulation (Karniski & Aronson 1985, Saleh et al 1996). The
kinetics of this pH-dependent formate transport suggested a mediated process of
H⁺/formate cotransport or OH⁻/formate exchange rather than simple nonionic
diffusion (Saleh et al 1996). These findings suggested a model by which formate
might facilitate NaCl entry across the apical membrane of proximal tubule cells
(Karniski & Aronson 1985), as illustrated in Fig. 1. The inward H⁺ gradient generated by apical membrane Na⁺/H⁺ exchange would drive filtered formate into the

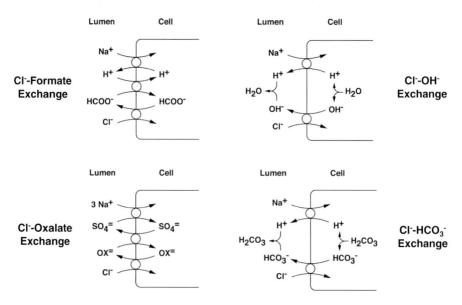

FIG. 1.   Models of apical membrane NaCl entry mediated by different modes of Cl⁻/base
exchange.

cell by H^+-coupled transport, shown in Fig. 1 as H^+/formate cotransport. The outward formate gradient would then drive Cl^- entry by Cl^-/formate exchange.

We similarly evaluated possible mechanisms to mediate lumen to cell transport of oxalate. Although we did not detect significant Na^+/oxalate cotransport, we did find that oxalate is capable of exchanging for sulfate in renal brush border vesicles (Kuo & Aronson 1996). These findings suggested a model by which oxalate might facilitate NaCl entry across the apical membrane of proximal tubule cells (Kuo & Aronson 1996), as also illustrated in Fig. 1. Uphill sulfate absorption by Na^+/ sulfate cotransport would generate an outward sulfate gradient that could then drive oxalate uptake by sulfate/oxalate exchange. The outward oxalate gradient would then drive Cl^- entry by Cl^-/oxalate exchange.

Although we did not detect appreciable Cl^-/OH^- or Cl^-/HCO_3^- exchange activity in renal brush border membrane vesicles, measurements of intracellular pH have indicated that Cl^-/OH^- or Cl^-/HCO_3^- exchange activity is present on the apical membrane of intact proximal tubule cells (Kurtz et al 1994, Sheu et al 1995), although there is nephron heterogeneity in this regard (Sheu et al 1995). As illustrated in Fig. 1, operation of Cl^-/OH^- or Cl^-/HCO_3^- exchange in parallel with Na^+/H^+ exchange is yet another potential mechanism for NaCl entry across the apical membrane of proximal tubule cells. This mechanism would not require the presence of formate or oxalate.

It was important to test the models of NaCl transport shown in Fig. 1 and to determine whether formate and oxalate actually promote transtubular NaCl reabsorption. Indeed, we found that addition of formate or oxalate strongly stimulated the rate of transtubular NaCl absorption when tubules were microperfused with a low HCO_3^-, high Cl^- solution as occurs physiologically in the late proximal tubule (Schild et al 1987, Wang et al 1992). Shown in Fig. 2, the increments in NaCl absorption induced by formate and oxalate were not additive and were completely inhibited by addition of an anion exchange inhibitor, DIDS, to the lumen perfusate or a Cl^- channel blocker, DPC, to the capillary perfusate (Wang et al 1992). These findings were consistent with the proposed roles of Cl^-/formate and Cl^-/oxalate exchange in mediating NaCl transport as shown in Fig. 1. Importantly, as also illustrated in Fig. 2, the baseline rate of NaCl absorption measured in the absence of formate and oxalate was completely insensitive to DIDS and DPC. This finding suggested that the transcellular Cl^- transport pathways operating in the presence of formate and oxalate did not operate in their absence, arguing against a significant role for Cl^-/OH^- or Cl^-/HCO_3^- exchange in mediating NaCl transport in the proximal tubule under these conditions.

We also conducted experiments to test the proposed mechanisms of formate and oxalate recycling that were illustrated in Fig. 1 (Wang et al 1996). Luminal application of the Na^+/H^+ exchange inhibitor EIPA abolished stimulation of Cl^- absorption by formate, but had no effect on stimulation by oxalate. Conversely,

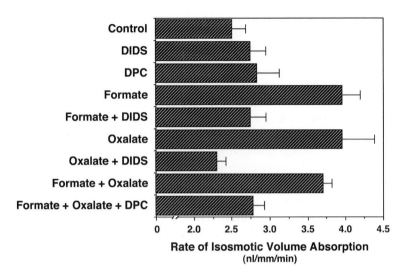

Lumen Perfusate: 145 Na$^+$, 140 Cl$^-$, 5 HCO$_3^-$, pH 6.7
Capillary Perfusate: 145 Na$^+$, 115 Cl$^-$, 25 HCO$_3^-$, pH 7.4

FIG. 2.   Effects of formate and oxalate on volume absorption (J_v) by rat proximal tubules microperfused *in situ*. Data from Wang et al (1992).

oxalate stimulation of Cl$^-$ absorption required the presence of sulfate in the perfusion solutions, but stimulation by formate was sulfate-independent. These observations strongly supported the models for NaCl reabsorption in Fig. 1 in which formate recycling is dependent on Na$^+$/H$^+$ exchange, and oxalate recycling occurs by Na$^+$/sulfate cotransport in parallel with sulfate-oxalate exchange. Further supporting these models were the findings that formate-stimulated Cl$^-$ absorption was abolished in NHE3 null mice, whereas oxalate-stimulated transport was unaffected by the absence of NHE3 (Wang et al 2001a).

Role of SLC26 anion exchangers in proximal tubule transport

The molecular identification of the apical membrane anion exchanger(s) mediating proximal tubule Cl$^-$ reabsorption had been elusive. An important advance toward this goal was the discovery by Karniski and co-workers that pendrin (SLC26A4) actually is a monovalent anion exchanger with the ability to mediate Cl$^-$/formate exchange (Scott et al 1999, Scott & Karniski 2000). We therefore evaluated pendrin as a candidate to mediate apical membrane Cl$^-$/formate exchange in the proximal tubule. We generated an anti-pendrin antibody to study immunocytochemical localization of pendrin in the kidney. We could detect no staining for pendrin in proximal tubule cells, but found pendrin expression on the apical membrane of a subpopulation of cells in the collecting tubule (Knauf et al 2001). A more detailed analysis of pendrin kidney expression was performed by Royaux et al (2001) who identified this cell population as non-α-intercalated cells mediating HCO_3^- secretion in the cortical collecting tubule. A significant functional role for pendrin in mediating proximal tubule Cl^- absorption was effectively ruled out by our studies with Karniski indicating that there is no reduction in either brush border membrane Cl^-/formate exchange or formate-stimulated NaCl absorption in pendrin null mice that had been generated by Everett and colleagues (Karniski et al 2002).

As a strategy to identify additional candidate anion exchangers capable of mediating Cl^- reabsorption in the proximal tubule, we screened the expressed sequence tag (EST) database for homologues of pendrin (Knauf et al 2001). We isolated a cDNA encoding a pendrin homologue expressed in kidney that appeared to be the mouse orthologue of human SLC26A6, a putative anion exchanger of then unknown function (Lohi et al 2000, Waldegger et al 2001). We raised antibodies against the encoded protein and demonstrated its expression in the kidney by Western blot analysis and on the brush border membrane of proximal tubule cells by immunocytochemistry (Knauf et al 2001).

We functionally expressed the transporter successfully in *Xenopus* oocytes (Knauf et al 2001). When we found that it can mediate DIDS-sensitive Cl^-/formate exchange we tentatively named it CFEX. But our subsequent studies demonstrated that CFEX (SLC26A6) can mediate additional anion exchange activities previously described in renal brush border membrane vesicles including electrogenic Cl^-/oxalate exchange, and oxalate/sulfate exchange (Jiang et al 2002). We (Jiang et al 2002) and other groups (Wang et al 2002, Xie et al 2002) also found that SLC26A6 can mediate Cl^-/HCO_3^- exchange and Cl^-/OH^- exchange. We found that affinity of SLC26A6 for oxalate was the highest among the substrates tested and that a concentration of oxalate in the range found in urine ($100\,\mu M$) can significantly convert the function of SLC26A6 from Cl^-/HCO_3^- exchange to Cl^-/oxalate exchange (Jiang et al 2002).

Taken together, these functional expression studies clearly indicated that SLC26A6 is at least capable of operating in all four of the Cl^-/base exchange modes previously illustrated in Fig. 1 that could contribute to NaCl entry across the apical membrane of proximal tubule cells. But a key issue was which if any of these modes of anion exchange are actually mediated by SLC26A6 in proximal tubule cells under physiological conditions. To address this issue, tubule microperfusion studies were performed in *Slc26a6* null mice generated by Soleimani's group (Wang et al 2005).

Proximal tubules were microperfused *in situ* with a low bicarbonate (5 mM), low pH (6.7) solution simulating conditions in the later portions of the proximal tubule after substantial secondary active reabsorption of bicarbonate has taken place. Under these conditions, the rate of volume absorption (J_v) is an index of 'isosmotic' NaCl absorption (Wang et al 1996). As shown in Fig. 3, the increment in volume absorption induced by the addition of oxalate to the perfusion solution was completely abolished in *Slc26a6* null mice. The increment in volume absorption induced by formate was also at least partially inhibited. Interestingly, the baseline rate of volume reabsorption measured in the absence of added formate and oxalate was not measurably decreased in the *Slc26a6* null mice.

The findings in Fig. 3 indicated that SLC26A6 primarily functions to mediate Cl$^-$ absorption occurring by Cl$^-$/oxalate exchange and probably Cl$^-$/formate exchange in the proximal tubule, but does not mediate Cl$^-$ absorption by Cl$^-$/HCO$_3^-$ or Cl$^-$/OH$^-$ exchange. These latter modes of Cl$^-$/base exchange, if operable, should have contributed to the baseline rate of NaCl absorption in tubules perfused with a low bicarbonate, low pH luminal solution. The lack of contribution of SLC26A6 to the baseline rate of NaCl transport is consistent with the observations shown earlier in Fig. 2 that the disulfonic stilbene DIDS, an inhibitor of SLC26A6 activity (Knauf et al 2001, Jiang et al 2002), completely abolished the

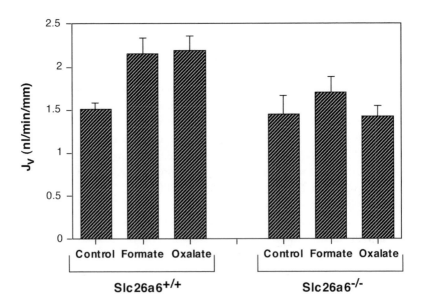

Lumen Perfusate: 145 Na$^+$, 140 Cl$^-$, 5 HCO$_3^-$, pH 6.7

FIG. 3.   Effects of formate and oxalate on volume absorption (J_v) by proximal tubules microperfused *in situ* in wild-type and SLC26A6 null mice. Data from Wang et al (2005).

increments in proximal tubule volume reabsorption induced by formate or oxalate but did not affect the baseline rate of transport (Wang et al 1992). Thus, although SLC26A6 had been shown in functional expression studies to be capable of mediating all of the modes of Cl⁻/base exchange illustrated in Fig. 1, in the intact proximal tubule under physiological conditions it appeared to primarily mediate Cl⁻/oxalate exchange and probably Cl⁻/formate exchange, but not Cl⁻/HCO₃⁻ or Cl⁻/OH⁻ exchange. Taken together, our results support the model of transcellular NaCl absorption shown in Fig. 4 in which SLC26A6, operating in parallel with Na⁺/H⁺ exchanger NHE3 and Na⁺/sulfate cotransporter NaSi, represents the principal pathway for entry of Cl⁻ across the apical membrane of proximal tubule cells.

The reasons why some modes of SLC26A6-mediated Cl⁻/base exchange might be favoured over others in a given epithelium are not yet known. One possible explanation derives from the prediction that the function of SLC26A6 should depend on the relative concentrations and affinities for substrates and inhibitors. For example, we found that oxalate is a relatively high affinity substrate for SLC26A6 and that para-aminohippurate, a prototypic organic anion secreted by proximal tubule cells, is a non-transported inhibitor (Jiang et al 2002). Perhaps the intracellular concentration of inhibitor anions like hippurates accumulated by proximal tubule cells is sufficient to block activity of SLC26A6. If the binding of

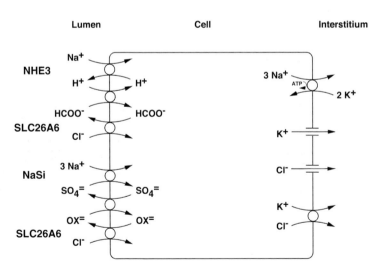

FIG. 4. Model for role of SLC26A6 in mediating transcellular NaCl absorption in the proximal tubule.

hippurates can be competitively released by the high affinity substrate oxalate but not by bicarbonate it might then explain why SLC26A6 functions principally as a Cl⁻/oxalate exchanger in the proximal tubule. Another possibility to explain tissue-specific differences in modes of transport mediated by SLC26A6 is that they result from tissue-specific differences in interactions with neighbouring membrane proteins. For example, NaCl absorption by Cl⁻/formate or Cl⁻/oxalate exchange may require clustering of SLC26A6 with the transporters mediating formate or oxalate recycling across the apical membrane as illustrated in the models of Fig. 1 and Fig. 4.

As an initial strategy to investigate the role of protein interactions in governing membrane expression or function of SLC26A6, we evaluated its interaction with the brush border scaffolding protein PDZK1. Sequence analysis of SLC26A6 had revealed the presence of a predicted PDZ-interaction motif at its extreme c-terminus (Lohi et al 2003), raising the possibility that it may interact with a PDZ domain-containing protein in the brush border. Indeed, by use of yeast two-hybrid and fusion protein interaction assays, Murer and colleagues identified PDZK1 as a potential binding partner for SLC26A6 (CFEX) (Gisler et al 2003). PDZK1 has four predicted PDZ domains and is expressed in pancreas, liver, gastrointestinal tract, adrenal cortex and kidney, where it is localized exclusively in the brush border of the proximal tubule (Kocher et al 1998).

We performed a series of pulldown assays from detergent-solubilized brush border membranes and confirmed that a fusion protein containing the C-terminal 185 amino acids of SLC26A6 was capable of interacting with native PDZK1. We then performed overlay assays to test the ability of SLC26A6 fusion proteins with or without truncation of the putative C-terminal PDZ-interaction motif to bind to a full-length PDZK1 fusion protein. These studies demonstrated that the interaction of SLC26A6 with PDZK1 was due to direct binding mediated through the putative PDZ-interaction motif located at the C-terminus of SLC26A6.

Finally, to determine the role of interaction with PDZK1 in governing the brush border membrane expression of SLC26A6, we assessed the abundance and membrane localization of SLC26A6 in PDZK1 null mice. Western blot analysis of renal membranes revealed that expression of CFEX was dramatically reduced in the kidneys of PDZK1 null mice. Immunofluorescence microscopy confirmed that expression of SLC26A6 on the brush border membrane of proximal tubule cells was markedly reduced in PDZK1 null mice. These findings indicated that the PDZ domain scaffolding protein PDZK1 plays an essential role in brush border membrane expression of SLC26A6 (Thomson et al 2005). We have recently observed that another SLC26 isoform, SLC26A7, is also expressed on the apical membrane of proximal tubule cells (Dudas et al 2006). Whether PDZK1 similarly affects the membrane expression of SLC26A7 in the proximal tubule remains to be determined.

Acknowledgements

Work in the author's laboratory has been supported by National Institutes of Health grants R01-DK37933 and PO1-DK17433.

References

Aronson PS, Giebisch G 1997 Mechanisms of chloride transport in the proximal tubule. Am J Physiol 273:F179–192

Biemesderfer D, Reilly RF, Exner M et al 1992 Immunocytochemical characterization of Na⁺-H⁺ exchanger isoform NHE-1 in rabbit kidney. Am J Physiol 263:F833–840

Biemesderfer D, Pizzonia J, Abu-Alfa A et al 1993 NHE3: A Na⁺/H⁺ exchanger isoform of renal brush border. Am J Physiol 265:F736–742

Biemesderfer D, Rutherford PA, Nagy T et al 1997 Monoclonal antibodies for high-resolution localization of NHE3 in adult and neonatal rat kidney. Am J Physiol 273:F289–299

Dudas PL, Mentone S, Greineder CF, Biemesderfer D, Aronson PS 2006 Immunolocalization of anion transporter Slc26a7 in mouse kidney. Am J Physiol Renal Physiol 290:F937–945

Gisler SM, Pribanic S, Bacic D et al 2003 PDZK1: I. A major scaffolder in brush borders of proximal tubular cells. Kidney Int 64:1733–1745

Goyal S, Vanden Heuvel G, Aronson PS 2003 Renal expression of novel Na⁺/H⁺ exchanger isoform NHE8. Am J Physiol Renal Physiol 284:F467–473

Goyal S, Mentone S, Aronson PS 2005 Immunolocalization of NHE8 in rat kidney. Am J Physiol Renal Physiol 288:F530–538

Jiang Z, Grichtchenko II, Boron WF, Aronson PS 2002 Specificity of anion exchange mediated by mouse Slc26a6. J Biol Chem 277:33963–33967

Karniski LP, Aronson PS 1985 Chloride/formate exchange with formic acid recycling: A mechanism of active chloride transport across epithelial membranes. Proc Natl Acad Sci USA 82:6362–6365

Karniski LP, Aronson PS 1987 Anion exchange pathways for Cl⁻ transport in rabbit renal microvillus membranes. Am J Physiol 253:F513–521

Karniski LP, Wang T, Everett LA et al 2002 Formate-stimulated NaCl absorption in the proximal tubule is independent of the pendrin protein. Am J Physiol Renal Physiol 283:F952–956

Kinsella JL, Aronson PS 1980 Properties of the Na⁺-H⁺ exchanger in renal microvillus membrane vesicles. Am J Physiol 238:F461–469

Kinsella JL, Aronson PS 1981a Amiloride inhibition of the Na⁺-H⁺ exchanger in renal microvillus membrane vesicles. Am J Physiol 241:F374–379

Kinsella JL, Aronson PS 1981b Interaction of NH₄⁺ and Li⁺ with the renal microvillus membrane Na⁺-H⁺ exchanger. Am J Physiol 241:C220–226

Knauf F, Yang CL, Thomson RB et al 2001 Identification of a chloride-formate exchanger expressed on the brush border membrane of renal proximal tubule cells. Proc Natl Acad Sci USA 98:9425–9430

Kocher O, Comella N, Tognazzi K, Brown LF 1998 Identification and partial characterization of PDZK1: A novel protein containing PDZ interaction domains. Lab Invest 78:117–125

Kuo SM, Aronson PS 1996 Pathways for oxalate transport in rabbit renal microvillus membrane vesicles. J Biol Chem 271:15491–15497

Kurtz I, Nagami G, Yanagawa N et al 1994 Mechanism of apical and basolateral Na⁺-independent Cl⁻/base exchange in the rabbit superficial proximal straight tubule. J Clin Invest 94:173–183

Lohi H, Kujala M, Kerkela E et al 2000 Mapping of five new putative anion transporter genes in human and characterization of SLC26A6, a candidate gene for pancreatic anion exchanger. Genomics 70:102–112

Lohi H, Lamprecht G, Markovich D et al 2003 Isoforms of SLC26A6 mediate anion transport and have functional PDZ interaction domains. Am J Physiol Cell Physiol 284: C769–779

Murer H, Hopfer U, Kinne R 1976 Sodium/proton antiport in brush-border-membrane vesicles isolated from rat small intestine and kidney. Biochem J 154:597–604

Royaux IE, Wall SM, Karniski LP et al 2001 Pendrin, encoded by the pendred syndrome gene, resides in the apical region of renal intercalated cells and mediates bicarbonate secretion. Proc Natl Acad Sci USA 98:4221–4226

Saleh AM, Rudnick H, Aronson PS 1996 Mechanism of H^+-coupled formate transport in rabbit renal microvillus membranes. Am J Physiol 271:F401–407

Schild L, Giebisch G, Karniski LP, Aronson PS 1987 Effect of formate on volume reabsorption in the rabbit proximal tubule. J Clin Invest 79:32–38

Schultheis PJ, Clarke LL, Meneton P et al 1998 Renal and intestinal absorptive defects in mice lacking the NHE3 Na^+/H^+ exchanger. Nat Genet 19:282–285

Scott DA, Karniski LP 2000 Human pendrin expressed in Xenopus laevis oocytes mediates chloride/formate exchange. Am J Physiol Cell Physiol 278:C207–211

Scott DA, Wang R, Kreman TM et al 1999 The pendred syndrome gene encodes a chloride-iodide transport protein. Nat Genet 21:440–443

Seifter JL, Knickelbein R, Aronson PS 1984 Absence of Cl-OH exchange and NaCl cotransport in rabbit renal microvillus membrane vesicles. Am J Physiol 247:F753–759

Sheu JN, Quigley R, Baum M 1995 Heterogeneity of chloride/base exchange in rabbit superficial and juxtamedullary proximal convoluted tubules. Am J Physiol 268:F847–853

Thomson RB, Wang T, Thomson BR et al 2005 Role of PDZK1 in membrane expression of renal brush border ion exchangers. Proc Natl Acad Sci USA 102:13331–13336

Waldegger S, Moschen I, Ramirez A et al 2001 Cloning and characterization of SLC26A6, a novel member of the solute carrier 26 gene family. Genomics 72:43–50

Wang T, Giebisch G, Aronson PS 1992 Effects of formate and oxalate on volume absorption in rat proximal tubule. Am J Physiol 263:F37–42

Wang T, Egbert AL Jr, Abbiati T et al 1996 Mechanisms of stimulation of proximal tubule chloride transport by formate and oxalate. Am J Physiol 271:F446–450

Wang T, Yang CL, Abbiati T et al 2001a Essential role of NHE3 in facilitating formate-dependent NaCl absorption in the proximal tubule. Am J Physiol Renal Physiol 281: F288–292

Wang T, Hropot M, Aronson PS, Giebisch G 2001b Role of NHE isoforms in mediating bicarbonate reabsorption along the nephron. Am J Physiol Renal Physiol 281:F1117–1122

Wang Z, Petrovic S, Mann E, Soleimani M 2002 Identification of an apical Cl^-/HCO_3^- exchanger in the small intestine. Am J Physiol Gastrointest Liver Physiol 282:G573–579

Wang Z, Wang T, Petrovic S et al 2005 Renal and intestinal transport defects in Slc26a6-null mice. Am J Physiol Cell Physiol 288:C957–965

Wu MS, Biemesderfer D, Giebisch G, Aronson PS 1996 Role of NHE3 in mediating renal brush border Na^+-H^+ exchange. Adaptation to metabolic acidosis. J Biol Chem 271:32749–32752

Xie Q, Welch R, Mercado A et al 2002 Molecular characterization of the murine Slc26a6 anion exchanger: functional comparison with Slc26a1. Am J Physiol Renal Physiol 283:F826–838

DISCUSSION

Markovich: Have you tested any other PDZ proteins to see whether they interact?

Aronson: No.

Soleimani: What is the renal phenotype of the PDZK1 knockout?

Aronson: We are trying to look at this. There aren't many labs that can microperfuse proximal tubules any more. Tong Wang at Yale has done a few studies. She started with oxalate-stimulated chloride transport, and this looked as if it was abolished. Also, Brent Thomson, who is the main person doing the work on PDZK1 in our lab, has preliminary data that Cl⁻/oxalate exchange in brush border membranes is greatly reduced in PDZK1 null mice.

Lee: What are the underlying molecular mechanisms of the decrease of the A6 expression in the PDZK1 knockout? Is it defects in protein sorting or in the plasma membrane half-life?

Aronson: We did things backwards. Most people would start with the association and go to transfected cells. We started with the end result and then worked backwards, to make sure there was an effect in the real animal. We did try this in oocytes, and this is an illustration of the difference between oocytes and renal proximal tubules. Coexpressing A6 with PDZK1 in oocytes made no difference, nor did truncating the C-terminus of A6. We now need to study this in a proximal tubule cell line. My guess is that it plays a role in retention, in the absence of which it is endocytosed and degraded. But this is just a postulate.

Mount: One issue with PDZK1, which is supported in CFTR, is that in some papers it seems that the NHERF1 or PDZK1 activate CFTR by tethering monomers together (Wang et al 2000, Raghuram et al 2001). It is surprising in this sense that they wouldn't function that way. I don't know how controversial this phenomenon is, the activation of channels by PDZ domain multimerization.

Aronson: That is possible, certainly. We haven't looked at NHERF1. The fact that NHE3 protein expression wasn't affected in the PDZK1 null animals doesn't mean that there is not an effect on its function. The NHERF1 knockout had perfectly normal localization of NHE3 (Shenolikar et al 2002), and NHERF1 was originally described as an NHE3-associated protein. PDZK1 had the most attention in the proximal tubule because it was found to be a major associated protein for NaPi-IIa. Yet in PDZK1 null animals there was no change in NaPi-II surface expression, at least on a control diet (Capuano et al 2005). This emphasizes that there might be multiple proteins that are interacting and competing with each other.

Alper: You have described regulation of NHE3 activity by GP330/600/megalin. Is it a stretch to say that megalin is part of a LRP family that includes the scavenger receptor? The scavenger receptor is grossly dysfunctional in the PDZK1 knockout mouse. Is megalin regulation of NHE3 altered?

Aronson: We haven't looked at that, and so far we haven't stained for megalin in the PDZK1 null mice. The phenotype that was originally found by Kocher in PDZK1 null animals was elevated cholesterol and high density lipoprotein (HDL) levels. These were attributed to loss of membrane expression of HDL receptor scavenger receptor class B type I (SR-BI) that interacts with PDZK1 in the liver (Kocher et al 2003). For those studying other SLC26s, PDZK1 is known to be expressed in liver, pancreas and intestine (Kocher et al 1998).

Stewart: Have you checked the A6 knockout for up-regulation of A7, given that they are both probably expressed in the same place?

Aronson: No.

Turner: Do you think the basal activity is just passive?

Aronson: Our assumption over the years had been that since the baseline rate of NaCl absorption measured in the absence of formate and oxalate was DIDS insensitive, and also insensitive to an inhibitor on the basolateral side that could abolish the oxalate and formate-stimulated transport, perhaps it was entirely passive and paracellular. This may be the case, particularly if A7 is not a chloride/bicarbonate exchanger. We were intrigued by the possibility that A7 had been reported to be significantly less DIDS sensitive than A6 (Petrovic et al 2003). Our assumption had been that if transtubular transport is DIDS insensitive that there is probably no anion exchange and it is passive. Perhaps this assumption is wrong if some members of this family have considerably less DIDS sensitivity. In particular, where these experiments are being done under physiological conditions with high substrate concentration, the DIDS sensitivity could be skewed even further. Our assumption has always been that it is passive, but now we are entertaining the idea that perhaps a DIDS insensitive chloride/base exchanger could play a role.

Turner: In your perfusion studies, do you always perfuse the same segment?

Aronson: These are proximal tubule loops that are accessible on the surface. This is *in situ* work, not in isolated perfused tubules.

Markovich: Do the A6 knockout mice develop oxalate kidney stones?

Soleimani: We have collaborated on this issue with Dr Marguerite Hatch at the University of Florida. We know that up to one year of age the A6 knockout animals do not develop kidney stones. What we know is that in wild-type animals there is a net secretion of oxalate, but in the knockout there is a complete reversal, resulting in net absorption of oxalate. This raises the question as to where this absorbed oxalate goes.

Markovich: Do you see a lot of oxalate in the urine?

Soleimani: We haven't looked at the crystals, but we are currently measuring this.

Alper: Has there been a mouse model of kidney stones? Aren't mice extremely resistant to oxalate nephrolithiasis?

Mount: There is calcium phosphate nephrolithiasis in mice but not oxalate.

Soleimani: We are examining the S3 segment of the proximal tubule in A6 knockout mice, as this segment is the main site of chloride reabsorption in the proximal tubule. Dr Aronson's lab is predominantly examining the S2 segment of the proximal tubule. There is a possibility that one can get a different phenotype in S3 vs. S2 segment.

Romero: Getting back to the animal models, I think they are quite useful. It is nice to see the field come back to real physiology. It is also worth mentioning that in the context of funding times getting difficult, Larry Karniski was the only person in the world who would have thought to test formate. This speaks to the fact that when people are trained in a particular arena they bring things with them to other fields that give unique insights. It is important to keep our eyes open, to use what we know from other systems for application in new systems, otherwise insights like this are going to be lost. 20 years down the road a very controversial observation that fitted well for the model of sodium chloride absorption is starting to come full circle, and we now have animal models that can be applied to disease.

Aronson: It is gratifying to see the molecular observation for some of the things we observed. But there is a larger question that we haven't tested yet: how much does this really contribute to proximal tubule reabsorption with native glomerular filtration going on?

Welsh: You can do these studies in the animals.

Aronson: Yes, we are on the waiting list for a micropuncture study.

Welsh: What about the whole organs?

Aronson: One problem with the proximal tubule is that it is easily buffered by the loop of Henle with respect to NaCl transport. In terms of having a major impact on Na^+ balance, it almost certainly doesn't. Manoocher's original description of the animals was that they had normal blood pressure. In the steady state they are going to come into Na^+ balance; the question is, at what blood pressure? This was normal. We monitored blood pressure on a low salt diet and it was also not different. We are currently evaluating effects of loop diuretics and thiazides: if we give such agents to block downstream NaCl reabsorption, there may be a bigger fall in blood pressure.

Mount: The corollary to that is that we don't know that up-regulation of A6 doesn't play a role in hypertension. There are data from Pedro Jose's lab suggesting that dopamine doesn't inhibit A6 or A6-like activity in the same way that it doesn't inhibit NHE3 activity in hypertensive rats (Pedrosa et al 2004). Does this explain why NaC1 doesn't collaborate? Does NaSi-1 or NaSi-2 interact with PDZ domain proteins?

Markovich: It does, apparently, according to a postdoc called Serge Gisler in Heini Murer's lab at the University of Zurich, Switzerland. But there isn't a consensus PDZ domain in the C-terminus. It seems to be somewhere else.

Aronson: It should be emphasized that NHE3 is associated with NHERF2 through an internal sequence rather than through its predicted C-terminal PDZ binding motif (Yun et al 1998). A protein that doesn't necessarily have a classical PDZ motif can still associate with a PDZ domain protein.

Mount: The whole clustering doesn't explain the collaboration of NHE3 for formate exchange, does it?

Aronson: We are trying to determine which domains of PDZK1 associate with SLC26A6 and NHE3. If it is one and the same domain, you could still have networks with multimers, but it would be nicer if different domains were involved.

Welsh: One of the things I thought particularly interesting was the lack of effect of knockout on Cl⁻ and bicarbonate. I don't know the significance of this. Perhaps it has to do with the ion concentrations or gradients.

Aronson: Manoocher Soleimani's lab has shown that we can make a distinction between chloride–bicarbonate exchange activity on the apical membrane and its contribution to transtubular transport.

Soleimani: In perfused S3 segment of the proximal tubule the apical chloride/bicarbonate exchange is decreased by more than 60% compared with the wild-type. This is tested by measuring the rate of intracellular pH alteration in response to changing the luminal chloride concentration. What it means in terms of net chloride absorption needs further studies. Peter Aronson's studies are focusing on the S2 segment.

Aronson: We are not ruling out that later on in the proximal tubule perhaps the chloride–bicarbonate exchange isn't working. But at least in the section of the proximal tubule being microperfused in our studies there is a mode, namely Cl^-/HCO_3^- exchange, that is perfectly demonstrable in an oocyte but doesn't seem to be participating in transport. In terms of explanations, we think oxalate has high affinity. It is possible that when oxalate is present, or other endogenous anions that might be in a proximal tubule, these anions are binding with high affinity and keeping bicarbonate off. For chloride/bicarbonate exchange to participate with Na^+/H^+ exchange to give transtubular transport, carbonic anhydrase needs to be in the complex and perhaps in this epithelium it is not in the complex.

Soleimani: What happened with the oxalate in the S3 segment?

Aronson: We haven't done this study.

Soleimani: You could look at volume absorption.

Aronson: Years ago when we did studies in rabbit, not mouse, we could stimulate reabsorption in S1 and S2 segments of proximal tubule by adding formate but we didn't study S3 (Schild et al 1987). Baum and colleagues have reported that there is heterogeneity in the proximal tubule and that certain segments have formate-independent chloride/base exchange that contributes to NaCl absorption (Sheu et al 1995).

Soleimani: Our studies show that intracellular pH is decreased in the S3 segment of the proximal tubule in A6 knockout mouse. One has to be careful that this happens in S3. We have looked at the apical Na^+/H^+ exchanger NHE3 activity in isolated perfused S3 segment and found it to be significantly reduced. I don't know whether NHE3 activity is also reduced in the S2 segment. It is possible that depending on the ion gradients or other, as yet to be discovered, properties a transporter can work in two different functional modes in two different segments in the kidney.

Muallem: I like the fact that you don't see much effect on HCO_3^- transport. But if you are ever going to learn anything from the *in vitro* system that is relevant to *in vivo* we need to agree what these transporters do. How can we get transporters to function in one way in the kidney and another in the pancreas? Do NHE3 and A6 co-immunoprecipitate? Do they exist in the same complex?

Aronson: We haven't been able to demonstrate that.

References

Capuano P, Bacic D, Stange G et al 2005 Expression and regulation of the renal Na/phosphate cotransporter NaPi-IIa in a mouse model deficient for the PDZ protein PDZK1. Pflugers Arch 449:392–402

Kocher O, Comella N, Tognazzi K Brown LF 1998 Identification and partial characterization of PDZK1: A novel protein containing PDZ interaction domains. Lab Invest 78:117–125

Kocher O, Yesilaltay A, Cirovic C et al 2003 Targeted disruption of the PDZK1 gene in mice causes tissue-specific depletion of the high density lipoprotein receptor scavenger receptor class B type I and altered lipoprotein metabolism. J Biol Chem 278:52820–52825

Pedrosa R, Jose PA, Soares-da-Silva P 2004 Defective D1-like receptor-mediated inhibition of the Cl⁻/HCO₃⁻ exchanger in immortalized SHR proximal tubular epithelial cells. Am J Physiol Renal Physiol 286:F1120–1126

Petrovic S, Ju X, Barone S et al 2003 Identification of a basolateral Cl⁻/HCO₃⁻ exchanger specific to gastric parietal cells. Am J Physiol Gastrointest Liver Physiol 284:G1093–1103

Raghuram V, Mak DD, Foskett JK 2001 Regulation of cystic fibrosis transmembrane conductance regulator single-channel gating by bivalent PDZ-domain-mediated interaction. Proc Natl Acad Sci USA 98:1300–1305

Shenolikar S, Voltz JW, Minkoff CM et al 2002 Targeted disruption of the mouse NHERF-1 gene promotes internalization of proximal tubule sodium-phosphate cotransporter type IIa and renal phosphate wasting. Proc Natl Acad Sci USA 99:11470–11475

Wang S, Yue H, Derin RB, Guggino WB, Li M 2000 Accessory protein facilitated CFTR-CFTR interaction, a molecular mechanism to potentiate the chloride channel activity. Cell 103: 169–179

SLC26 transporters and the inhibitory control of pancreatic ductal bicarbonate secretion

Péter Hegyi*†, Zoltán Rakonczay Jr.*†, László Tiszlavicz‡, András Varró§, András Tóth§, Gábor Rácz¶, Gábor Varga¶, Michael A. Gray* and Barry E. Argent*[1]

*Institute for Cell and Molecular Biosciences, University of Newcastle upon Tyne, NE2 4HH, UK, †First Department of Medicine, ‡Department of Pathology and §Department of Pharmacology and Pharmacotherapy, Faculty of Medicine, University of Szeged, H-6720, Szeged, Hungary, and ¶Molecular Oral Biology Research Group, Department of Oral Biology, Hungarian Academy of Sciences and Semmelweis University, H-1089 Budapest, Hungary

Abstract. SLC26 anion exchangers (probably SLC26A3 and SLC26A6) are expressed on the apical membrane of pancreatic duct cells and play a key role in HCO_3^- secretion; a process that is inhibited by the neuropeptide, substance P (SP). SP had no effect on basolateral HCO_3^- transporters in the duct cell or on CFTR Cl^- channels, but inhibited a Cl^--dependent HCO_3^- efflux step on the apical membrane. In microperfused ducts, luminal H_2DIDS (0.5 mM) caused intracellular pH to alkalinize (consistent with inhibition of HCO_3^- efflux) and, like SP, inhibited HCO_3^- secretion. SP did not reduce HCO_3^- secretion further when H_2DIDS was applied to the duct lumen, suggesting that SP and H_2DIDS inhibit the same transporter on the apical membrane. As SLC26A6 is DIDS-sensitive, this isoform is the likely target for SP. The inhibitory effect of SP was mimicked by phorbol 12,13-dibutyrate (PDBu), an activator of protein kinase C (PKC). Moreover, bisindolylmaleimide, a blocker of PKC, relieved the inhibitory effect of both SP and PDBu on HCO_3^- secretion. Western blot analysis revealed that guinea pig pancreatic ducts express the α, β_1, δ, ε, η, θ, ζ and μ isoforms of PKC. We conclude that PKC is a negative regulator of SLC26 activity in pancreatic duct cells.

2006 Epithelial anion transport in health and disease: the role of the SLC26 transporters family. Wiley, Chichester (Novartis Foundation Symposium 273) p 164–176

SLC26 anion exchangers play a key role in the mechanism by which pancreatic ducts secrete a fluid containing up to 140 mM $NaHCO_3$ (Fig. 1) (for reviews see Argent et al 2006, Steward et al 2005). SLC26A3 (down regulated in adenoma, DRA) and SLC26A6 (putative anion transporter 1, PAT1) are the most likely

---

[1]This paper was presented at the symposium by Barry E. Argent, to whom correspondence should be addressed.

candidates for the apical anion exchanger in the duct cell (Fig. 1). Importantly, SLC26A6 has been localized to the apical membrane of human pancreatic duct cells by immunocytochemistry (Lohi et al 2000).

How the SLC26 exchangers and apical Cl⁻ channels interact to produce a high HCO_3^- secretion is controversial (Argent et al 2006, Steward et al 2005). One hypothesis is that HCO_3^- is secreted on the SLC26 exchanger until the luminal concentration reaches about 70 mM, after which the additional HCO_3^- required to raise the luminal concentration to ~140 mM is transported via either cystic fibrosis transmembrane conductance regulator (CFTR) or Ca^{2+}-activated Cl⁻ channels (CACC), depending on whether the cAMP or the $[Ca^{2+}]_i$ intracellular signalling pathway has been activated (Sohma et al 2000, Ishiguro et al 2002). A second hypothesis is that all of the HCO_3^- is secreted on the exchangers (Ko et al 2004). In this case CFTR and CACC act to dissipate the intracellular Cl⁻ that accumulates as the exchangers cycle. The SLC26 exchangers are activated by cyclic AMP-dependent phosphorylation of CFTR (Ko et al 2002), via an interaction between the R-domain of CFTR and the STAS domain of the exchanger (Gray 2004,

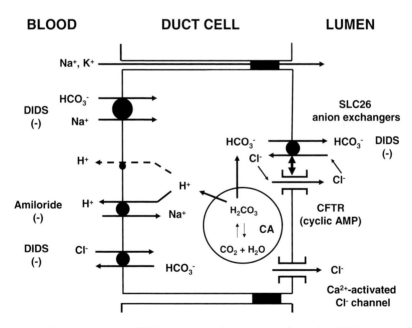

FIG. 1. Cellular mechanism of HCO_3^- secretion by pancreatic duct cells. HCO_3^- ions are first accumulated across the basolateral membrane and then exit across the apical membrane by a mechanism that involves Cl⁻ channels and SLC26 anion exchangers. The basolateral Cl⁻/HCO_3^- exchanger (probably AE2), which should oppose HCO_3^- secretion, is probably down-regulated following stimulation. CA, carbonic anhydrase; DIDS, diisothiocyanatostilbene-disulfonic acid; (−), inhibitor. For reviews see Argent et al (2006), Steward et al (2005).

Ko et al 2004). A rise in $[Ca^{2+}]_i$ also activates SLC26 exchangers, provided the cells express wild-type CFTR (Namkung et al 2003). There is evidence that SLC26A3 and A6 are electrogenic (Ko et al 2002, Xie et al 2002), although this finding is controversial (Chernova et al 2003). However, with electrogenic exchangers it is possible to envisage a secretory model in which HCO_3^- transport occurs via SLC26A6 (2 HCO_3^-/1 Cl^-) in the proximal ducts where luminal $[Cl^-]$ should be high and via SLC26A3 (1 HCO_3^-/2 Cl^-) in the more distal ducts in which $[Cl^-]$ will be lower (Ko et al 2004). However, calculations based on the electrochemical gradients for Cl^- and HCO_3^- across the luminal membrane of guinea-pig duct cells suggest that the maximum luminal HCO_3^- concentration that could be achieved by this model is rather less than the 140 mM found in pancreatic juice (Steward et al 2005).

Figure 2 summarizes the control of pancreatic ductal HCO_3^- secretion. Here we focus on inhibitory pathways, which may be physiologically important in terms of

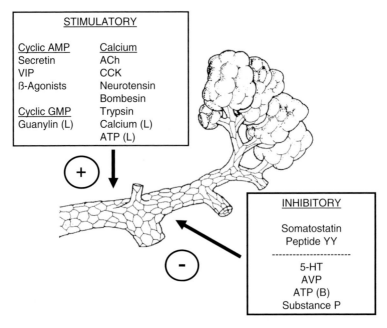

FIG. 2.   Regulation of pancreatic ductal HCO_3^- secretion. Stimulation of ductal HCO_3^- secretion can occur by activation of cyclic AMP, cyclic GMP and Ca^{2+} intracellular signalling pathways. Inhibitory control includes somatostatin and peptide YY (both of which probably work by interfering with the hormonal/neural mechanisms activating secretion) plus a number of other agents that have a direct effect on the ductal epithelium. L and B, receptors located on the luminal and basolateral sides of the duct cell respectively.

limiting the hydrostatic pressure that develops within the lumen of the duct and in terms of switching off pancreatic secretion after a meal. The neuropeptide SP has been identified in periductal nerves of the pancreas (Hegyi et al 2005), and has been shown to inhibit fluid secretion from the intact gland *in vivo* (Konturek et al 1981, Katoh et al 1984, Iwatsuki et al 1986) and *in vitro* (Kirkwood et al 1999). Furthermore, we have shown that SP inhibits HCO_3^- secretion from isolated pancreatic ducts, providing evidence for a direct effect of SP on the ductal epithelium (Ashton et al 1990, Hegyi et al 2003, 2005). The inhibitory effect of SP on isolated ducts is dose-dependent and reversed by spantide, a SP receptor antagonist (Ashton et al 1990, Hegyi et al 2003). The aim of this study was to identify the cellular mechanism by which SP exerts its inhibitory effect on pancreatic ductal HCO_3^- secretion.

Experimental procedures

The experiments were performed on small inter- and intralobular ducts isolated from the guinea-pig pancreas (Argent et al 1986). The ducts were maintained in culture overnight, during which time their ends seal and they inflate due to continued secretion into the closed luminal space (Argent et al 1986). HCO_3^- secretion in these sealed ducts was measured using changes in intracellular pH (pH_i) (Hegyi et al 2003, 2004, 2005). The 'inhibitor stop' method involves suddenly stopping HCO_3^- accumulation by the basolateral base loaders with amiloride (0.2 mM) and diisothiocyanatostilbene-disulfonic acid (DIDS, 0.1 mM) and then following the decline in pH_i as HCO_3^- secretion continues across the apical membrane (Szalmay et al 2001). The 'alkali load' method involves alkali loading cells by exposure to NH_4Cl (20 mM) and then following the recovery of pH_i. In some experiments the sealed ends of the ducts were cut off and the ducts microperfused allowing access to both the basolateral and luminal membranes (Hegyi et al 2005).

Results and discussion

SP inhibits an anion exchanger on the apical membrane of the duct cell

Initial studies on sealed ducts ruled out an effect of SP on the basolateral $NaHCO_3$ cotransporters and Na^+/H^+ exchangers and indicated that SP inhibits a Cl^--dependent HCO_3^- transport step on the apical membrane of the duct cell (Hegyi et al 2003). As SP had no effect on whole-cell CFTR currents, we concluded that the peptide inhibits an SLC26 anion exchanger (Hegyi et al 2003). Support for this idea came from studies on microperfused ducts. As with sealed ducts, SP caused a marked inhibition of both basal and secretin-stimulated HCO_3^- secretion

Inhibitor stop method

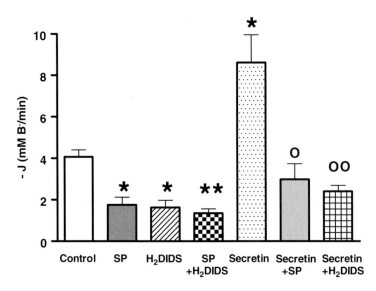

FIG. 3.   Effect of secretin, substance P and H_2DIDS on HCO_3^- secretion in microperfused ducts. Microperfused ducts were exposed to basolateral 0.2 mM amiloride and 0.5 mM H_2DIDS for 5 minutes ('inhibitor stop' method). The initial rates of intracellular acidification were measured and, knowing the buffering capacity of the cell, converted to a base flux, $-J(B^-)$ (Hegyi et al 2003, 2005). 1% albumin (control, and the vehicle for secretin and SP), 20 nM SP and/or 10 nM secretin were administered basolaterally with/without luminal H_2DIDS (0.5 mM) for 10 minutes before the test. Each column is the mean ± SEM for 5 ducts, except for the secretin + H_2DIDS experiments where n was 4. *$P < 0.05$ vs. the control, **$P < 0.001$ vs. the control; $^oP < 0.05$ vs. secretin, $^{oo}P \leq 0.05$ vs. H_2DIDS alone and secretin alone (ANOVA). (Hegyi et al 2005, used with permission from The American Physiological Society.)

(Fig. 3). Furthermore, exposing the apical membrane to H_2DIDS caused pH_i to alkalinize (consistent with inhibition of a HCO_3^- exit step on the apical membrane), and also reduced HCO_3^- secretion to a similar extent to the inhibition observed with SP (Fig. 3). As extracellular DIDS has no effect on CFTR (Gray et al 1993), we concluded that DIDS inhibits HCO_3^- secretion by inhibiting the apical anion exchanger. Importantly, SP did not further inhibit HCO_3^- secretion in the presence of H_2DIDS, consistent with SP and the disulfonic stilbene inhibiting the same transporter (Fig. 3). As SLC26A6 is DIDS sensitive (Ko et al 2002, Wang et al 2002), whereas SLC26A3 is only weakly inhibited by DIDS (Ko et al 2002, Chernova et al 2003), we conclude that A6 is the important isoform in guinea-pig ducts and that this transporter is inhibited following exposure to SP (Hegyi et al

2005). Furthermore, as H_2DIDS and SP virtually abolished the stimulatory effect of secretin on HCO_3^- secretion (Fig. 3), our data support a secretory model in which the SLC26 anion exchanger is the major pathway for HCO_3^- transport across the apical membrane of the duct cell.

Involvement of protein kinase C in the inhibitory effect of SP

Previous work in our laboratory showed that SP had no effect on intracellular cAMP concentration in isolated ducts (Ashton et al 1993), and also inhibited fluid secretion in response to dibutyryl cAMP (Ashton et al 1990). These data indicated that the inhibitory effect of SP must occur downstream in the intracellular signalling pathway from the generation of cAMP. SP interacts with tachykinin receptors, which are coupled via the G_q/G11 family of G proteins to the Ca^{2+} and protein kinase C (PKC) intracellular signalling pathways. We considered that the involvement of intracellular Ca^{2+} in the inhibitory effect of SP was unlikely as the peptide has either no effect on $[Ca^{2+}]_i$ (Evans et al 1990, Hug et al 1994), or causes only a small increase at high doses (Hug et al 1994). As it was known that activation of PKC with phorbol esters inhibits pancreatic secretion *in vivo* (Salido et al 1990), we tested the possibility that PKC was involved in the inhibitory effect of SP observed in isolated ducts. Activation of PKC with phorbol 12,13-dibutyrate (PDBu) inhibited HCO_3^- secretion from isolated ducts (Hegyi et al 2005) (Fig. 4). In contrast, the inactive phorbol ester, PDD, had no effect (Hegyi et al 2005). Furthermore, bisindolylmaleimide (BIS), a highly selective, cell-permeable, PKC inhibitor, caused a dose-dependent reversal of the inhibitory effects of both SP and PDBu on HCO_3^- secretion (Hegyi et al 2005) (Fig. 5). Finally, using Western blotting we showed that guinea-pig pancreatic ducts expressed $PKC\alpha$ and β_1 of the classic isoforms, $PKC\delta$, $-\varepsilon$, $-\eta$ and $-\theta$ of the novel isoforms and $PKC\zeta$ and $-\mu$ of the atypical isoforms (Hegyi et al 2005).

Inhibitory effects of PKC on Cl^-/HCO_3^- and Cl^-/OH^- exchange processes have previously been reported in hepatocytes (Benedetti et al 1994) and in the human Caco-2 intestinal cell line (Saksena et al 2002, 2005). In Caco-2 cells the effect of PKC on Cl^-/OH^- exchange was blocked by a specific inhibitor of the novel, Ca^{2+}-independent, $PKC\varepsilon$ (Saksena et al 2002). The same group has recently shown that c-Src and $PKC\delta$ are involved in the inhibition of Cl^-/OH^- exchange in Caco-2 cells by 5-hydroxytryptamine (Saksena et al 2005). So how might PKC activation down-regulate SLC26 exchangers? One possibility is that PKC phosphorylates the exchanger itself or an accessory protein (e.g. CFTR). Another possibility, for which some evidence exists, is that PKC stimulates endocytosis so reducing the number of transporters in the apical membrane. Activation of PKC in MDCK and Caco-2 cells has been shown to increase apical membrane endocytosis (Holm et al 1995), and PKC-mediated internalization of the human amino acid transporter hCAT-1

Alkali load method

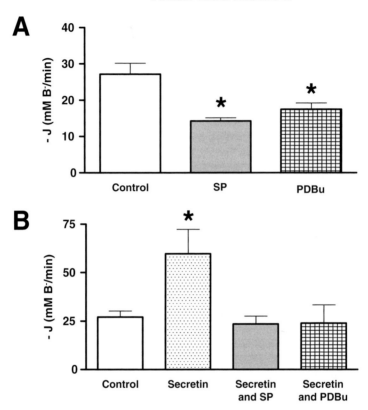

FIG. 4. Effect of protein kinase C activation on pancreatic ductal HCO_3^- secretion measured in sealed ducts using the 'alkali load' method. (A) Basal secretion. (B) Secretin-stimulated secretion. The additions, alone or in combination, were: 1% albumin (control, and the vehicle for secretin and SP), 20 nM SP, 100 nM phorbol 12,13-dibutyrate and 10 nM secretin. Base efflux ($-/[B^-]$) was calculated as described in the legend to Fig. 3. Each column is the mean ± SEM for 6–8 ducts. *$P < 0.05$ vs. the control (ANOVA). (Used with permission from The American Physiological Society, Hegyi et al 2005.)

underlies the inhibitory effect of PKC on amino acid transport in Caco-2 cells (Rotmann et al 2004).

In conclusion, our data suggest that SP and PKC are negative regulators of SLC26 anion exchangers in pancreatic duct cells. The PKC isoform that is activated by SP, the temporal and spatial aspects of PKC translocation in the duct cell, and whether PKC stimulates apical endocytosis are questions that remain to be answered.

Alkali load method

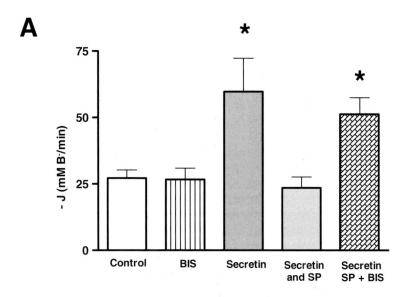

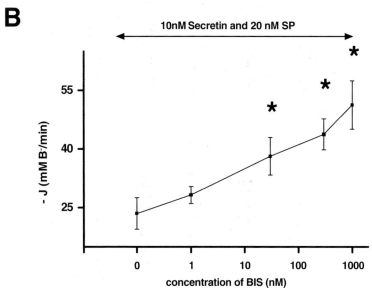

FIG. 5.   Effect of protein kinase C inhibition on pancreatic ductal HCO_3^- secretion measured using the 'alkali load' method. (A) Effect of 1000 nM bisindolylmaleimide (BIS) on HCO_3^- secretion in the presence of secretin and SP. The additions, alone or in combination, were: 1% albumin (control, and the vehicle for secretin and SP), 20 nM SP, 1000 nM bisindolylmaleimide I (BIS) and 10 nM secretin. Base efflux $(-/[B^-])$ was calculated as described in the legend to Fig. 3. Each column is the mean ± SEM of data for 6 ducts. (B) Dose–response curve for the effect of BIS on HCO_3^- secretion in the presence of secretin and SP. Each point is the mean ± SEM for five ducts. *$P < 0.05$ vs. the control (ANOVA). (Used with permission from The American Physiological Society, Hegyi et al 2005.)

Acknowledgements

Funded by a Wellcome Trust International Research Development Award (Grant No. 022618) and a Hungarian Scientific Research Fund Postdoctoral (Grant No. D42188) to PH, and by a Wellcome Trust Travelling Fellowship (Grant No. 069470) to ZR.

References

Argent BE, Arkle S, Cullen MJ, Green R 1986 Morphological, biochemical and secretory studies on rat pancreatic ducts maintained in tissue culture. Q J Exp Physiol 71:633–648

Argent BE, Gray MA, Steward MC, Case RM 2006 Cell physiology of pancreatic ducts. In: Johnson LR (eds) Physiology of the gastrointestinal tract, 4th edn. Elsevier, San Diego, in press

Ashton N, Argent BE, Green R 1990 Effect of vasoactive intestinal peptide, bombesin and substance P on fluid secretion by isolated rat pancreatic ducts. J Physiol (Lond) 427:471–482

Ashton N, Evans RL, Elliott AC, Green R, Argent BE 1993 Regulation of fluid secretion and intracellular messengers in isolated rat pancreatic ducts by acetylcholine. J Physiol (Lond) 471:549–562

Benedetti A, Strazzabosco M, Ng OC, Boyer JL 1994 Regulation of activity and apical targeting of the Cl^- HCO_3^- exchanger in rat hepatocytes. Proc Natl Acad Sci USA 91:792–796

Chernova MN, Jiang L, Shmukler BE et al 2003 Acute regulation of the SLC26A3 congenital chloride diarrhoea anion exchanger (DRA) expressed in Xenopus oocytes. J Physiol (Lond) 549:3–19

Evans RL, Ashton N, Elliot AC 1990 Intracellular free $[Ca^{2+}]$ in rat pancreatic ducts is increased by acetylcholine but not by bombesin or substance P. J Physiol (Lond) 429:90P

Gray MA 2004 Bicarbonate secretion: it takes two to tango. Nat Cell Biol 6:292–294

Gray MA, Plant S, Argent BE 1993 cAMP-regulated whole cell chloride currents in pancreatic duct cells. Am J Physiol 264:C591–602

Hegyi P, Gray MA, Argent BE 2003 Substance P inhibits bicarbonate secretion from guinea pig pancreatic ducts by modulating an anion exchanger. Am J Physiol 285:C268–276

Hegyi P, Rakonczay Z, Gray MA, Argent BE 2004 Measurement of intracellular pH in pancreatic duct cells—a new method for calibrating the fluorescence data. Pancreas 28: 427–434

Hegyi P, Rakonczay Jr Z, Tiszlavicz L et al 2005 Protein kinase C mediates the inhibitory effect of substance P on HCO_3^- secretion from guinea-pig pancreatic ducts. Am J Physiol 288: C1030–1041

Holm PK, Eker P, Sandvig K, Vandeurs B 1995 Phorbol-myristate acetate selectively stimulates apical endocytosis via protein-kinase-C in polarized MDCK cells. Exp Cell Res 217:157–168

Hug H, Pahl C, Novak I 1994 Effect of ATP, carbachol and other agonists on intracellular calcium activity and membrane voltage of pancreatic ducts. Pflügers Arch 426:412–418

Ishiguro H, Steward MC, Sohma Y et al 2002 Membrane potential and bicarbonate secretion in isolated interlobular ducts from guinea-pig pancreas. J Gen Physiol 120:617–628

Iwatsuki N, Yamgishi F, Chiba S 1986 Effects of substance P on pancreatic exocrine secretion stimulated by secretin, dopamine and cholecystokinin in dogs. Clin Exp Pharmacol Physiol 13:663–670

Katoh K, Murai K, Nonoyama T 1984 Effects of substance-P on fluid and amylase secretion in exocrine pancreas of rat and mouse. Res Vet Sci 36:147–152

Kirkwood KS, Kim EH, He XD et al 1999 Substance P inhibits pancreatic exocrine secretion via a neural mechanism. Am J Physiol 277:G314–320

Ko SBH, Shcheynikov N, Choi JY et al 2002 A molecular mechanism for aberrant CFTR-dependent HCO_3^- transport in cystic fibrosis. EMBO J 21:5662–5672

Ko SBH, Zeng WZ, Dorwart MR et al 2004 Gating of CFTR by the STAS domain of SLC26 transporters. Nat Cell Biol 6:343–350

Konturek SJ, Jaworek J, Tasler J, Cieszkowski M, Pawlik W 1981 Effect of substance P and its C-terminal hexapeptide on gastric and pancreatic secretion in the dog. Am J Physiol 241: G74–81

Lohi H, Kujala M, Kerkela E, Saarialho-Kere U, Kestila M, Kere J 2000 Mapping of five new putative anion transporter genes in human and characterization of SLC26A6, a candidate gene for pancreatic anion exchanger. Genomics 70:102–112

Namkung W, Lee JA, Ahn WI et al 2003 Ca^{2+} activates cystic fibrosis transmembrane conductance regulator- and Cl^--dependent HCO_3^- transport in pancreatic duct cells. J Biol Chem 278:200–207

Rotmann A, Strand D, Martine U, Closs EI 2004 Protein kinase C activation promotes the internalization of the human cationic amino acid transporter hCAT-1: a new regulatory mechanism for hCAT-1 activity. J Biol Chem 279:54185–54192

Saksena S, Gill RK, Syed IA et al 2002 Inhibition of apical Cl^-/OH^- exchange activity in Caco-2 cells by phorbol esters is mediated by $PKC\varepsilon$. Am J Physiol 283:C1492–1500

Saksena S, Gill RK, Tyagi S et al 2005 Involvement of c-Src and $PKC\delta$ in the inhibition of Cl^-/OH^- exchange activity in Caco-2 cells by serotonin. J Biol Chem 280:11859–11868

Salido GM, Francis LP, Camello PJ, Singh J, Madrid JA, Pariente JA 1990 Effects of phorbol esters and secretin on pancreatic juice secretion in the anaesthetized rat. Gen Pharmacol 21:465–469

Sohma Y, Gray MA, Imai Y, Argent BE 2000 HCO_3^- transport in a mathematical model of the pancreatic ductal epithelium. J Membr Biol 176:77–100

Steward MC, Ishiguro H, Case RM 2005 Mechanisms of bicarbonate secretion in the pancreatic duct. Annu Rev Physiol 67:377–409

Szalmay G, Varga G, Kajiyama F et al 2001 Bicarbonate and fluid secretion by cholecystokinin, bombesin and acetylcholine in isolated guinea-pig pancreatic ducts. J Physiol (Lond) 535.3: 795–807

Wang Z, Petrovic S, Mann E, Soleimani M 2002 Identification of an apical Cl^-/HCO_3^- exchanger in the small intestine. Am J Physiol 282:G573–579

Xie Q, Welch R, Mercado A, Romero MF, Mount DB 2002 Molecular characterization of the murine Slc26a6 anion exchanger: functional comparison with Slc26a1. Am J Physiol 283: F826–838

DISCUSSION

Case: Have you superimposed any acetazolamide on these experiments to see whether there were any additional effects?

Argent: No.

Knipper: Now we have at least two hints that endocytosis might be involved. In Peter Aronson's experiments, the PDZK1 knockout and the missing SLC26A6 (CFEX) may also point to this. Do you have any evidence that the phenotype of those knockouts that exist have lysosomal defects, such as defects in lysosomal membrane proteins such as LIMP?

Argent: No.

Aronson: For what it is worth, Hatim Hassan in my lab has found that activation of protein kinase C causes endocytosis of A6 in oocytes resulting in inhibition of transport activity, whereas it had no effect on pendrin.

Argent: This is interesting: there might be a specific effect of PKC on A6. A6 expression has been detected in the apical membrane of human pancreatic duct cells using immunocytochemistry, so A6 is the strongest candidate for involvement in ductal HCO_3^- secretion.

Markovich: Is that a direct effect by PKC?

Aronson: Hassan then screened a bunch of PKC inhibitors and found it was blocked by rottlerin, which is a relatively specific PKCδ inhibitor. There was one putative PKCδ site which was mutated but didn't affect the inhibition. We also looked at phosphorylation and couldn't detect a change. My guess is that it is an indirect effect.

Argent: What is the timescale of the PKC effect?

Aronson: 15–30 min.

Argent: We exposed the ducts to substance P for 10 min.

Case: Why did you choose 10 min?

Argent: It was the time we were using for secretin stimulation, and we observed an effect of substance P after a 10 min preincubation.

Ishiguro: Obviously apical HCO_3^- transport via anion exchange is dependent on the gradients of HCO_3^- and Cl^- across the apical membrane. I want to ask about the data from the alkaline load method by NH_4^+ application. During the alkaline load, is the HCO_3^- concentration in the cell high?

Argent: I would assume that the intracellular HCO_3^- concentration is high because the pH is high.

Ihsiguro: So perhaps there is some outward gradient of HCO_3^- and inward gradient of Cl^- across the apical membrane. If there is an anion exchanger in the apical membrane, it must work for HCO_3^- efflux in this situation. Many years ago I tried alkaline load with a different method. We superfused the isolated pancreatic ducts with a HCO_3^--CO_2-buffered solution and then changed to a HCO_3^--CO_2-free solution. We saw a big alkalinization and then recovery from alkaline load. When I removed bath Cl^- at the same time, intracellular pH went up and didn't come down again.

Argent: That would point to a Cl^--dependent HCO_3^- efflux process, which could be an SLC26 anion exchanger.

Ishiguro: An apical anion exchanger would work for HCO_3^- efflux during the alkaline load (a special experimental condition) but not necessarily in a physiological condition. In physiological conditions, intracellular pH never goes up above 7.4 even with high HCO_3^- (125–145 mM) in the lumen after secretin stimulation. The gradients of HCO_3^- and Cl^- across the apical membrane are not large enough for any anion exchanger to work for HCO_3^- efflux.

Argent: I appreciate your point about the alkali load method being unphysiological, and that is why we repeated the experiments using the inhibitor stop method as well. Using that technique the intracellular pH and HCO_3^- concentration should be physiological at the start of the experiment. We got the same results. So regardless of whether the intracellular HCO_3^- concentration is normal (inhibitor stop method) or high (alkali load method), substance P inhibits a Cl^- dependent HCO_3^- efflux mechanism. I also appreciate the fact that your careful measurements of the Cl^- and HCO_3^- electrochemical gradients across the luminal membrane of the duct cell indicate that they are not large enough to drive secretion of ~140 mM HCO_3^- on an anion exchanger. Nevertheless, in microperfusion experiments, we found that luminal H_2DIDS inhibited HCO_3^- secretion consistent with SLC26A6, rather than CFTR, being the major pathway for HCO_3^- secretion.

Alper: I wanted to follow up on what happens to Na^+/K^+ ATPase and membrane capacitance during the 15–30 min period with phorbol ester.

Aronson: We have been slow about publishing these results for that reason. For us, pendrin was a control which didn't change under the same conditions.

Muallem: Does SP activate the WNK4 kinase?

Aronson: I don't know. We have collaborated with Rick Lifton's lab, and they found that WNK4 was present in pancreatic duct at the site of A6 (Kahle et al 2004). Coexpressing WNK4 with A6 in oocytes markedly inhibits A6 activity (Kahle et al 2004).

Muallem: Stimulation of that kinase markedly induces endocytosis of A6. This might explain its inhibition by SP.

Argent: Is there any evidence that A3 and A6 are actually phosphorylated and that phosphorylation can modulate activity?

Markovich: I don't think this has been studied.

Argent: When we ran A3 and A6 through the programs for looking at phosphorylation sites there were six or seven PKC consensus sites.

Markovich: Whether they use them or not is unknown.

Argent: That is why I was asking the question. Is direct phosphorylation going to be a regulatory phenomenon, or are we always going to be talking about regulation via accessory proteins (if I can include CFTR in that category!)?

Seidler: The alkaline load method that you describe is also used for assessing the activity of NKCCs when you take the fall in intracellular pH after ammonium chloride exposure that is sensitive to loop diuretics. Do you have NKCC1 expression in your pancreatic ducts?

Argent: No, NKCC1 is not present in guinea-pig duct cells. However, NKCC1 is expressed on the basolateral membrane of rat duct cells and this probably accounts for the rat having a low bicarbonate concentration, about 70 mM, in its pancreatic juice. It seems that rat duct cells secrete a mixture of NaCl and $NaHCO_3$, whereas the duct cells of guinea-pigs, cats, dogs and humans secrete only $NaHCO_3$ up to a concentration of about 140 mM.

Reference

Kahle KT, Gimenez I, Hassan H et al 2004 WNK4 regulates apical and basolateral Cl⁻ flux in
    extrarenal epithelia. Proc Natl Acad Sci USA 101:2064–2069

Regulatory interaction between CFTR and the SLC26 transporters

Nikolay Shcheynikov*, Shigeru B. H. Ko*†[2], Weizhong Zeng*, Joo Young Choi*, Michael R. Dorwart*, Philip J. Thomas*[2] and Shmuel Muallem*[1]

Department of Physiology, University of Texas Southwestern Medical Center at Dallas, Dallas, TX 75390-9040, USA and †Division of Gastroenterology, Department of Medicine, Nagoya University Graduate School of Medicine, Nagoya 466-8550, Japan

Abstract. Most epithelia that express CFTR secrete fluid rich in HCO_3^- and poor in Cl^- that is generated by a CFTR-dependent Cl^- absorption and HCO_3^- secretion process that when aberrant leads to human diseases such as cystic fibrosis and congenital chloride diarrhoea. Epithelial Cl^- absorption and HCO_3^- secretion require expression of CFTR and other Cl^- and HCO_3^- transporters in the luminal membrane of the secreting cells. Recent advances in understanding this critical epithelial function revealed that the luminal Cl^- and HCO_3^- transporters are members of the SLC26 family. Characterization of several members of the family reveals that all characterized thus far are electrogenic with an isoform specific Cl^-/HCO_3^- transport stoichiometry. *In vivo* these transporters exist in a transporting complex with CFTR. The SLC26 transporters and CFTR are recruited to the complex by binding to scaffolds containing PDZ domains. Upon stimulation and PKA-dependent phosphorylation of CFTR R domain, the R domain binds to the SLC26 transporter STAS domain. Interaction of the R and STAS domains results in a marked and mutual activation of CFTR and the SLC26 transporters. The significance of this mode of regulation to epithelial Cl^- absorption and HCO_3^- secretion is obvious.

2006 Epithelial anion transport in health and disease: the role of the SLC26 transporters family. Wiley, Chichester (Novartis Foundation Symposium 273) p 177–192

Cl^- absorption and HCO_3^- secretion are among the most important functions of secretory epithelia, including the respiratory, reproductive and gastrointestinal systems that are vital for the health and survival of the tissues (Hadorn et al 1968, Kunzelmann & Mall 2002, Melvin et al 2005, Sokol 2001, Steward et al 2005, Wilschanski & Durie 1998). The critical importance of these activities is reflected most prominently in cystic fibrosis (CF), in which aberrant Cl^- absorption and HCO_3^- secretion play a central role in epithelial tissue destruction (Sokol 2001,

---

[1]This paper was presented at the symposium by Shmuel Muallem, to whom correspondence should be addressed.
[2]Equal contributors to work presented in this article.

Wilschanski & Durie 1998). Cl^- absorption and HCO_3^- secretion predominate in the ductal systems of secretory glands or their equivalents (Melvin et al 2005, Steward et al 2005) and surface epithelial cells in the gastrointenstinal (GI) tract (Kunzelmann & Mall 2002). The reciprocal transport of the two anions is tightly coupled, suggesting mutual dependence of the transport. However, in spite of the physiological and pathological importance of these processes, the mechanism and control of epithelial Cl^- absorption and HCO_3^- secretion remain obscure.

Epithelial Cl^- and HCO_3^- transport show considerable tissue specificity in the contribution of the electrogenic and electroneutral components to the transport and to the Cl^- and HCO_3^- gradients generated by the tissue. Probably the steepest gradient is generated by the pancreatic duct, which modifies the pancreatic juice to contain 20 mM Cl^- and 140 mM HCO_3^- (Steward et al 2005). Elegant recent work showed that the airway glands can also secrete a fluid containing more than 80 mM HCO_3^- (Irokawa et al 2004). The discovery that the pancreatic juice of CF patients is devoid of HCO_3^- and is high in Cl^- (Hadorn et al 1968) suggests that the function of CFTR is associated with HCO_3^- secretion. The fact that CFTR functions as a cAMP-regulated Cl^- channel (Sheppard & Welsh 1999) that can also conduct HCO_3^- (Poulsen et al 1994), led to the suggestion that CFTR mediates most HCO_3^- transport across the luminal membrane (LM) of epithelia. CFTR-dependent HCO_3^- current was found in several epithelia (Spiegel et al 2003, Wang et al 2003). However, the findings that some mutations in CFTR associated with CF inhibit CFTR-dependent HCO_3^- more than Cl^- transport (Choi et al 2001) and that CFTR is a relatively poor HCO_3^- transporter compared to Cl^- (Linsdell et al 1997) indicates that CFTR-dependent Cl^- absorption and HCO_3^- secretion involves several transporters that communicate with CFTR. Furthermore, CFTR activity is regulated by external Cl^- (Cl^-)(Wright et al 2004), which affects HCO_3^- transport by CFTR (Shcheynikov et al 2004). This is illustrated in Fig. 1.

Figure 1 shows the change in membrane potential and intracellular pH (pH$_i$) in *Xenopus* oocytes expressing CFTR. Removal of Cl^- resulted in the expected depolarization but with a delayed activation of HCO_3^- influx. HCO_3^- influx was observed only after the switch in membrane potential (first shaded area in Fig. 1A), suggesting delayed activation of HCO_3^- conductance by CFTR that resulted in the reduction of the membrane potential. Furthermore, re-addition of Cl^- hyperpolarized the cells, indicating activated CFTR, but failed to induce HCO_3^- efflux (second shaded area in Figure 1A). The reason for the delayed activation of HCO_3^- influx and lack of HCO_3^- efflux by activated CFTR became apparent when the effect of a gradual change in Cl^- on activation of HCO_3^- transport was examined. Fig. 1B shows that reducing Cl^- to below 20 mM was required to observe the switch in membrane potential (change in conductance) and increased HCO_3^- transport by CFTR. These and similar observations lead to the conclusion that Cl^- regulates HCO_3^- conductance of CFTR and that Cl^- has to be reduced to below 20 mM before HCO_3^- transport by CFTR becomes physiologically relevant and that addi-

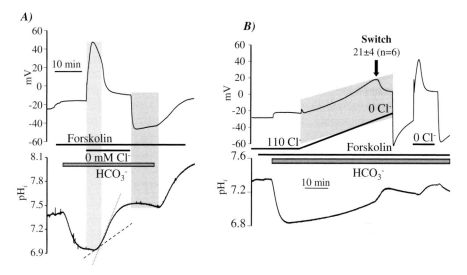

FIG. 1.   Effect of Cl⁻ on CFTR HCO₃⁻ permeability. *Xenopus* oocytes expressing CFTR were stimulated with Forskolin and bathed in HCO₃⁻-buffered media. Where indicated, the oocytes were exposed to Cl⁻-free media (A) or perfused with media in which the Cl⁻ concentration was gradually decreased (B). Membrane potential and pH$_i$ were measured with microelectrodes. (Adapted from Shcheynikov et al 2004.)

tional luminal Cl⁻ and HCO₃⁻ transporters participate in epithelial Cl⁻ absorption and HCO₃⁻ secretion (Shcheynikov et al 2004).

For a long time members of the SLC4 family of Cl⁻/HCO₃⁻ exchanger were considered as the potential Cl⁻ and HCO₃⁻ transporters at the luminal membrane (LM) of epithelia. However, members of the family were found to be expressed only in the basolateral membrane (BLM) of most epithelial cells. A breakthrough in understanding epithelial Cl⁻ and HCO₃⁻ transport was made by the discovery of the new family of Cl⁻ and HCO₃⁻ transporters, the SLC26 family. The family consists of ten genes, each of which may have several splice variants (see Mount & Romero 2004 and review by Kere 2006, this volume). Mutations in several members are linked to human diseases (Everett et al 1997, Makela et al 2002, Superti-Furga et al 1996) that arise from defective anion transport, including defective Cl⁻ and HCO₃⁻ transport. Several members are expressed in restricted tissues (Everett et al 1997, Lohi et al 2002), whereas others are wide-spread (Ko et al 2002). The excitement in the discovery of the SLC26 transporters stems from the finding that many members of the family are expressed in the LM of epithelial cells (Mount & Romero 2004). Characterization of anion transport by three members of the family, SLC26A3 (Melvin et al 1999), SLC26A4 (Soleimani et al 2001) and SLC26A6 (Ko et al 2002), revealed that all function as Cl⁻/HCO₃⁻ exchangers, whereas SLC26A7 is a H⁺-regulated Cl⁻ channel (Kim et al 2005). SLC26A6 can also transport oxalate (Jiang et al 2002) and oxalate transport appears to be the major function of SLC26A6 in the kidney (Wang et al 2005).

The potential role of the SLC26 transporters in luminal Cl^- absorption and HCO_3^- secretion prompted us and others to determine their expression profile and characterize their function and interaction with CFTR. An exploratory RT-PCR assay showed that the pancreatic and submandibular gland ducts express SLC26A2, SLC26A3, SLC26A6 and SLC26A11 and the airway epithelia express all of these transporters and SLC26A9, indicating expression of multiple SLC26 transporters isoforms in each cell. Additional key recent findings demonstrated that the SLC26 transporters are electrogenic transporters (Ko et al 2002, Xie et al 2002) that exhibit isoform specific Cl^-/HCO_3^- stoichiometry and are markedly (five–sixfold) activated by CFTR (Ko et al 2002). Some of these properties are discussed below but full details can be found in Ko et al (2002, 2004). Figure 2 shows the electrogenic properties of SLC26A3 and SLC26A6 expressed in *Xenopus* oocytes. Measurement of membrane potential revealed that removal of Cl^- depolarized the membrane potential of cells expressing SLC26A3 and bathed in HEPES-buffered media and that incubation in HCO_3^--buffered media reduced the depolarization (Fig. 2A). Current measurements showed that HCO_3^- shifted the I/V relationships to more depolarized voltages (Fig. 2B,C), indicating higher coupling of Cl^- and HCO_3^- than Cl^- and OH^- transport. This behaviour is consistent with a Cl^-/HCO_3^- stoichiometry higher than one. Indeed, in recent work to be published elsewhere in which we measured Cl^- and HCO_3^- transport in the same cells we found a Cl^-/HCO_3^- stoichiometry close to 2.

The results obtained with SLC26A6 are shown in Fig. 2D–F. Remarkably, removal of Cl^- hyperpolarized the membrane potential of cells expressing SLC26A6 and HCO_3^- increased the rate of membrane potential changes due to Cl^- removal and addition (Fig. 2D). The I/V curves in the presence and absence of HCO_3^- are shown in Fig. 2E, F. Similar changes in membrane potential of oocytes expressing SLC26A6 were reported by Xie et al (2002). On the other hand, Chernova et al (2005) expressed the human and mouse orthologues of SLC26A6 and concluded that both function as electroneutral Cl^-/HCO_3^- exchangers. A noteworthy problem with the studies of Alper and colleagues is that pH_i was measured with BCECF (2',7'-bis-[2-carboxyethyl]-5-[and-6]-carboxyfluorescein) in the large *Xenopus* oocytes. This method may not be suitable for true pH_i changes in *Xenopus* oocytes since in the oocytes BCECF reports the pH only underneath the plasma membrane, which can be very different from that in the bulk cytosol and is thus prone to artefacts. Indeed, Chernova el at (2005) reported that while the human and mouse SLC26A6 orthologues show identical 'Cl^-/HCO_3^- exchange activity' the mouse orthologue showed robust while the human orthologue showed no $^{36}Cl^-$ fluxes. This highlights the caution with which the BCECF technique should be used to measure true pH_i changes and transport function in *Xenopus*

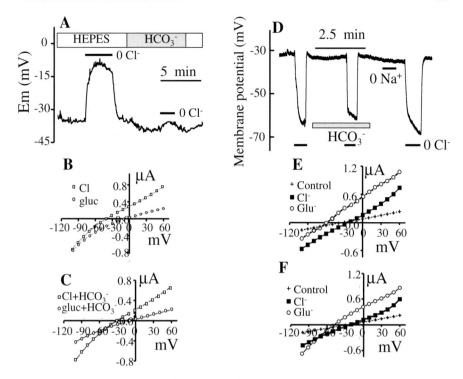

FIG. 2. Electrogenicity of the SLC26A3 and SLC26A6 transporters. Oocytes expressing SLC26A3 (A–C) or SLC26A6 (D–F) were used to measure membrane potential (A, D) or the voltage dependence of the current in the absence (B, E) or presence of HCO_3^- (C, E). (Adapted from Ko et al 2002.)

oocytes. Simultaneous Cl^- and pH_i measurements disclosed that Cl^- and HCO_3^- transport by SLC26A6 are strongly affected by the membrane potential and the HCO_3^-/Cl^- transport stoichiometry is close to 2. Therefore, it is clear that all SLC26 transporters examined so far are electrogenic and show isoform specific modes of transport and Cl^-/HCO_3^- transport stoichiometry.

CFTR expressing cells require CFTR to secrete alkaline fluid that is poor in Cl^- and rich in HCO_3^-. If the SLC26 transporters participate in this activity it would be expected that CFTR and the SLC26 transporters will exist in the same Cl^- and HCO_3^- transporting complex and will affect the activity of each other. To examine such regulatory interactions we followed the SLC26 transporters' activity by measuring Cl^-/OH^- exchange in HEK293 cells expressing SLC26 transporters alone or together with CFTR. Cl^-/OH^- exchange activity was measured since the SLC26 transporters but not CFTR mediate Cl^-/OH^- exchange (Ko et al 2002). Figures 3A–F show that CFTR prominently activates SLC26A3, SLC26A4 and SLC26A6.

Notably, stimulation of CFTR by cAMP was required for activation of the SLC26 transporters by CFTR. The reciprocal experiments are shown in Figs 3G, H. In these experiments HEK293 cells co-expressing CFTR and SLC26A3 (Fig. 3) or SLC26A6 (Ko et al 2004) were used to measure the cAMP-activated Cl⁻ current mediated by CFTR. Figure 3 shows that SLC26A3 increased CFTR current by about twofold. Analysis of surface expression of CFTR by biotinylation indicated that the increased current was not due to increased expression of CFTR in the plasma membrane and single channel analysis in cell-attached patches revealed that the SLC26 transporters increased CFTR NP by reducing the mean close time (Ko et al 2004).

A search for the mechanism by which CFTR and SLC26 transporters mutually activate each other revealed multiple interactions between the proteins. CFTR

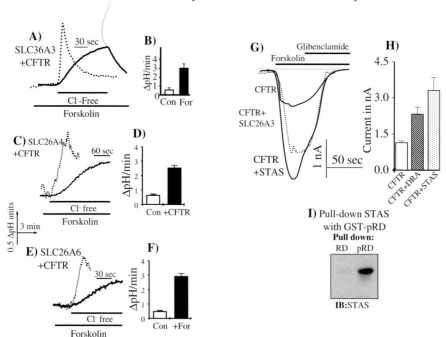

FIG. 3.   Activation of SLC26A3, SLC26A4 and SLC26A6 by CFTR (A–F) and of CFTR by SLC26A3 and its STAS domain (G–I). HEK293 cells expressing SLC26A3 (A, B), SLC26A4 (C, D) or SLC26A6 (E, F) without (solid traces) or with CFTR (doted traces) bathed in HEPES-buffered media were used to measure Cl⁻/OH⁻ exchange activity by the SLC26 transporters. Note the three–fivefold activation of the SLC26 transporters by CFTR (Adapted from Ko et al 2002.) In (G, H) HEK293 cells expressing CFTR alone, CFTR and SLC26A3 or CFTR and the SLC26A3-STAS domain were used to measure CFTR-mediated Cl⁻ current. In (I) recombinant unphosphorylated (RD) or phosphorylated (pRD) R domain were used to pull down the STAS domain. (Adapted from Ko et al 2004, with permission from Nature, www. nature.com.)

(Wang et al 1998) and most SLC26 transporters (Mount & Romero 2004) have a PDZ domain binding motif in their C-terminus. Deletion of either motif resulted in dissociation of the CFTR–SLC26A3 complexes and loss of activation of the SLC26A3 by CFTR. However, upon overexpression of the proteins the interaction and activation were restored, leading to the discovery that the SLC26 transporters' STAS domain interacts with the CFTR R domain (Ko et al 2004). Furthermore, as shown in Fig. 3G,H the isolated STAS domain was even more effective than the SLC26 transporters in activating CFTR. Binding assays with expressed (Ko et al 2004) or purified recombinant proteins (Fig. 3I) showed that phosphorylation of the R domain markedly increased its binding to SLC26A3 and the STAS domain.

That the STAS domain was sufficient for activation of CFTR raised the question of whether the CFTR R domain is sufficient for activation of the SLC26 transporters. Expression of the R domain alone failed to activate Cl^-/OH^- exchange by SLC26A3. Suspecting that this could be because the soluble R domain failed to target to the plasma membrane and bind to SLC26A3 we attached the R domain to either half of CFTR to aid in its targeting to the plasma membrane. The constructs are the same and were prepared as reported before (Yue et al 2000). Figure 4 shows that the first (TM1) or second (T2N2) transmembrane sectors of CFTR devoid of the R domain failed to activate SLC26A3. However, when the R domain was attached to these constructs (T1N1**R** and **R**T2N2) they activated SLC26A3 Cl^-/OH^- exchange activity by 90–170%. These findings suggest that the R domain may be sufficient for activation of the SLC26 transporters.

Based on these finding it is possible to suggest the model illustrated in Fig. 5 for the interaction and mutual regulation of CFTR and the SLC26 transporters. In this model, the transporters bind with their PDZ ligands to a common scaffolding protein that possesses at least two PDZ domains. The binding serves to recruit the CFTR and SLC26 transporters to a Cl^- and HCO_3^- transporting complex. In the complex the STAS domain can interact with the R domain. The interaction takes place or is markedly enhanced by PKA-mediated phosphorylation of the R domain. Whether phosphorylation of the STAS domain is also required for its binding to the R domain remains to be determined. Binding of the R and STAS domains results in the mutual activation of the two transporters and stimulation of epithelial Cl^- absorption and HCO_3^- secretion. The SLC26 transporters that mediate activation of CFTR and the Cl^- and HCO_3^- transport depends on the repertoire of SLC26 transporters expressed in each cell type. The fact that cells express multiple SLC26 transporters suggests that different members of the SLC26 transporters have specific and perhaps redundant roles in epithelial Cl^- absorption and HCO_3^- secretion. This may be inferred from the mild phenotype observed in mice from which the gene coding for the ubiquitous SLC26A6 was deleted (Wang et al 2005). The continued effort to understand the transport function of each

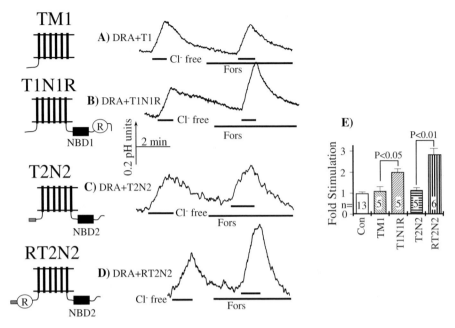

FIG. 4. The CFTR R domain is sufficient to activate SLC26A3. HEK293 cells were transfected with SLC26A3 (DRA) and the CFTR constructs TM1, T1N1R, T2N2 or RT2N2 and used to measure activation of SLC26A3 by the plasma membrane targeted R domain. The CFTR constructs were prepared according to Yue et al 2000.

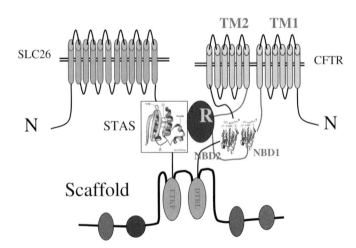

FIG. 5. A model depicting interaction between CFTR and SLC26A6 transporters with the aid of PDZ domains containing scaffolding protein.

SLC26 transporter and the use of mutant mice will undoubtedly provide important information to better understand the physiological role of this intriguing family of Cl⁻ and HCO₃⁻ transporters in epithelial Cl⁻ absorption and HCO₃⁻ secretion and other activities.

References

Chernova MN, Jiang L, Friedman DJ et al 2005 Functional comparison of mouse slc26a6 anion exchanger with human SLC26A6 polypeptide variants: Differences in anion selectivity, regulation, and electrogenicity. J Biol Chem 280:8564–8580

Choi JY, Muallem D, Kiselyov K, Lee MG, Thomas PJ, Muallem S 2001 Aberrant CFTR-dependent HCO₃⁻ transport in mutations associated with cystic fibrosis. Nature 410:94–97

Everett LA, Glaser B, Beck JC et al 1997 Pendred syndrome is caused by mutations in a putative sulphate transporter gene (PDS). Nat Genet 17:411–422

Hadorn B, Johansen PG, Anderson CM 1968 Pancreozymin secretin test of exocrine pancreatic function in cystic fibrosis and the significance of the result for the pathogenesis of the disease. Can Med J 98:377–384

Irokawa T, Krouse ME, Joo NS, Wu JV, Wine JJ 2004 A 'virtual gland' method for quantifying epithelial fluid secretion. Am J Physiol Lung Cell Mol Physiol 287:L784–793

Jiang Z, Grichtchenko II, Boron WF, Aronson PS 2002 Specificity of anion exchange mediated by mouse Slc26a6. J Biol Chem 277:33963–33967

Kere J 2006 Overview of the SLC26 family and associated diseases. In: Epithelial anion transport in health and disease: the role of the SLC26 transporter family. Wiley, Chichester (Novartis Found Symp 273) p 2–18

Kim KH, Shcheynikov N, Wang Y, Muallem S 2005 SLC26A7 is a Cl- channel regulated by intracellular pH. J Biol Chem 280:6463–6470

Ko SB, Shcheynikov N, Choi JY et al 2002 A molecular mechanism for aberrant CFTR-dependent HCO₃⁻ transport in cystic fibrosis. EMBO J 21:5662–5672

Ko SB, Zeng W, Dorwart MR 2004 Gating of CFTR by the STAS domain of SLC26 transporters. Nat Cell Biol 6:343–350

Kunzelmann K, Mall M 2002 Electrolyte transport in the mammalian colon: mechanisms and implications for disease. Physiol Rev 82:245–289

Linsdell P, Tabcharani JA, Rommens et al 1997 Permeability of wild-type and mutant cystic fibrosis transmembrane conductance regulator chloride channels to polyatomic anions. J Gen Physiol 110:355–364

Lohi H, Kujala M, Makela S et al 2002 Functional characterization of three novel tissue-specific anion exchangers SLC26A7, -A8, and -A9. J Biol Chem 277:14246–14254

Makela S, Kere J, Holmberg C, Hoglund P 2002 SLC26A3 mutations in congenital chloride diarrhea. Hum Mutat 20:425–438

Melvin JE, Park K, Richardson L, Schultheis PJ, Shull GE 1999 Mouse down-regulated in adenoma (DRA) is an intestinal Cl⁻/HCO₃⁻ exchanger and is up-regulated in colon of mice lacking the NHE3 Na⁺/H⁺ exchanger. J Biol Chem 274:22855–22861

Melvin JE, Yule D, Shuttleworth T, Begenisich T 2005 Regulation of fluid and electrolyte secretion in salivary gland acinar cells. Annu Rev Physiol 67:445–469

Mount DB, Romero MF 2004 The SLC26 gene family of multifunctional anion exchangers. Pflugers Arch 447:710–721

Poulsen JH, Fischer H, Illek B, Machen TE 1994 Bicarbonate conductance and pH regulatory capability of cystic fibrosis transmembrane conductance regulator. Proc Natl Acad Sci USA 91:5340–5344

Shcheynikov N, Kim KH, Kim KM et al 2004 Dynamic control of cystic fibrosis trans-membrane conductance regulator Cl^-/HCO_3^- selectivity by external Cl^-. J Biol Chem 279:21857–21865

Sheppard DN, Welsh MJ 1999 Structure and function of the CFTR chloride channel. Physiol Rev 79:S23–45

Sokol RZ 2001 Infertility in men with cystic fibrosis. Curr Opin Pulm Med 7:421–426

Soleimani M, Greenley T, Petrovic et al 2001 Pendrin: and apical $Cl^-/OH^-/HCO3^-$ exchanger in the kidney cortex. Am J Physiol 280:F356–364

Spiegel S, Phillipper M, Rossmann H, Riederer B, Gregor M, Seidler U 2003 Independence of apical Cl^-/HCO_3^- exchange and anion conductance in duodenal HCO3- secretion. Am J Physiol Gastrointest Liver Physiol 285:G887–897

Steward MC, Ishiguro H, Case RM 2005 Mechanism of bicarbonate secretion in the pancreatic duct. Annu Rev Physiol 67:377–409

Superti-Furga A, Hastbacka J, Wilcox WR et al 1996 Achondrogenesis type IB is caused by mutations in the diastrophic dysplasia sulphate transporter gene. Nat Genet 12:100–102

Wang S, Raab RW, Schatz PJ, Guggino WB, Li M 1998 Peptide binding consensus of the NHE-RF-PDZ1 domain matches the C-terminal sequence of cystic fibrosis transmembrane conductance regulator (CFTR). FEBS Lett 427:103–108

Wang XF, Zhou CX, Shi QX et al 2003 Involvement of CFTR in uterine bicarbonate secretion and the fertilizing capacity of sperm. Nat Cell Biol 5:902–906

Wang Z, Wang T, Petrovic S et al 2005 Kidney and intestine transport defects in Slc26a6 null mice. Am J Physiol Cell Physiol 288:C957–965

Wilschanski M, Durie PR 1998 Pathology of pancreatic and intestinal disorders in cystic fibrosis. J R Soc Med 91(suppl 34):40–49

Wright AM, Gong X, Verdon B et al 2004 Novel regulation of cystic fibrosis transmembrane conductance regulator (CFTR) channel gating by external chloride. J Biol Chem 279:41658–41663

Xie Q, Welch R, Mercado A, Romero MF, Mount DB 2002 Molecular characterization of the murine Slc26a6 anion exchanger: functional comparison with Slc26a1. Am J Physiol 283:F826–838

Yue H, Devidas S, Guggino WB 2000 The two halves of CFTR form a dual-pore ion channel. J Biol Chem 275:10030–10034

DISCUSSION

Welsh: Just as you could stimulate CFTR with the isolated STAS domain, could you stimulate SLC26 with the isolated R domain? Did you show that?

Muallem: No, the R domain does not stimulate the transporter by itself. The only way we can get it to stimulate the transporters is if we hook it up to a membrane-targeting domain. When we hook it up to the N- or the C-terminal halves of CFTR it does activate the SLC26 transporters.

Lee: How does the STAS domain activate CFTR?

Thomas: We have data addressing this directly. Earlier on in our work we expressed the STAS as a fusion with MPB (Ko et al 2004). This is not well enough behaved biochemically for structural studies. We have now developed better expression systems. We have expressed the human SLC26A3 STAS domain from position 509–741. We get several milligrams per litre and we can concentrate this protein to a reasonably high level, as is necessary for these kinds of structural studies. The circular dichroism (CD) spectrum indicates that it is alpha helical with some beta structure. This is what we would expect if it is structurally homologous with the anti-sigma factor (Aravind & Koonin 2000). We have also performed ultracentrifugation experiments and shown that it is a monomer. 1D NMR spectra of unlabelled protein collected by Jennifer Baker in Julie Forman-Kay's lab in Toronto show that this is a well folded protein (personal communication). The presence of peaks corresponding to methylene hydrogens packed against aromatics are indicative of there being a hydrophobic core to the protein.

Julie also has produced ^{15}N-labelled R domain and she has assigned all the peaks (personal communication). When the STAS domain I have just described is mixed with the phosphorylated R domain, they see about 20% of the peaks are dramatically reduced in intensity, including phosphorylated S670. There is some indication that there are some specific low affinity interactions, and we are beginning to get an idea of exactly which residues are involved. We have a detailed model for how this might activate CFTR, but currently I do not know how the interaction leads to enhanced SLC26 activity. This model builds on work on a methanococcal protein, homologous with the CFTR NBDs, we performed in collaboration with John Hunt at Columbia. We mutated the catalytic base responsible for ATP hydrolysis. This mutant protein will bind ATP but it won't hydrolyse it (Moody et al 2002). When it binds ATP, the ATP binds at the interface and causes a dimerization of these two ATP binding cassettes (Moody et al 2002, Smith et al 2002). David Gadsby's recent work has shown that CFTR behaves similarly and this dimeric state probably corresponds to the open state of the CFTR channel (Vergani et al 2005). However, when we look at the structure of the murine NBD (Lewis et al 2004), there is a helical extension that is in the way of this interface. At the end of this helical extension is residue S670. Other data suggest that this helical extension is dynamic. It is in and out of this position in different crystal structures (Lewis et al 2005). Thus, one potential model could be that one conformation of the helical extension interferes with NBD dimerization, locking the channel in a closed state. When position S670 is phosphorylated, it still dynamically flips in and out, but now the STAS can bind to it and stabilize it in a position that does not occlude the interface, thereby allowing the channel to open. Mike Welsh, you published data a couple of years ago demonstrating that, contrary to expectation, it is not the phosphorylated R domain that stabilizes the open state, but rather that the non-phosphorylated R domain inhibits its formation (Winter &Welsh 1997). This structural model is consistent with those functional data.

Welsh: Do you see other sites, too?

Thomas: The Forman-Kay group is vigorously pursuing this specific question. I should also point out that both the R domain and the NBD constructs contain the same helix, so it is important to note that there is an overlap between the two domains we are looking at. S670 is in both.

Muallem: This explains very well the incredible activation of CFTR by the STAS domain that we have seen. We can keep CFTR open for most of the time. The channel has a hard time remaining closed.

Welsh: How does the R domain structure change?

Thomas: The R domain does not have regular secondary structure nor are the dispersions in the NMR spectra indicative of an ordered globular protein; as you know it is a highly dynamic, disordered protein (Ostedgaard et al 2000). Julie's model is that this is important for getting a non-linear response (Finerty et al 2005). This is a highly cooperative interaction between multiple weak sites giving a steep response as one titrates these phosphorylation sites. When one gets to a specific critical number of interactions you get a persistent interaction that will affect the channels. This may be why it was so confusing to determine which individual sites are important for regulating gating of the channel.

Alper: Was her assignment done by individual ^{15}N amino acid labelling?

Thomas: No, this was uniformly ^{15}N-labelled protein, and then she does the standard coupling experiments to complete the assignments. This is difficult and it has taken several years to get to this point.

Gray: Have you looked at either ATP binding or hydrolysis in the presence of the STAS domain?

Thomas: The problem is that we don't have a good model for NBD2 of CFTR. This is what we would like to do; we'd like to see whether we can evaluate the dimerization model using biochemical methods.

Gray: What about using your model of the MJ0796 methanococcal ABC protein to investigate this point?

Thomas: There is no reason to suspect that this protein would interact with the STAS domain.

Soleimani: A major focus of the presentations in this meeting, and one area of discrepancy, has been what happens *in vivo* versus *in vitro*. This work has been done in HEK cells. Have you done studies to test the model, for example, in the duodenum? This goes back to the issue of chloride–bicarbonate exchange: this depends on where you look at it. How does this *in vitro* study translate to *in vivo*?

Muallem: What Min Goo Lee found in the SLC26A6-/- mouse pancreas is very much what I would expect to see. Min Goo sees a reduction in the cAMP-stimulated Cl^-/HCO_3^- exchange. My interpretation of this is that as long as CFTR is unchanged, once A6 is present you need SLA26A6(A6) to be activated by CFTR which mediates some of the Cl^- and HCO_3^- transport. This Cl^- and HCO_3^- transport is very important for the proximal part of the duct, the site of most ductal HCO_3^- secretion. You will then get a reduction of stimulated Cl^-/HCO_3^- exchange activity. One of the reasons you might not see effects on the final HCO_3^- secretion by the duct is because SLC26A3(A3) or an A3-like protein will be the one that will eventually compensate for A6 and will absorb more of the Cl⁻, secreting more HCO_3^-. The final secretion will be the same but the regional activity will be different. This is dependent on what happens to CFTR. If CFTR is activated you will probably not see much of an effect on the final HCO_3^- secretion unless you delete almost all of the SLC26 transporters, because once Cl⁻ is reduced below 20 mM the Cl⁻ inhibitory effect is removed from CFTR. We do not know at present by how much it has to be reduced in the duct cells to remove the inhibitory Cl⁻ effect on CFTR.

Gray: You start getting significant effects on CFTR activity (gating) when external Cl⁻ falls below 120 mM (Wright et al 2004).

Muallem: What I am emphasizing is that there are two different effects here. Extracellular Cl⁻ has a marked inhibitory effect on CFTR function. We looked at it as an effect on HCO_3^- transport (Shcheynikov et al 2004); Mike looks at it while measuring Cl⁻ channel activity (Wright et al 2004). Assuming that this is the same site, when we start reducing chloride below 80–40 mM, now CFTR becomes a decent HCO_3^- conductive pathway. In our model, the CFTR will clamp the HCO_3^- at 140 mM. If the membrane depolarizes we get more HCO_3^- secretion than we want and CFTR will clamp it at 140 mM.

Argent: Your secretory model requires that A6 is expressed at the top end of the duct system and that A3 is expressed at the bottom end. It is difficult to see how the model could work if A3 and A6 are both expressed evenly down the ductal system. If A6 and A3 are in the same cell then anion secretion on the exchangers is likely to be electroneutral.

Muallem: We don't have good A3 antibodies. They work reasonably well with expressed protein, but we are not getting good immunolocalization.

Alper: You had convincing chloride apparent affinity regulation by imposed voltage holding potential. Do you see regulation of DIDS affinity?

Muallem: These experiments were done only last week! When the paper showing that the current was DIDS insensitive came out, it became important to look at the current. In the original paper (Ko et al 2002) we showed that the current is indeed DIDS inhibitable. Again, when we started measuring simultaneously pH and Cl⁻ in the same cell we decided that we should look at this hyperpolarization that we see. The hyperpolarization was inhibited by DIDS. In your trace the hyperpolarization is pretty evident when you remove external chloride.

Alper: When you compare nitrate–bicarbonate exchange with nitrate currents, are they entirely the same? They seem different.

Muallem: No, this is still puzzling to us. When we add NO_3^- it seems that the entire transport is uncoupled. The only thing that the protein (A3) does is appear as a NO_3^- channel. This is almost heresy: taking a coupled transporter, then adding NO_3^-, and once you get enough NO_3^- into the oocyte it doesn't move anything but NO_3^-. In this manner A3 functions as an electronic, coupled Cl^-/HCO_3^- transporter but in the absence of Cl^- and presence of NO_3^- it functions as a NO_3^- channel. Although there is a precedence for such a behaviour of transporters we are not ready to commit to such a conclusion.

Alper: But it is conductive. Why not postulate the induction of a native nitrate channel as you drop intracellular chloride?

Muallem: We never see anything like that in the cells that don't have A3. A6 is a poor NO_3^- transporter.

Ashmore: Why is it non-linear, if it is a nitrate–nitrate channel?

Muallem: We think it might be a voltage-regulated channel.

Case: If you replaced your plasma chloride by nitrate, what would happen to pancreatic secretion?

Muallem: It should cease.

Case: It doesn't. In the isolated gland fluid secretion is almost as great in the presence of nitrate as it is with chloride (Case et al 1979).

Muallem: I would say A3 doesn't move HCO_3^- any more. You might get some transport with A6. A6 will mediate HCO_3^-/NO_3^- transport; it is just a lousy substrate compared with Cl^-.

Romero: We had data from years ago that got buried in Christopher M. Sciortino's PhD thesis, from experiments on *Drosophila* Na^+-dependent chloride–bicarbonate exchanger. We had interesting nitrate results. When we did chloride replacements, the effects with nitrate and nitrite were dependent on whether or not they were done in a bicarbonate buffer. In a non-bicarbonate buffer using nitrate electrodes we demonstrated large nitrate fluxes when the solution was spiked with nitrate. We were looking at the pH effects of doing nitrate in bicarbonate and non-bicarbonate buffers. In the presence of bicarbonate we saw an unanticipated result with nitrate in that pH went up rather than down. It turns out that in a non-bicarbonate situation nitrate could easily replace the halide. In the presence of bicarbonate nitrate didn't replace the halide that was being transported, but replaced the bicarbonate. It seemed to have a better affinity for the bicarbonate replacement than it did the nitrate. This fits with some of your old nitrate data from pancreatic duct cells. In terms of anion binding sites or transport sites, with nitrate it depends on what the conditions are. If the conditions are varied, especially in terms of bicarbonate, then the recognition sites look different. Nitrate looks like bicarbonate in 3D space rather than chloride.

Quinton: I am trying to understand how you get so much current out of a nitrate exchanger if it is tightly coupled. In my younger days we used to talk about 'slippage'. Does this have any role here? Is something slipping through instead of being exchanged?

Muallem: We are generating a huge current. Slippage will not translate into a current of this magnitude. It is difficult conceptually to understand how coupled transporters all of a sudden switch and start behaving like a channel, but for the most part this looks like a NO_3^- channel. HEK cells are the easiest place to see it. There is a beautiful current, seen with NO_3^- and only with A3. We never see it with A6, for example.

Seidler: In the colonic surface there is a high concentration of DRA and NHE3 expression, with low expression of CFTR. If you assume that DRA is electrogenic in that tissue as well, you would have more hydrogen secretion than bicarbonate secretion during resting state electroneutral salt absorption, if NHE3 is working with DRA in a 1:1 fashion to absorb salt. I do find that the colonic pH in the lumen can go down in the proximal column. I can understand this, but I can't understand how this mechanism would work: what would it need to keep going if it is an electrogenic process on the one hand and an electroneutral process on the other?

Muallem: I don't know enough about the colon transporters to answer this fully. What you need is a mechanism that will supply you with Cl^-. If you have this, you would secrete HCO_3^-.

Seidler: There is low to zero bicarbonate secretion taking place in the proximal colon during absorption. Would an apical potassium conductance be a candidate to maintain electroneutrality?

Muallem: That is a perfect way to regulate these transporters. You want them to be electrogenic if you want them to be regulated by the membrane potential. In some cells you would like to activate them, so you depolarize the membrane potential; in others you want the transporters to work less. In this case you would expect hyperpolarization of the membrane potential. These transporters need to be considered in the context of where they work and the other transporters that they interact with. Also important in this case is the identity of the transporters that sit next to them. The microenvironment at the mouth of those transporters will be responsible for much of this regulation.

Seidler: A K^+ conductance has been described in the apical membrane in colonocytes, whose molecular identity and physiological function has not yet been worked out. So it might be a candidate.

Muallem: If it will hyperpolarize, then yes.

Romero: What is the voltage across the apical membrane of colonocytes? The colon is a bizarre environment. I wouldn't worry about potassium, but I would worry about ammonium having a significant effect.

Seidler: You can take the isolated colon and look at sodium/chloride absorption and bicarbonate secretion when there is no ammonium present. One thing that has been found is a lumenal potassium conductance in the proximal colon. At least in the proximal colon, where there is no sodium conductance, its function so far is a little mysterious.

Romero: Is it still present in the distal colon?

Seidler: In the distal colon there is electrogenic sodium uptake by ENaC, and a potassium conductance is needed to maintain electroneutrality. There is also an H/K ATPase to recycle the potassium that is lost. In the proximal colon there is no ENaC yet there is potassium conductance.

Welsh: How is the R domain regulating A3?

Muallem: We don't know that it is. We know that the STAS domain is needed because when this is mutated most mutants are not targeted to the plasma membrane and are retained in the ER/Golgi.

References

Case RM, Hotz J, Hutson D, Scratcherd T, Wynne RDA 1979 Electrolyte secretion by the isolated cat pancreas during replacement of extracellular bicarbonate by organic anions and chloride by inorganic anions. J Physiol 286:563–576

Aravind L, Koonin EV 2000 The STAS domain—link between anion transporters and antisigma-factor antagonists. Curr Biol 10:R53–55

Finerty PJ Jr, Mittermaier AK, Muhandiram R, Kay LE, Forman-Kay JD 2005 NMR dynamics-derived insights into the binding properties of a peptide interacting with an SH2 domain. Biochemistry 44:694–703

Ko SB, Zeng W, Dorwart MR et al 2004 Gating of CFTR by the STAS domain of SLC26 transporters. Nat Cell Biol 6:343–350

Lewis HA, Buchanan SG, Burley SK et al 2004 Structure of nucleotide-binding domain 1 of the cystic fibrosis transmembrane conductance regulator. EMBO J 23:282–293

Lewis HA, Zhao X, Wang C et al 2005 Impact of the deltaF508 mutation in first nucleotide-binding domain of human cystic fibrosis transmembrane conductance regulator on domain folding and structure. J Biol Chem 280:1346–1353

Moody JE, Millen L, Binns D, Hunt JF, Thomas PJ 2002 Cooperative, ATP-dependent association of the nucleotide binding cassettes during the catalytic cycle of ATP-binding cassette transporters. J Biol Chem 277:21111–21114

Ostedgaard LS, Baldursson O, Vermeer DW, Welsh MJ, Robertson AD 2000 A functional R domain from cystic fibrosis transmembrane conductance regulator is predominantly unstructured in solution. Proc Natl Acad Sci USA 97:5657–5662

Shcheynikov N, Kim KH, Kim KM et al 2004 Dynamic control of CFTR Cl$^-$/HCO$_3^-$ selectivity by external Cl. J Biol Chem 279:21857–21865

Smith PC, Karpowich N, Millen L et al 2002 ATP binding to the motor domain from an ABC transporter drives formation of a nucleotide sandwich dimer. Mol Cell 10:139–149

Vergani P, Lockless SW, Nairn AC, Gadsby DC 2005 CFTR channel opening by ATP-driven tight dimerization of its nucleotide-binding domains. Nature 433:876–880

Winter MC, Welsh MJ 1997 Stimulation of CFTR activity by its phosphorylated R domain. Nature 389:294–296

Wright AM, Gong X, Verdon B et al 2004 Novel regulation of cystic fibrosis transmembrane conductance regulator (CFTR) channel gating by external chloride. J Biol Chem 279:41658–41663

Insights from a transgenic mouse model on the role of SLC26A2 in health and disease

Antonella Forlino*, Benedetta Gualeni*, Fabio Pecora*, Sara Della Torre*, Rocco Piazza*, Cecilia Tiveron†, Laura Tatangelo‡, Andrea Superti-Furga§, Giuseppe Cetta and Antonio Rossi*[1]

*Dipartimento di Biochimica 'Alessandro Castellani', Università di Pavia, I-27100 Pavia, Italy, †Foundation EBRI Rita Levi-Montalcini, 00143 Rome, Italy, ‡EMBL, I-00016 Monterotondo, Italy and §Centre for Pediatrics and Adolescent Medicine, Freiburg University Hospital, D-79106, Freiburg, Germany

Abstract. Mutations in the *SLC26A2* cause a family of recessive chondrodysplasias that includes in order of decreasing severity achondrogenesis 1B, atelosteogenesis 2, diastrophic dysplasia and recessive multiple epiphyseal dysplasia. The gene encodes for a widely distributed sulfate/chloride antiporter of the cell membrane whose function is crucial for the uptake of inorganic sulfate that is needed for proteoglycan sulfation. To investigate the mechanisms leading to skeletal dysplasia, we generated a transgenic mouse with a mutation in *Slc26a2* causing a partial loss of function of the sulfate transporter. Homozygous mutant mice were characterized by skeletal dysplasia with chondrocytes of irregular size, delay in the formation of the secondary ossification centre and osteoporosis of long bones. Impaired sulfate uptake was demonstrated in chondrocytes, osteoblasts and fibroblasts, but proteoglycan undersulfation was detected only in cartilage. The similarity with human diastrophic dysplasia makes this mouse a model to explore pathogenetic and therapeutic aspects of SLC26A2-related disorders.

2006 Epithelial anion transport in health and disease: the role of the SLC26 transporters family. Wiley, Chichester (Novartis Foundation Symposium 273) p 193–212

Because of its anionic charge, extracellular sulfate is transported into the cytoplasm by specific anion transporters. SLC26A2 (also known as diastrophic dysplasia sulfate transporter, or DTDST) is a member of the SLC26 (solute carrier 26) anion transporter family and acts as a Na$^+$-independent sulfate/chloride antiporter of the cell membrane. The isolation of the gene coding for this sulfate transporter was the result of a long-term project aimed at elucidating the molecular basis of

---

[1]This paper was presented at the symposium by Antonio Rossi, to whom correspondence should be addressed.

diastrophic dysplasia, a recessive skeletal dysplasia particularly frequent in the Finnish population (Hästbacka et al 1994), but present worldwide (Rossi & Superti-Furga 2001). Other members of the SLC26 family have been identified: SLC26A1, SLC26A3, SLC26A4 (previously known as SAT1, CLD or DRA and PDS, respectively), SLC26A5–A9 (Bissig et al 1994, Schweinfest et al 1993, Scott et al 1999, Ludwig et al 2001, Zheng et al 2000, Lohi et al 2000, 2002, Waldegger et al 2001), SLC26A10 (sequence accession no. NM 133489) and SLC26A11(Vincourt et al 2003). Their expression is limited to specific tissues, suggesting that they play specific anion transport activities; on the contrary, SLC26A2 is expressed in many tissues (Hästbacka et al 1994). Based on multiple sequence alignment comparisons with other sulfate transporters (SLC26A1 and SLC26A3), SLC26A2 show evidence for 12 transmembrane domains in corresponding locations and an N- and C-terminal cytoplasmic domain (Fig. 1A). In the C-terminal domain an hydrophobic region named STAS domain (sulfate transporter and anti-sigma antagonist) present also in SLC26A3 and SLC26A4 may indicate that phosphorylation is involved in regulation of anion transport (Aravind & Koonin 2000). Even if SLC26A2 is ubiquitously expressed, the main phenotypic consequences of its functional impairment are restricted to bone and joints (Superti-Furga 2002).

Mutations in the *SLC26A2* gene are associated with a family of recessively inherited chondrodysplasias including, in order of increasing severity, a recessive form of multiple epiphyseal dysplasia (rMED) (Superti-Furga et al 1999), diastrophic dysplasia (DTD) (Hästbacka et al 1994), atelosteogenesis type 2 (Hästbacka et al 1996) and achondrogenesis 1B (Superti-Furga et al 1996). Achondrogenesis 1B and atelosteogenesis type 2 are lethal before or shortly after birth; patients with DTD survive with major physical impairments, whereas the clinical phenotype in rMED is relatively mild (Ballhausen et al 2003). The mutations identified so far in SLC26A2 chondrodysplasias include single amino acid substitutions or deletions, premature stop codons and reduced mRNA levels caused by splice-site mutations (Rossi & Superti-Furga 2001).

The main biochemical consequence of reduced intracellular sulfate levels caused by sulfate transport impairment is cartilage proteoglycan undersulfation. Thus, cartilage from SLC26A2 patients has a reduced total sulfate content (Superti-Furga et al 1996), contains undersulfated glycosaminoglycans, as shown by chondroitin sulfate disaccharide analysis, and stains poorly with toluidine blue or alcian blue (Rossi et al 1996, 1997, 1998). On the basis of clinical, molecular and biochemical studies performed so far, a correlation has been traced between the different clinical phenotypes, the nature of the mutation in the SLC26A2, the residual activity of the sulfate transporter and cartilage proteoglycan undersulfation (Rossi et al 1998).

The major involvement of cartilage in the human phenotypes of SLC26A2 disorders demonstrates the crucial role of sulfate activation in the maturation of cartilage extracellular matrix macromolecules and the consequences of defects in this

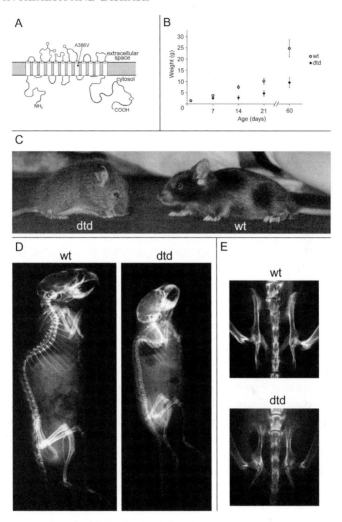

FIG. 1. (A) Predicted transmembrane structure of the SLC26A2. The coding sequence of the *SLC26A2* predicts a protein of 739 amino acids. Two and one potential *N*-glycosylation sites are present in the second (residues 199 and 205) and fourth (residue 357) extracellular domains, respectively. The rhomb in the C-terminal part indicates the STAS domain (residues 568–719). The A386V substitution which has been knocked-in is indicated. (B) Body weight of mutant (dtd) and wild-type (wt) mice from P1–P60. The body weight confirms delayed growth of mutant animals resulting in a reduction of approximately 60% at P60. The average values with the standard deviation are shown ($n = 10$ for each age point). (C) Mutant (dtd) and wild-type (wt) mice at P60. Affected mice show reduced growth, and a severe thoracic kyphosis causing a characteristic posture with flexed head. (D) X-rays at P60. In the mutant animal, shown at a higher magnification than the wild-type, shortening of tubular bones, severe thoracic kyphosis, bite overclosure and hip dysplasia is observed. (E) The pelvis is shown at higher magnification to illustrate hip dysplasia.

pathway on the architecture, composition, and mechanical properties of cartilage as well as on skeletogenesis. However, a detailed picture of the molecular and cellular events that could account for the specific defects in cartilage and bone matrix resulting in dwarfism, progressive joint disease, and malformed skeleton is far from complete. In order to provide new insight on the role of SLC26A2 and of macromolecular sulfation in the development and homeostasis of the skeleton, we have generated the first mouse strain with a mutation in the *Slc26a2* gene causing a partial loss of function of the sulfate transporter.

Materials and methods

Generation of transgenic mice

A gene targeting vector containing 8.3 kb of *Slc26a2* DNA sequence was designed in order to knock-in the C1184T transition in the murine *Slc26a2* locus causing the A386V substitution in the protein product. Targeted embryonic stem cells were injected in BDF1 × C57Bl/6J mouse blastocysts according to standard procedures (Ramirez-Solis et al 1993) and the resulting chimeric male mice were mated with C57Bl/6J females to yield heterozygous animals.

Cell cultures and sulfate uptake assays

Chondrocyte and skin fibroblast cultures were established from pups at postnatal day (P)1 and were cultured in Dulbecco's modified Eagle's medium (DMEM) with 10% fetal calf serum (FCS) and antibiotics at 37 °C in 5% CO_2. Chondrocytes were released from epiphyseal cartilage by digestion with collagenase A (Roche) as described previously (Rossi et al 1996). Osteoblasts were recovered from the femoral diaphyses of mice at P60 as previously described (Robey & Termine 1985) and cultured in DMEM/Ham's F-12K (1:1) supplemented with 25 μg/ml ascorbic acid and 10% FCS.

For sulfate uptake assays the same protocol was used for either chondrocytes or osteoblasts in primary culture or for skin fibroblasts at passage 4. Sulfate uptake was performed as described previously (Rossi et al 1996) using low ionic strength buffer (1 mM $MgCl_2$, 300 mM sucrose, 10 mM Tris-Hepes, pH 7.5) containing either 50 μM or 250 μM Na_2SO_4. In preparation for the assay, cells were washed with pre-warmed sulfate-free low ionic strength buffer and pre-incubated for 2 min in the same buffer at 37 °C. Cells were then incubated for 1 min at 37 °C in low ionic strength buffer containing different concentrations of Na_2SO_4 and a constant concentration (0.1 μM, corresponding to 150 μCi/ml) of carrier-free $Na_2[^{35}S]O_4$ (PerkinElmer Life Sciences). Following incubation, the uptake medium was removed and cells were washed with ice-cold medium containing

100 mM sucrose, 100 mM NaNO$_3$, 1 mM MgCl$_2$ and 10 mM Tris-Hepes, pH 7.5. Cells were then lysed in 2% SDS and assayed for radioactivity and protein content.

Chondroitin sulfate disaccharide analysis

Disaccharide analysis was performed in skin, cartilage and bone in pups at P1–60. Skin was excised and freed from subcutaneous fat. Cartilage was obtained from the femoral heads by dissection under the microscope. Bone was obtained from the diaphysis of the femora after removal of periosteum and bone marrow by flushing with phosphate buffer saline.

To recover glycosaminoglycans (GAGs), the carbohydrate part of proteoglycans, tissue specimens were digested with papain and released GAGs recovered by precipitation with 1% cetylpyridinium chloride. Hyaluronic acid was removed by digestion with *Streptomyces* hyaluronidase followed by ultrafiltration with Ultrafree-0.5 Centrifugal filter units (Millipore). Purified GAGs were digested with chondroitinase ABC and ACII (Seikagaku Corp.) (Rossi et al 1997) and released disaccharides were fractionated with a Supelcosil LC-SAX1 (Supelco) HPLC column (4.6 × 250 mm) using a gradient of 5–400 mM KH$_2$PO$_4$, pH 4.5, at room temperature, and the elution profile was measured at 232 nm (Rossi et al 1998).

Results and discussion

Transgenic mice harbouring an A386V substitution in the eighth transmembrane domain of Slc26a2 were generated by homologous recombination in embryonic stem cells. This mutation was detected in the homozygous state in a patient with a non-lethal form of DTD characterized by short stature, cleft palate, deformity of the external ear and 'hitchhiker' thumb deformity (Rossi & Superti-Furga 2001).

Beside the known pathogenic amino acid substitution, the intronic *neomycine* cassette in the mutant allele contributed to the hypomorphic phenotype by inducing abnormal splicing (data not shown). In spite of the dual pathogenetic mechanism, the model was not rejected because (i) most individuals affected by SLC26A2 dysplasias are not homozygote, but rather compound heterozygote for two of the many different known mutations, and compound heterozygosity for a null allele and a structural allele is frequent; (ii) all functional studies, in particular the sulfate uptake assay, confirmed the central role of sulfate transport inactivation in the pathogenesis of the observed phenotypic changes.

Mice heterozygous for the *Slc26a2* mutation did not show any phenotypic changes, as is the case for carriers of *SLC26A2* mutations in humans.

The phenotype of homozygous animals was visually inspected from birth to two months of age and skeletal defects were studied by radiographic analysis. At birth mutant pups were alive and slightly shorter than wild-type littermates. However, during the first days of life, mutant animals became easily identifiable because their growth was reduced. At P14 thoracic kyphosis was observed at visual inspection which became progressively more pronounced until P60 (Fig. 1C). Delayed growth resulted in progressive body weight reduction to approximately 60% at P60 (Fig. 1B) (P < 0.0001).

X-rays at P60 recapitulate the skeletal defects: shortening of tubular bones, severe thoracic kyphosis and abnormal thoracic cage with reduced volume, bite overclosure (overgrowth of the mandible relative to the maxilla, resulting in lower incisors being placed anteriorly to upper incisors) and hip dysplasia with pelvic deformity (Fig. 1D, E). Most homozygous mutant mice died before 6 months of age; the direct cause of death could be respiratory complications related to kyphosis, although we cannot exclude neurological problems or malnutrition due to dental malocclusion. Considering the species differences between humans and mice, the first strain of mice we obtained showed a phenotype that closely reproduced DTD, being characterized by dwarfism, skeletal deformities, craniofacial abnormalities and joint contractures.

Histological studies of cartilage tissue sections also demonstrated similarities to human SLC26A2 chondrodysplasia patients. Sections of the tibial and femoral epiphysis from animals from P1 to P60 were stained with toluidine blue, a cationic dye with affinity to sulfated, polyanionic proteoglycans, or with H&E. In mutant animals cartilage matrix stained less intensely with toluidine blue compared to age-matched wild-type littermates as already observed in patients with SLC26A2 disorders due to proteoglycan undersulfation (Fig. 2) (Hästbacka et al 1996, Superti-Furga 1994). The formation of the secondary ossification centre was studied from P7–P21; in the wild-type the second ossification centre was well developed at P14, while in mutant animals at the same age its formation was still at the beginning, suggesting a delay in chondrocyte differentiation (Fig. 2).

We hypothesized that tubular bone shortening and reduced growth might be produced by changes at the growth plate. The overall architecture of the growth plate at P21 was preserved, with resting, proliferative and hypertrophic zones well defined. A marked change in the growth plate was observed at P60: at low magnification it appeared as a straight line in wild-type animals, whereas in mutant animals it was irregularly delineated and in some regions appeared to be disrupted, with bony fusion of the epiphysis to the metaphysis (Fig. 3A). At higher magnification few chondrocyte columns in the proliferative zone were evident (Fig. 3B). Long bones were osteoporotic: epiphyseal and metaphyseal trabecular structures were thinner, fewer and poorly connected to each other compared to age-matched controls (Fig. 3C).

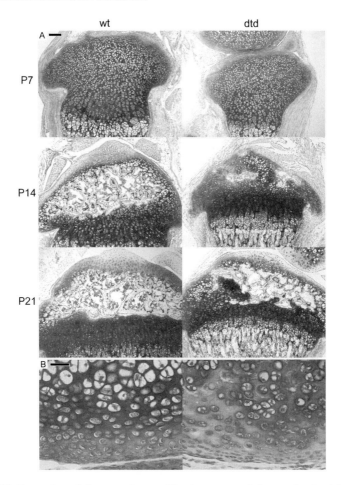

FIG. 2. (A) Formation of the secondary ossification centre of the proximal epiphysis from the tibia in wild-type (wt) and mutant animals (dtd). At P14 the secondary ossification centre is well developed in wt animals, but in the dtd its formation is delayed suggesting a defect in chondrocyte differentiation. Sections are stained with toluidine blue. (B) Decreased toluidine blue staining of cartilage from mutant animals at P14. In mutant animals the cartilage matrix from the distal epiphysis of the femur is less intensely stained with toluidine blue compared to wild-type animals with some areas staining less than other areas. In addition chondrocytes are smaller and their size is more variable compared to wild-type littermates. Bars: A = 1000 μm; B = 300 μm.

The activity of the sulfate transporter was tested in fibroblasts and in primary cultures of chondrocytes and osteoblasts by sulfate uptake assays. In chondrocytes and fibroblasts from homozygous animals the uptake was reduced compared to wild-type or heterozygous animals (Fig. 4). These data validate our mouse model

wt   dtd

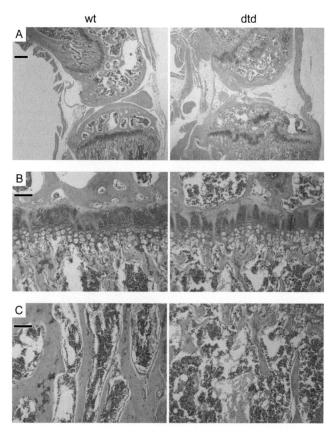

FIG. 3.   (A) H&E staining of the knees of animals at P60. In mutant animals (dtd) the growth plate is irregularly delineated and in some regions is disrupted with bony fusion of the epiphysis to the metaphysis. In addition, articular cartilage in mutant animals is degenerated and fragmented with sinovia hypertrophy. (B) Enlarged view of the growth plate. In dtd animals sections of the tibial growth plate do not show significant differences in growth plate architecture, except for a possible rarefaction of chondrocytes and a few chondrocyte columns in mutant animals. (C) Enlarged view of bone tissue. In bone tissue trabecular structures are thinner and poorly connected to each other compared to age matched controls. Bars: A = 3000 μm; B,C = 700 μm.

FIG. 4.   Functional assay of Slc26a2 in different cell lines. Chondrocytes and osteoblasts in primary culture and skin fibroblasts from wild-type (wt) and mutant animals (dtd) were analysed for uptake of inorganic sulfate. In the three cell lines considered, sulfate uptake is impaired in dtd animals demonstrating that the chondrodysplasia phenotype is caused by targeting of *Slc26a2*. The activity of the sulfate transporter in the heterozygote is within normal values confirming the lack of any phenotypic change in these animals (data not shown).

Fibroblasts

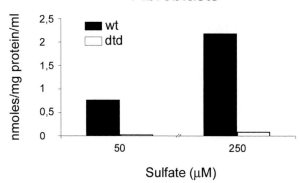

Sulfate (μM)

Chondrocytes

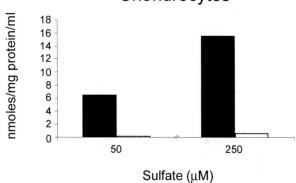

Sulfate (μM)

Osteoblasts

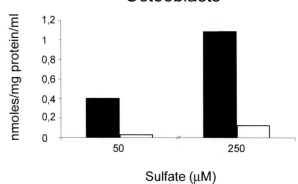

Sulfate (μM)

since in patients with SLC26A2 chondrodysplasias sulfate uptake impairment has been observed in both cell types (Rossi et al 1996). The same assay performed in osteoblasts from mutant animals showed sulfate uptake impairment suggesting that also in this cell type, a large proportion of the intracellular sulfate pool is probably mediated by Slc26a2 (Fig. 4).

Sulfation of proteoglycans was then studied in cartilage, bone and skin of the mouse strain. It has been demonstrated (with the exception of rMED) that cartilage chondroitin sulfate proteoglycans are undersulfated in SLC26A2 chondrodysplasias (Rossi et al 1998), but, even if tissue abnormalities are restricted to cartilage, no information is available regarding proteoglycan sulfation in other tissues such as skin or bone where the SLC26A2 is expressed. In addition, sulfation was measured at different ages from P1–P60 to trace the course of proteoglycan sulfation during post-embryonic development. Chondroitin sulfate proteoglycan sulfation was measured as the amount of mono-sulfated disaccharides (ΔDi-4S and ΔDi-6S) relative to non-sulfated disaccharide (ΔDi-0S) and mono-sulfated disaccharides. HPLC separation of cartilage samples demonstrated that in mutant animals the relative amount of sulfated disaccharides was lower compared to wild-types, providing direct evidence of reduced chondroitin sulfate proteoglycan sulfation (Fig. 5). Interestingly, in mutant animals the lowest values of proteoglycan sulfation were observed at birth (about 68% of mono-sulfated disaccharide versus 90.5% in controls, $P < 0.0001$), while proteoglycan sulfation increased with age, even if not within normal values. Proteoglycans obtained from the femoral diaphysis of mutant animals were slightly undersulfated. Chondroitin/dermatan sulfate disaccharide analysis of skin from mutant animals showed that proteoglycan sulfation was comparable to that of wild-type animals consistent with the lack of any phenotypic alteration of skin in the mouse strain as well as in DTD patients (Fig. 5).

→

FIG. 5.   Sulfation of chondroitin sulfate proteoglycans in skin, cartilage and bone of animals at different ages from P1–P60. The degree of proteoglycan sulfation has been determined by HPLC on the basis of the amount of mono-sulfated disaccharides (ΔDi-4S and ΔDi-6S) relative to non-sulfated (ΔDi-0S) and mono-sulfated disaccharides (ΔDi-4S and ΔDi-6S). Analysis of cartilage from the femoral head demonstrates that in mutant animals (dtd) the relative amount of sulfated disaccharides is lower compared to wild-type (wt) littermates, providing direct evidence of reduced chondroitin sulfate proteoglycan sulfation ($P < 0.05$). Interestingly, in mutant animals the highest values of proteoglycan undersulfation are observed at birth, while sulfation increases with age. Proteoglycans obtained from the femoral diaphysis of mutant animals are slightly undersulfated and, as already observed in cartilage, there is an increase in sulfation with age ($P < 0.05$ at P7 and P14, at the other age point differences were not statistically significant). Bone sulfation analysis in mutant animals at P1 has not been possible because of the reduced amount of bone available (n.d., not determined). Sulfation of skin proteoglycans in mutant animals is within normal values at all age points considered. Bars represent the average values with the standard deviation ($n = 4$) for wild-type and mutant animals.

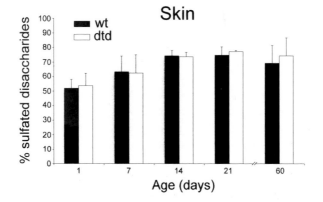

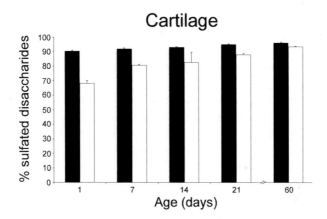

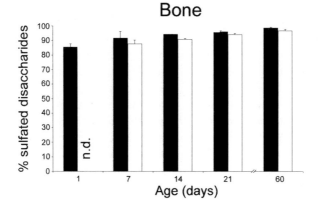

In SLC26A2 disorders the restriction of tissue alterations to cartilage has been linked to the high amount of proteoglycans synthesized by chondrocytes and thus to the ensuing high sulfate requirements. Previous studies performed in fibroblasts from SLC26A2 patients demonstrated that the degree of proteoglycan sulfation depends also on the rate of proteoglycan synthesis (Rossi et al 1998). Since the intracellular sulfate pool is limited due to a reduced sulfate uptake per time unit, its depletion depends on proteoglycan synthesis rate and on proteoglycan species. Thus, the tendency to more normal proteoglycan sulfation observed in mutant mice over time may be explained by a slower rate of proteoglycan synthesis in older mice. An additional factor that may affect proteoglycan sulfation comes from the formation of the secondary ossification centre which can provide sulfate and other sulfur compounds through the blood vessels penetrating the epiphyses and through the degradation of cartilage matrix with lysosomal catabolism of glycosaminogly-can chains (Harper et al 1993).

A third factor modulating proteoglycan sulfation is intracellular sulfate recruit-ment from the oxidation of thiols (e.g. cysteine and methionine) which is signifi-cant in fibroblasts and chondrocytes from SLC26A2 patients (Rossi et al 1997, 2003).

The lack of any phenotype in skin (normal proteoglycan sulfation) is in agree-ment with the lower amount and type of proteoglycans present in this tissue com-pared to cartilage. Thus, in fibroblasts the intracellular sulfate pool remains adequate to allow for sulfation of skin proteoglycans.

Despite the improvement in cartilage proteoglycan sulfation with ageing, the histological and clinical data demonstrate that the development of skeletal abnor-malities is progressive, suggesting that other factors or sulfation of other macro-molecules are involved in the pathogenesis of SLC26A2 disorders. One possible explanation is that degenerative changes in cartilage are cumulative, and a late, partial recovery in proteoglycan sulfation may not be enough to restore tissue tro-phism. Another possible pathogenic mechanism for the progression of skeletal changes may involve heparan sulfate proteoglycans; these macromolecules are relevant for efficient fibroblast growth factor (FGF) signalling which is crucial for the regulation of chondrocyte proliferation and differentiation (Ornitz & Marie 2002). Sulfation of heparan sulfate proteoglycans has never been studied in SLC26A2 disorders; however, it is likely that modulating the overall sulfation status and/or the sulfation at specific positions along the glycosaminoglycan chains will influence the availability of critical ligands and the optimal presentation of signalling molecules to their receptors.

Conclusions

In this work we have generated a mouse strain with a partial loss of function of *Slc26a2* causing a non-lethal chondrodysplasia phenotype; the biochemical, histological and morphological studies validate this strain as a useful animal model of DTD to explore pathogenetic and therapeutic aspects of SLC26A2 disorders. The non-lethal phenotype allow us to study the influence of proteoglycan undersulfation on skeleton growth and maintenance, and therefore to trace the progression of the DTD phenotype. No pharmacological treatment of DTD has been available so far; the demonstration that cysteine derived compounds can increase the intracellular sulfate pool in chondrocytes, suggests that pharmacological treatments might be tried in order to increase proteoglycan sulfation in the patients (Rossi et al 2003). No data are available *in vivo* regarding the contribution of this alternative pathway of sulfate recruitment to the sulfate pool and the effective dose required to increase proteoglycan sulfation in cartilage. The answers might be provided by the animal model before treatment of DTD patients might be considered.

Acknowledgments

We thank Angelo Gallanti for expert cell culturing and Professor Laura Pozzi for the use of AB1-ES cells. This work was supported by grants from the Telethon-Italy (grant no. D.083 and no. B.055), MIUR PRIN grant no. 2003052778 and Consorzio Interuniversitario per le Biotecnologie 2003.

References

Aravind L, Koonin EV 2000 The STAS domain—a link between anion transporters and antisigma-factor antagonists [letter]. Curr Biol 10:R53–55

Ballhausen D, Bonafe L, Terhal P et al 2003 Recessive multiple epiphyseal dysplasia (rMED): phenotype delineation in eighteen homozygotes for DTDST mutation R279W. J Med Genet 40:65–71

Bissig M, Hagenbuch B, Stieger B, Koller T, Meier PJ 1994 Functional expression cloning of the canalicular sulfate transport system of rat hepatocytes. J Biol Chem 269:3017–3021

Harper GS, Rozaklis T, Bielicki J, Hopwood JJ 1993 Lysosomal sulfate efflux following glycosaminoglycan degradation: measurements in enzyme-supplemented Maroteaux-Lamy syndrome fibroblasts and isolated lysosomes. Glycoconj J 10:407–415

Hastbacka J, de la Chapelle A, Mahtani MM et al 1994 The diastrophic dysplasia gene encodes a novel sulfate transporter: positional cloning by fine-structure linkage disequilibrium mapping. Cell 78:1073–1087

Hästbacka J, Superti-Furga A, Wilcox WR et al 1996 Atelosteogenesis type II is caused by mutations in the diastrophic dysplasia sulfate-transporter gene (DTDST): evidence for a phenotypic series involving three chondrodysplasias. Am J Hum Genet 58:255–262

Lohi H, Kujala M, Kerkela E et al 2000 Mapping of five new putative anion transporter genes in human and characterization of SLC26A6, a candidate gene for pancreatic anion exchanger. Genomics 70:102–112

Lohi H, Kujala M, Makela S et al 2002 Functional characterization of three novel tissue-specific anion exchangers SLC26A7, -A8, and -A9. J Biol Chem 277:14246–14254

Ludwig J, Oliver D, Frank G et al 2001 Reciprocal electromechanical properties of rat prestin: the motor molecule from rat outer hair cells. Proc Natl Acad Sci USA 98: 4178–4183

Ornitz DM, Marie PJ 2002 FGF signaling pathways in endochondral and intramembranous bone development and human genetic disease. Genes Dev 16:1446–1465

Ramirez-Solis R, Davis AC, Bradley A 1993 Gene targeting in embryonic stem cells. Methods Enzymol 225:855–878

Robey PG, Termine JD 1985 Human bone cells *in vitro*. Calcif Tissue Int 37:453–460

Rossi A, Superti-Furga A 2001 Mutations in the diastrophic dysplasia sulfate transporter (DTDST) gene (SLC26A2): 22 novel mutations, mutation review, associated skeletal phenotypes, and diagnostic relevance. Hum Mutat 17:159–171 (Erratum in Hum Mutat 18:82)

Rossi A, Bonaventure J, Delezoide AL, Cetta G, Superti-Furga A 1996 Undersulfation of proteoglycans synthesized by chondrocytes from a patient with achondrogenesis type 1B homozygous for an L483P substitution in the diastrophic dysplasia sulfate transporter. J Biol Chem 271:18456–18464

Rossi A, Bonaventure J, Delezoide AL, Superti-Furga A, Cetta G 1997 Undersulfation of cartilage proteoglycans *ex vivo* and increased contribution of amino acid sulfur to sulfation *in vitro* in McAlister dysplasia/atelosteogenesis type 2. Eur J Biochem 248:741–747

Rossi A, Kaitila I, Wilcox WR et al 1998 Proteoglycan sulfation in cartilage and cell cultures from patients with sulfate transporter chondrodysplasias: relationship to clinical severity and indications on the role of intracellular sulfate production. Matrix Biol 17:361–369

Rossi A, Cetta A, Piazza R et al 2003 *In vitro* proteoglycan sulfation derived from sulfhydryl compounds in sulfate transporter chondrodysplasias. Pediatr Pathol Molec Med 22: 311–321

Schweinfest CW, Henderson KW, Suster S, Kondoh N, Papas TS 1993 Identification of a colon mucosa gene that is down-regulated in colon adenomas and adenocarcinomas. Proc Natl Acad Sci USA 90:4166–4170

Scott DA, Wang R, Kreman TM, Sheffield VC, Karnishki LP 1999 The Pendred syndrome gene encodes a chloride-iodide transport protein. Nat Genet 21:440–443

Superti-Furga A 1994 A defect in the metabolic activation of sulfate in a patient with achondrogenesis type IB. Am J Hum Genet 55:1137–1145

Superti-Furga A 2002 Skeletal dysplasias related to defects in sulfate metabolism. In: Steinmann B, Royce P (eds) Connective tissue and its heritable disorders. Wiley-Liss, New York p 939–960

Superti-Furga A, Hästbacka J, Wilcox WR et al 1996 Achondrogenesis type IB is caused by mutations in the diastrophic dysplasia sulphate transporter gene. Nat Genet 12:100–102

Superti-Furga A, Neumann L, Riebel T et al 1999 Recessively inherited multiple epiphyseal dysplasia with normal stature, club foot, and double layered patella caused by a DTDST mutation. J Med Genet 36:621–624

Vincourt JB, Jullien D, Amalric F, Girard JP 2003 Molecular and functional characterization of SLC26A11, a sodium-independent sulfate transporter from high endothelial venules. Faseb J 17:890–892

Waldegger S, Moschen I, Ramirez A et al 2001 Cloning and characterization of SLC26A6, a novel member of the solute carrier 26 gene family. Genomics 72:43–50

Zheng J, Shen W, He DZ et al 2000 Prestin is the motor protein of cochlear outer hair cells. Nature 405:149–155

DISCUSSION

Welsh: Could you feed patients with SLC26A2 disorders with high cysteine diets? There are transgenic plants that over-produce a variety of specific amino acids: do these exist for cysteine? If so, a trial might be justified.

Rossi: The demonstration that cysteine and cysteine-derived compounds can increase the intracellular sulfate pool in patients with SLC26A2 disorders suggests that pharmacological trials might be tried in order to increase proteoglycan sulfation in the patients.

Cysteine is present in low amounts in plasma and inside the cell, because it is highly insoluble. Moreover cysteine is likely to be used mainly for protein synthesis. For these reasons we prefer to use cysteine derivatives such as *N*-acetylcysteine. They are much more soluble and inside the cell their final destination is to be catabolized in order to produce sulfate. We focus on *N*-acetylcysteine because its pharmacokinetics are well known, but we cannot exclude that other cysteine derivatives might be effective.

Lee: Have you any preliminary data on the treatment of this defect in your mice?

Rossi: We are injecting the mice in this period and unfortunately we have no results yet. As a first step to study the efficacy of the treatment, we plan to measure sulfation of cartilage proteoglycans in order to check whether a recovery in sulfation has been obtained in mutant animals injected with the drug.

Markovich: Did you measure blood sulfate levels?

Rossi: No we did not, and to my knowledge nobody has looked at sulfate concentration in blood from the patients.

Markovich: Have you looked at the role of proteoglycans of the brain?

Rossi: No we have not and unfortunately no data have been reported regarding sulfation of brain macromolecules in patients with DTD.

Markovich: Do humans with DTD have neurological problems?

Rossi: Mental development and intelligence are normal in DTD patients. Neurological complications may occur, particularly in the cervical region. Broadening of the cervical spine and cervical kyphosis may be seen in radiographs from newborns. In most cases it ameliorates with age, but in some cases severe cervical kyphosis may lead to spinal cord compression.

Quinton: Do humans or mice with this defect have any problems with respiratory infections? If so, do you know anything about the sulfated mucins?

Rossi: There are no data regarding mucin sulfation. DTD patients in the neonatal period may have respiratory complications such as pneumonia, but it may be caused by respiratory insufficiency because of the small rib cage and tracheal instability and collapsibility.

Quinton: That's the mechanics. Do they have any problems with clearance?

Rossi: No, they don't.

Thomas: Can you summarize the fate of the mutant protein in these animal models?

Rossi: We do not know the fate of the mutant Dtdst protein in cell from our mouse model. Studies on the fate of mutant SLC26A2 proteins have been performed by Karniski (2004) in transfected HEK-293 cells; in particular he has studied six SLC26A2 mutations already detected in patients. Four mutations, A715V, C653S, Q454P and R279W, were expressed on the plasma membrane of the cell, but in reduced amounts when compared to wild-type SLC26A2. The ΔV340 mutant was not detected suggesting that the mutant protein was poorly expressed or rapidly degraded; interestingly, the G678V mutation was not expressed on the plasma membrane, but was trapped within the cytoplasm.

Soleimani: There is a lot of A2 in the gastrointestinal (GI) tract. Are there any other phenotypes other than the chondrocyte problem? And since A2 is more similar to A1 than the other SLC26 members, and it is presumably basolateral, the same membrane domain as A1, is there any up-regulation of A1 in this model?

Rossi: DTD patients do not have any problems in the GI tract and our mouse model has not shown any phenotype linked to the intestine. We have no data as to whether the defect in the SLC26A2 might influence the expression of SLC26A1.

Markovich: No one has looked at the localization of A2 in the tissues that it is expressed in. Coming back to mucins, if serum sulfate levels are low in this mouse model you would expect a change in mucins. We see this in a sodium sulfate transporter knockout mouse which has extremely low serum sulfate levels. The mucin layer in the GI tract is completely destroyed. We are testing whether these mice are more prone to infection.

Alper: Is there any evidence for direct binding or complex formation between either A1 or A2 with synthases or sulfate transferases? Is there a sulfate metabolon which assembles around these?

Markovich: No one has tried it.

Welsh: What is so special about the chondrocyte? I can think of many proteins that require sulfation. One example is dystroglycan. This is the whole basis for its binding. You could imagine muscle phenotypes or brain phenotypes but you don't see them.

Rossi: The restriction of tissue alterations to cartilage has been linked to the high amount of proteoglycans synthesized by chondrocytes and to the high sulfation level of cartilage proteoglycans compared to other tissues such as skin which is normal. Aggrecan, the main proteoglycan in cartilage, is a huge proteoglycan with several glycosaminoglycan chains which have to be sulfated. In skin other proteoglycans are present such as decorin, which is a small leucine-rich proteoglycan with only one glycosaminoglycan chain. Thus in decorin only one glycosaminoglycan has to be sulfated, but in aggrecan there are 30–40 glycosaminoglycan chains that have to be sulfated. In conclusion the cartilage phenotype might be linked to the high sulfate requirements of this tissue. An additional factor that might affect cartilage proteoglycan sulfation comes from the poor vascularization of cartilage compared to other tissues which may impair diffusion of sulfate from blood plasma to chondrocytes.

Alper: What happens to adult haematopoeisis in the surviving animals with the trabecular malformation?

Rossi: We have no data regarding haematopoiesis in our mouse model.

Knipper: I have a question concerning the reduced proliferation rate you see in chondrocytes. This is of great interest for tissue engineering. In the ear/nose field the reconstruction of these cartilages suffers from the low proliferation rate of chondrocytes.

Rossi: In mutant animals we have observed reduced chondrocyte proliferation *in vitro* compared to wild-types indicating that reduced bone growth in SLC26A2 disorders might be linked to reduced chondrocyte proliferation.

In tissue engineering chondrocyte proliferation is the main problem because it occurs at a low rate. If you culture chondrocytes in order to make them proliferate they dedifferentiate; if you want to keep the chondrocytic phenotype they don't grow. Unfortunately, I do not think that the extent of chondrocyte proliferation for tissue engineering purposes might be linked directly to the extent of proteoglycan sulfation.

Markovich: Were you surprised that sulfate uptake in skin fibroblasts was reduced whereas the proteoglycan levels were the same?

Rossi: No, because I also observed the sulfate uptake defects in fibroblasts from SLC26A2 patients and in spite of this observation skin was normal. Normal sulfation of skin proteoglycans and thus the lack of any phenotype in skin is in agreement with the amount and type of proteoglycans present in this tissue when compared to cartilage. The amount of proteoglycans synthesized in skin is lower compared to cartilage, and in addition, versican, decorin and biglycan present in skin bear fewer glycosaminoglycan chains compared to cartilage macromolecules. Thus, in fibroblasts, the intracellular sulfate pool remains adequate to allow for sulfation of skin proteoglycans.

Romero: Has anyone looked at carbonic anhydrases in chondrocytes? Are any of the 14 or so known localized to cartilage?

Rossi: I have no data regarding carbonic anhydrases.

Romero: In teleost fish, for example, carbonic anhydrase coupling to sulfate transport is very important for the kidney and the GI tract to be able to secrete sulfate. But one can imagine a scenario in which either the patients or your knockout animals may have aberrant carbonic anhydrase distributions. You might be able to overcome some of the problems with sulfate metabolism by banging at the carbonic anhydrase aspect.

Aronson: If it functions mainly as a sulfate–chloride exchanger you wouldn't expect carbonic anhydrase to have a role.

Stewart: I heard a talk by Dr Robert Wilkins, at University Laboratory of Physiology, University of Oxford, who works on bovine chondrocytes, that suggested chondrocytes don't have any carbonic anhydrase.

Muallem: Are there any other SLC transporters in chondrocytes?

Rossi: I don't think that other sulfate transporters are expressed.

Muallem: Perhaps that is the only one expressed in chondrocytes, so when it is knocked out there is a severe phenotype.

Welsh: What about the heterozygotes? Do you tell the parents to stop running?

Rossi: The parents of patients with SLC26A2 disorders are normal and they do not have skeletal problems. Also in our mouse strain, heterozygous animals are phenotypically normal. Cartilage proteoglycan sulfation is within normal values as well as sulfate uptake in chondrocytes in primary culture.

Welsh: Do their joints wear out prematurely?

Rossi: No they do not; carriers of SLC26A2 mutations don't have any skeletal problem.

Markovich: So one copy is enough.

Alper: Has there been any screening of families with hereditary transmission of scoliosis and no other joint problems with respect to polymorphisms?

Rossi: Such studies have not been performed to date; however some observations suggest that it would be worth screening families with common joint or skeletal problems for SLC26A2 polymorphisms. Recessive multiple epiphyseal dysplasia, the last entity added to the SLC26A2 chondrodysplasia family, is a mild condition with normal or mild short stature, joint pain, hand and/or foot deformities and scoliosis. Most patients are homozygous for the R279W substitution in the SLC26A2. Since this condition is mild and not easily identifiable it may be present at higher frequencies in the population.

Alper: Recruitment by scoliosis or kyphosis might be a good way of finding these.

Welsh: Is there a strong genetic link in kyphosis?

Alper: There is a familial predisposition in some cases.

Mount: Can you disconnect for us the structural effects of the lack of sulfation and the developmental effects? What is your gut feeling? It has always struck me as more of a developmental abnormality than a structural abnormality. Is there bony degeneration over time in these patients?

Rossi: It is quite difficult to disconnect the structural effect of the lack of sulfation from the developmental effects. We can only make a hypothesis: in articular cartilage the lack of sulfation mainly affects the mechanical properties of the extracellular matrix and thus the lack of sulfation largely causes structural effects, while in the growth plate the effect of proteoglycan undersulfation may affect chondrocyte proliferation and maturation and is thus mainly a developmental effect.

To my knowledge no data have been reported regarding bone degeneration in DTD patients, but on the basis of the observations we got in the DTD mouse strain it might be worth also looking at bones in the patients.

Knipper: If this is the case, do you understand why some tissues which obviously also undergo chondrification before becoming ossified, such as the cochlea and nose, aren't affected?

Rossi: Unfortunately, we have no explanation why the cochlea and the nose are not affected. Patients do not show hearing loss and the nose is not shortened or stubby as in other chondrodysplasias. Within the first weeks of life a peculiar process affects the ear pinnae which is still unexplained: the pinnae develop acute inflammation and swelling with fluctuation or cystic changes. The consequence is a thickened and sometimes deformed pinna, which may eventually calcify.

Knipper: Even the capsule of the cochlea undergoes a chondrification first and then an ossification.

Markovich: Maybe they don't require as much sulfate.

Quinton: It could be a matter of timing. At some point in development there may be an incredible demand for sulfate.

Rossi: This could explain the increase of proteoglycan sulfation with age. In our mouse model chondroitin sulfate proteoglycans were undersulfated compared to wild-type littermates (as is observed in DTD), but interestingly, there was a progressive increase in proteoglycan sulfation with age from P1–P60. In SLC26A2 disorders the restriction of tissue alterations to cartilage has been linked to the high amount of proteoglycans synthesized by chondrocytes and thus to the ensuing high sulfate requirements. Since the intracellular sulfate pool is limited due to a reduced sulfate uptake per time unit, its depletion depends on proteoglycan synthesis rate and on proteoglycan species. Thus, the tendency to more normal proteoglycan sulfation observed in mutant mice over time may be explained by slower rate of proteoglycan synthesis in older mice. An additional factor that may affect proteoglycan sulfation comes from the formation of the secondary ossification center which can provide sulfate and other sulfur compounds through the blood vessels penetrating the epiphyses and through the degradation of cartilage matrix with lysosomal catabolism of glycosaminoglycan chains.

Mount: Is there any way you could rescue this *in utero*? Could the mothers be sulfate supplemented? This gives you a chance to manipulate sulfate at a period where you could probably grossly increase it *in utero*.

Markovich: Pregnant women and children already have an elevated serum sulfate level and their diet is already high in sulfate. If you supplement even more you might get diarrhoea.

Mount: If you give mice intraperitoneal sulfate, what happens?

Markovich: They'll excrete it via their kidneys.

Alper: A7 and A11 were first cloned out of endothelial cells. Heparin sulfate proteoglycans weren't among your assays. Is there any evidence for some kind of prothrombotic predisposition?

Rossi: Targeted experiments on blood coagulation have never been performed. However, coagulation in the patients as well as in mutant animals seems normal.

Alper: Have you done any heparan sulfate proteoglycan analysis?

Rossi: Sulfation of heparan sulfate proteoglycans has never been studied in SLC26A2 disorders because these macromolecules are present in low amounts. However it is certainly crucial to measure their sulfation since heparan sulfate proteoglycans are relevant for efficient FGF signalling which is important for the regulation of chondrocyte proliferation and differentiation. Moreover, they affect expression of other growth factors (i.e. Ihh) involved in endochondral bone formation. Several lines of evidence demonstrate that sulfation at specific positions along the glycosaminoglycan chains play crucial roles in the formation of biologically active domain structures. Thus, it is likely that modulating the overall sulfation status and/or the sulfation at specific positions along the glycosaminoglycan chains will influence the availability of critical ligands and the optimal presentation of signalling molecules to their receptor.

Reference

Karniski LP 2004 Functional expression and cellular distribution of diastrophic dysplasia sulfate transporter (DTDST) gene mutations in HEK cells. Hum Mol Genet 13: 2165–2171

New insights into the role of pendrin (SLC26A4) in inner ear fluid homeostasis

Lorraine A. Everett

Audiovestibular Genomics Group, The Wellcome Trust Sanger Institute, Wellcome Trust Genome Campus, Hinxton, Cambridge CB10 1SA, UK

Abstract. For over 100 years after the first description of the disorder, the molecular pathology underlying the deafness and thyroid pathology in Pendred syndrome (PS) remained unknown. In 1997, early progress towards understanding the molecular basis of the disorder was made when we identified the PS gene and found it to belong to the SLC26 family of anion transporters. The realization that an anion transporter was responsible for these clinical features soon highlighted a potential role for pendrin in thyroid hormone biosynthesis. The role of pendrin in deafness, however, remained unclear. Our determination of its expression pattern in the inner ear along with the development of a mouse with a targeted disruption of the *Slc26a4* gene has revealed that *Slc26a4* is expressed in areas of the endolymphatic compartment known to play a role in endolymph reabsorption and that absence of this protein leads to a profound prenatal endolymphatic hydrops and destruction of many of the epithelial cells surrounding the scala media. The precise mechanisms underlying endolymph reabsorption in the inner ear are not yet known; these studies, however, provide some of the groundwork for allowing the future delineation of these processes.

2006 Epithelial anion transport in health and disease: the role of the SLC26 transporters family. Wiley, Chichester (Novartis Foundation Symposium 273) p 213–230

Pendred syndrome (OMIM No. 274600) is an autosomal recessive disorder characterized by an early-onset sensorineural hearing loss combined with goitre, a swelling of the thyroid gland. The hearing loss in Pendred syndrome (PS) is usually prelingual and profound, although it is occasionally later in onset, with progression of the loss sometimes associated with head trauma. It is also associated with radiologically detectable malformations of the inner ear, namely a Mondini malformation of the cochlea and an enlarged vestibular aqueduct (Johnsen et al 1987, Mondini 1791). The goitre is associated with a positive perchlorate discharge test, described below, and is also very variable in its presentation; it often occurs around puberty but shows both intra- and interfamilial variability and may be completely absent in some individuals. Although the goitre can sometimes grow to a considerable size, individuals with PS are generally euthyroid, although hypothyroidism is sometimes seen later in life.

In 1997, just over one hundred years after Vaughan Pendred's original description of the disorder (Pendred 1896), we positionally cloned the gene responsible for Pendred syndrome (Everett et al 1997). Upon identification, the encoded protein of this gene (which we named pendrin) was seen to have database homology to a large number of plant sulphate transporters. Since 1997 however, many more homologous proteins have been identified; the gene family is now known as SLC26 and contains members capable of transporting a broad range of anions. The homologues of this gene family span all known taxa and a phylogenetic tree of all paralogues from a representative set of fully sequenced model organisms is depicted in Fig. 1. In mammals, there have been several late duplication events, occurring after the speciation event separating fish from other vertebrates; this (along with the discrete tissue expression pattern seen in many of the paralogues) suggests perhaps that some of these transporters will have mammal-specific or tissue-specific roles, rather than a ubiquitous role crucial for cellular survival.

Soon after the identification of *SLC26A4*, ion uptake studies in heterologous expression systems (Scott & Karniski 2000, Scott et al 1999) revealed that pendrin functioned as a sodium-independent transporter with a broad anion transport specificity and was capable of transporting a number of different anions including chloride, bromide, formate, nitrate and iodide. This demonstration of pendrin-

---→

FIG. 1.   Neighbour-joining phylogenetic tree of all SLC26 homologues from seven species. The species chosen for analysis represent selected model organisms from a wide taxonomic span whose full genomic sequence is available; this is to ensure that (i) all paralogues from each organism are included and (ii) the tree is easy to visualize due to a lack of sequences from similar organisms. This tree shows that the gene duplication leading to the different paralogues in each organism has generally occurred late, after the ancestral speciation events separating each of these taxa; on the whole, each model organism has its own set of paralogues. When comparing the two vertebrate lineages shown (*Fugu rubripes* and human), there are some direct orthologues in each species, but there is also evidence for gene duplication and gene loss in both mammals and fish. Methods: The ClustalW algorithm and ClustalX interface were used to perform the multiple sequence alignment and tree construction. To ensure the inclusion of every paralogue that exists in each organism, including those previously unreported, a TBLASTN search of each genomic sequence was performed using the pendrin protein sequence. For any previously unreported proteins detected, the predicted protein sequence was generated by a combination of GENCSAN output and correction by the manual piecing of TBLASTN-detected homologous segments. Sequences are annotated with either the protein name or some form of Accession ID such that the precise sequence used can be identified by the reader. The *Escherichia coli* sequence was used as the outgroup to root the tree. Although one can expect the automated multiple sequence alignment stage of the process to introduce errors and to thus cause the tree to differ slightly from the true evolutionary history of these genes, similar results are obtained from utilization both of full sequences and only the most highly conserved segments of these genes (which minimizes alignment errors) and the tree likely represents a reasonable approximation of the history of this gene family.

mediated iodide uptake immediately suggested that, in the thyroid, pendrin functioned as an apical porter of iodide from the thyrocyte to the follicular lumen (Fig. 2). This also explained the basis of the positive perchlorate discharge test observed in PS (Morgans & Trotter 1958), whereby radioactive iodide taken up by the thyroid is discharged from the thyroid again after administration of perchlorate. To briefly explain the biochemical basis of this phenomenon, one should state that in normal individuals, any iodide taken up by the thyroid rapidly crosses the apical membrane to enter the follicular lumen. Here it is rapidly oxidized by the membrane-bound thyroid peroxidase and then bound to the tyrosine residues in thyroglobulin, the precursor of thyroid hormones (Fig. 2). In individuals with PS, it

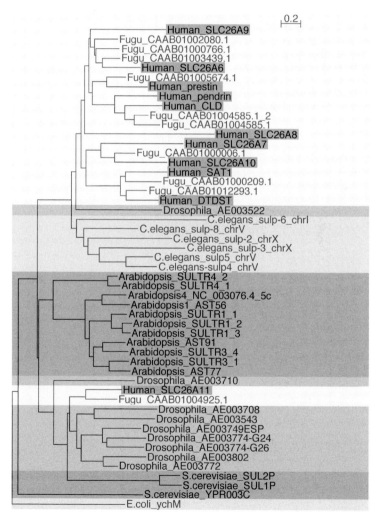

would appear that iodide transport across the apical membrane is retarded and iodide is thus trapped intracellularly. When perchlorate is administered, this competitively inhibits uptake of iodide by the sodium–iodide symporter (NIS) and the unbound intracellular iodide thus rapidly diffuses back across the basolateral membrane (down its high concentration gradient). It is therefore returned to the systemic circulation, causing a decrease in radioactivity over the thyroid and, as defined by this, a positive test result. The immunolocalization of pendrin to the apical membrane (Royaux et al 2000) further implicated pendrin's role in apical iodide transport and therefore, although the precise mechanism of transport is unknown—some studies suggest that pendrin functions more like an ion channel at this location (Yoshida et al 2002), the function of pendrin in thyroid hormone biosynthesis has been at least partly defined.

SLC26A4 and pendrin in the inner ear

While gene identification and determination of anion specificity alone led to a hypothesis on the role of pendrin in the thyroid, its role in the inner ear remained more elusive. Although some speculated on the idea that hypothyroidism was responsible for the hearing loss, this seemed unlikely since the deafness associated with congenital hypothyroidism is (i) mild, (ii) not associated with the structural defects of the inner ear (Uziel et al 1981), and (iii) PS patients are generally euthyroid. We therefore sought evidence that pendrin was directly involved in the auditory process in the inner ear.

First, several studies detected the presence of *Slc26a4* and pendrin in mouse inner ears (Dou et al 2004, Everett et al 1999, Royaux et al 2003, Yoshino et al 2004). Collectively, all of these show that, in the cochlea, pendrin is located in the cells of the external sulcus and spiral prominence as well as the external sulcus root cells; in the vestibule, pendrin is located in the transitional cells surrounding all of the sensory epithelia, and in the endolymphatic duct and sac there is pendrin immunostaining at the apical membrane of the mitochondrial-rich cells. Temporally, expression of *Slc26a4* is seen from embryonic day 13 (E13) onwards (Everett et al 1999). Interestingly, although the precise role of these cells in reabsorption of endolymph has not been determined, it would appear that these pendrin-positive cells in the cochlea are key players not only in anion absorption but in the reabsorption of Na^+ and K^+ through non-selective cation channels (Marcus & Chiba 1999) and BK channels (Chiba & Marcus 2000).

Simultaneously, we also developed a mouse with a targeted disruption of the *Slc26a4* gene (Everett et al 2001). Phenotypically, these mice are deaf, with a complete lack of auditory brainstem response waveforms at levels as high as 100 dB SPL (sound pressure level) in young adult mice. They also display signs of vestibular dysfunction, such as circling, head-tilting, and head bobbing (for video clips

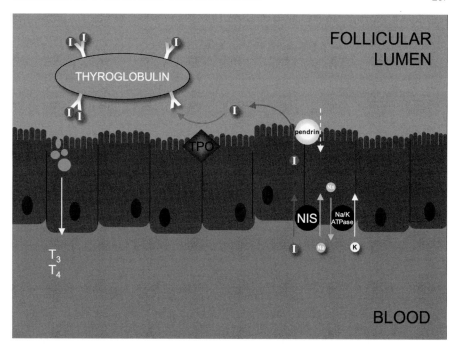

FIG. 2. Role of pendrin in thyroid hormone biosynthesis. In the thyroid, pendrin appears to mediate the transport of iodide across the apical membrane into the follicular lumen. Here it is rapidly oxidized by thyroperoxidase and bound to the tyrosine residues of thyroglobulin, the large globular precursor of thyroid hormones.

of *Slc26a4$^{-/-}$* mice, see *http://research.nhgri.nih.gov/publications/pdsko/*) and perform poorly on rotarod testing. Examination of the ears of these mice reveals profound dilatation of the endolymphatic compartment (Fig. 3A and 3B). This begins at around E15.5, which perhaps as one might expect, is shortly after the time that *Slc26a4* expression begins. Ultrastructural examination by electron and confocal microscopy shows abnormalities and loss of the sensory hair cells in cochlea (Fig. 3C and 3D) and vestibule, thus explaining the deafness and imbalance. The utricular and saccular maculae are often devoid of otoconia but, when they are present, they are often giant in nature (Fig. 3E and 3F). The abnormalities in these mice, however, are not limited to the sensory epithelia. Many of the other epithelial cells around the endolymphatic compartment take on an abnormal shape and lose their characteristic regular cobblestone appearance (Royaux et al 2003). These morphological abnormalities are clearly seen in the marginal, intermediate, and basal cell layers of the stria vascularis as well as in the cells of the endolymphatic sac. The total thickness of the stria vascularis is also significantly reduced, as is the thickness of the whole spiral ligament.

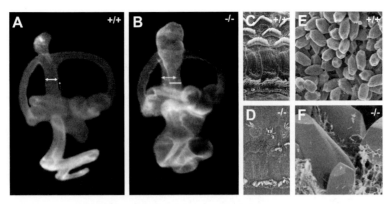

FIG. 3.   Inner ear abnormalities seen in *Slc26a4^{-/-}* mice (modified from Everett et al 2001). (A, B) At postnatal day 1 (P1), a marked dilatation (hydrops) of the endolymphatic compartment is seen. The widening of the endolymphatic duct is particularly striking: compare the non-arrowed bars in each of the two panels (the common crus is marked with an arrowed bar because of the overlap of these two structures). (C, D) On scanning electron microscopy, degeneration of the cochlear sensory hair cells is seen in young adult mice. (E, F) Otoconia are often scarce or absent in *Slc26a4^{-/-}* mice; when they are present, they are often giant in nature. Degeneration of the underlying otoconial membrane is also seen.

The severe abnormalities seen in these epithelial cells led us to wonder if they had ever been structurally normal, and therefore whether we were seeing degeneration or a deranged development of these areas. In the sensory cells at least, it appeared that hair cells of both the cochlea and the utricular and saccular maculae appeared completely normal until postnatal day 10 (Everett et al 2001), at which point degeneration began. It therefore seemed that two different pathological events were taking place, the first of which caused a profound swelling of the endolymphatic compartment from the time of onset of *Slc26a4* expression, and a second that caused some form of assault to a wide variety of epithelial cells bordering the endolymphatic compartment, beginning at postnatal day 10 (which, interestingly, is the time that endolymphatic electrolyte composition becomes mature and the endocochlear potential is set up).

In an attempt to define the nature of the assault to these epithelial cells, Royaux et al (2003) measured the composition of the endolymphatic fluid in the scala media. They determined that the high concentration of K^+ in this fluid was largely maintained; however a complete loss of the endocochlear potential was seen, and the usual negative endocochlear potential (EP) induced by anoxia also failed to appear, presumably due to the loss of sensory hair cells. On further analysis they went on to determine that the loss of the endocochlear potential was due to the loss of KCNJ10 from the intermediate cells of the stria vascularis (Wangemann et al 2004), rather than from a direct effect of the loss of pendrin.

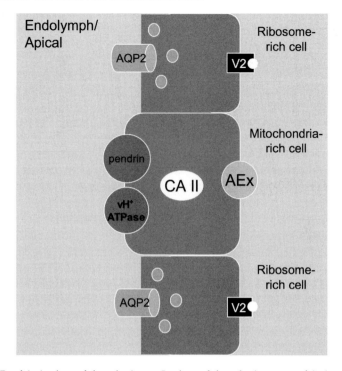

FIG. 4.   Pendrin in the endolymphatic sac. In the endolymphatic sac, pendrin immunolocalizes to the mitochondria-rich cells, which protrude slightly from the epithelial surface. Various proteins involved in electrolyte transport are known to localize to these cells, although the mechanisms in which they likely synergistically act remain to be delineated. It is interesting to note the striking similarity between the localization of proteins in these cells and those in the renal cortical collecting duct and connecting tubule. It would appear that the mitochondria-rich cells are analogous to the renal non-A non-B intercalated cells. This renal/sac similarity is not limited to these cells. The ribosome-rich cells contain AQP2 in their apical membrane and vasopressin receptors in the basolateral membrane, paralleling the principal cells of the renal cortical collecting duct and connecting tubule. The presence of these two proteins in this location has important implications for fluid homeostasis here since it suggests that these cells allow a vasopressin-mediated apical efflux of water (out of the cell and into the endolymph) when systemic fluid status requires the secretion of vasopressin; it would seem that this may have important implications for the treatment of disorders involving hydropic changes in this compartment.

Discussion

Results from numerous studies on the expression pattern/immunolocalization of Slc26a4/pendrin in the inner ear as well as those documenting the defects in $Slc26a4^{-/-}$ mice clearly show that pendrin directly participates in fluid homeostasis in the inner ear. This refutes the hypothesis that all inner ear defects are secondary to prenatal hypothyroidism. More importantly, the profound endolymphatic hydrops seen in *Slc26a4*$^{-/-}$ mice (beginning at the time of the onset of *Slc26a4* expression) shows that *Slc26a4* and the cells expressing it are responsible for the absorption of electrolytes from the endolymph (and thus reduction of the volume of the endolymphatic compartment). The clear visualization of this hydrops also likely explains the malformations of the inner ear in individuals with PS. While the cochlear Mondini malformation and the enlarged vestibular aqueduct were initially viewed as arrests in development, it now seems likely that both result from the dilatation of the endolymphatic compartment of the inner ear.

It is interesting to note that *Slc26a4* is expressed in several discrete areas in the endolymphatic compartment of the inner ear and this may help explain the findings of early research studies that sought to identify areas responsible for the reabsorption of endolymph. These studies showed that obliteration of the endolymphatic duct and sac induced an endolymphatic hydrops (Kimura & Schuknecht 1965) and this led to the suggestion that this structure was the area responsible for the endolymph reabsorption. Of course, this would require a longitudinal flow of endolymph from one area to the other, and from the apical turn of the cochlea to the basal turn. Later studies, in which markers were injected (Salt et al 1986) or iontophoresed (Salt & Thalmann 1988) into the scala media, however, suggested that the longitudinal flow rate in the cochlea is very low, and at a level incompatible with the idea of cochlear endolymph reabsorption occurring elsewhere. The expression of *Slc26a4* in both the cochlea and in the endolymphatic duct and sac perhaps helps to explain these two paradoxical findings; in a normally functioning inner ear, reabsorption of endolymph occurs throughout the endolymphatic compartment; in the ear with an obliterated sac, however, the other regions are unable to compensate for the loss of this large area of absorptive epithelium, excess vestibular endolymph flows through into the cochlea, and a uniform hydrops throughout the endolymphatic compartment develops.

The mechanisms underlying fluid homeostasis in the inner ear and the maintenance of the particularly usually electrolytic composition of the endolymph are currently still poorly understood. It has been shown that the stria vascularis and vestibular dark cells are responsible for the generation of the unusually high potassium concentration in the endolymph as well as, in the cochlea, its positive potential of +80 mV, the EP; however, the mechanisms of cationic absorption as well as anionic homeostasis have yet to be elucidated. Since chloride is the major anion in the endolymph, one can speculate that pendrin is involved in a chloride uptake mechanism that serves to reduce the volume of the endolymph. It also seems possible that pendrin plays a role in the regulation of pH in the endolymph. This notion is especially plausible given the otoconial irregularities seen in *Slc26a4*$^{-/-}$ mice. These abnormalities (which include the complete loss of otoconia or the formation of giant otoconia) have previously been observed after administration of acetazolamide (Kido et al 1991), a carbonic anhydrase inhibitor that has been shown to decrease the pH of fluid in the endolymphatic duct and sac (Tsujikawa et al 1993). They occur, presumably, because of the effect of altered acid-base balance on the maintenance of anionic sites on otoconin-90 onto which the calcite matrix of calcium carbonate is seeded.

Further evidence for a role in acid–base balance is provided by pendrin's role in the kidney, where it has been shown to be the long sought bicarbonate transporter of the B-intercalated cells of the cortical collecting duct (Royaux et al 2001, Wall et al 2003, Wall 2006, this volume). It is interesting to note the many similarities seen between the expression pattern of various solute carriers in these two different locations; clearly the development of these two areas is under the similar transcriptional control mechanisms during development, although it seems one should be cautious in making any assumptions that the cells in each location have a similar function. Nevertheless, in continuing the comparison, pendrin-positive/ mitochondrial-rich cells in the endolymphatic duct and sac seem to have most similarity to the non-A non-B cells of the cortical collecting duct (CCD) and connecting tubule (CNT), since both pendrin and vH$^+$ATPase localize to the apical membrane (Dou et al 2004, Stankovic et al 1997), with one of the SLC4 anion transporters located at the basolateral membrane (Stankovic et al 1997) and carbonic anhydrase (CA II) in the cytoplasm (Dou et al 2004). Unfortunately, the function of non-A non-B cells in the kidney is not clear, as they would not appear to be responsible for the secretion of either base or alkali. The ribosome-rich cells in the endolymphatic duct and sac would appear to also have similarity to cells of the CCD and CNT, in this case the principal cells. These have been shown to express AQP2 in apical membrane vesicles (Beitz et al 1999) although, due to the hypertonic nature of the endolymph with respect to perilymph or plasma (Konishi et al 1984), it would appear that mobilization of AQP2 to the apical membrane— mediated by activation of vasopressin receptors on the basolateral membrane of these cells (Beitz et al 1999, Kumagami et al 1998)—should cause an efflux of water; this hypothesis is strengthened by the fact that administration of vasopressin causes an endolymphatic hydrops (Takeda et al 2000). It seems important to note at this stage that water entry into the endolymph at this site is mediated by vasopressin; one could hypothesize that this could have important implications for the treatment of endolymphatic hydrops, since it would suggest that endolymphatic volume can be partially regulated by systemic fluid status.

Although the development of $Slc26a4^{-/-}$ mice has undoubtedly revealed new insights into the role of pendrin in the inner ear, it would seem that one major problem with this model is the fact that secondary degenerative changes are seen throughout the epithelia surrounding the endolymphatic compartment. The loss of the endocochlear potential, for example, is not surprising given the secondary degeneration of the stria vascularis and does not result directly from the loss of pendrin. This stria vascularis is responsible for the secretion of K$^+$ into the endolymphatic compartment, and the mechanism underlying this has been elucidated

(Fig. 5). KCNJ10, in the intermediate cells, is crucial to this process, and is responsible for the high intrastrial potential (and thus the endocochlear potential). Indeed, although the lack of *Slc26a4* and pendrin during embryonic development leads to an initial hydrops, one can speculate that the loss of KCNJ10 in these mice may prevent further expansion of the endolymphatic compartment, since the secretory epithelium becomes as damaged as the absorptive epithelium. Certainly, mice with targeted disruptions of the transporters and channels active in this secretory process, i.e. KCNQ1 (Casimiro et al 2001, Lee et al 2000), KCNE1 (Vetter et al 1996), NKCC1 (Dixon et al 1999) and KCNJ10 (Marcus et al 2002) all show varying degrees of collapse of the endolymphatic compartment, the reverse of that seen in *Slc26a4*$^{-/-}$ mice.

Future studies

At this stage, it is clear that we are still in only the early stages of elucidating the mechanisms underlying fluid and electrolyte homeostasis in the inner ear even if

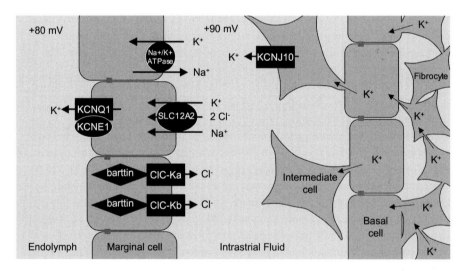

FIG. 5.   Mechanism of secretion of K$^+$ in the stria vascularis. The Na$^+$/K$^+$ATPase located in the basolateral membrane of the strial marginal cells is responsible for the active pumping of K$^+$ out of the intrastrial space. The presence of high K$^+$ in intermediate cells, generated by the recycling of K$^+$ through the basal cells and fibrocytes, causes efflux of K$^+$ into the intrastrial space down this concentration gradient, setting up a positive DC potential here. This in turn promotes the secretion of K$^+$ into the endolymph through the strial marginal cells. In *Slc26a4*$^{-/-}$ mice, all three layers of this structure are visibly abnormal, and loss of the KCNJ10 protein has been demonstrated. The loss of the endocochlear potential can therefore be attributed to this secondary effect.

such knowledge will likely be rapidly acquired within the next decade. One must ask, however, whether full characterization of these mechanisms must be achieved before performing studies aiming to discover methods of correcting defects in these processes. As a result of the studies described above, it would appear that PS is fairly unique amongst the causes of childhood deafness as it may be amenable to treatment by pharmacotherapeutic means. Children with PS often have useable hearing in early childhood, and children with large vestibular aqueduct syndrome (LVAS), a disorder often associated with one mutation in *SLC26A4*—as well as one or more unknown genetic alteration(s) elsewhere (Pryor et al 2005)—the progression of hearing loss is even slower. It seems unlikely that hearing loss could be prevented altogether, but if its onset could be slowed, it could allow a child to develop language skills before a profound hearing loss occurred; this would forever provide them with better communication skills, or a better outcome after cochlear implantation. At present, it is not known whether the hearing loss in PS and LVAS is simply due to the trauma caused to the endolymphatic compartment due to increased pressure changes transmitted from the intracranial cavity through the widened vestibular aqueduct, or whether it is due to disturbances in the electro-chemical composition of the endolymph. If it were the former, it would seem that the hearing loss could only be prevented if therapeutic agents with good transpla-cental transfer were administered to the mother and this would clearly appear to have a limited therapeutic role. If the progression of hearing loss is due to distur-bances of endolymphatic composition or volume however, then it seems that a variety of drugs could be tested to see if the natural history of the disease process could be altered. Many drugs have been shown to alter the electrolytic composition of the endolymph, especially diuretics. Examples in this class include canreoate, an aldosterone antagonist that decreases the endolymphatic sac potential (Mori et al 1991) and OPC-31260, a vasopressin antagonist that decreases the degree of experimental hydrops in animal models (Takeda et al 2003). Furthermore, since secretion of endolymph appears to be mediated by β-adrenergic stimulation (Fauser et al 2004), even the use of drugs such as β-adrenergic antagonists may be of benefit. Finally, if degeneration of the epithelia is due to free radical/oxidative damage, then free radical scavengers such as high-dose allopurinol may be of use. These studies appear to be the logical next step in trying to alter the disease process in PS and *Slc26a4*$^{-/-}$ mice and it would seem important to perform these studies in the near future.

References

Beitz E, Kumagami H, Krippeit-Drews P, Ruppersberg JP, Schultz JE 1999 Expression pattern of aquaporin water channels in the inner ear of the rat. The molecular basis for a water regu-lation system in the endolymphatic sac. Hear Res 132:76–84

Casimiro MC, Knollmann BC, Ebert SN et al 2001 Targeted disruption of the Kcnq1 gene produces a mouse model of Jervell and Lange-Nielsen Syndrome. Proc Natl Acad Sci USA 98:2526–2531

Chiba T, Marcus DC 2000 Nonselective cation and BK channels in apical membrane of outer sulcus epithelial cells. J Membr Biol 174:167–179

Dixon MJ, Gazzard J, Chaudhry SS, Sampson N, Schulte BA, Steel KP 1999 Mutation of the Na-K-Cl co-transporter gene Slc12a2 results in deafness in mice. Hum Mol Genet 8:1579–1584

Dou H, Xu J, Wang Z et al 2004 Co-expression of pendrin, vacuolar H^+-ATPase alpha4-subunit and carbonic anhydrase II in epithelial cells of the murine endolymphatic sac. J Histochem Cytochem 52:1377–1384

Everett LA, Belyantseva IA, Noben-Trauth K et al 2001 Targeted disruption of mouse Pds provides insight about the inner-ear defects encountered in Pendred syndrome. Hum Mol Genet 10:153–161

Everett LA, Glaser B, Beck JC et al 1997 Pendred syndrome is caused by mutations in a putative sulphate transporter gene (PDS). Nat Genet 17:411–422

Everett LA, Morsli H, Wu DK, Green ED 1999 Expression pattern of the mouse ortholog of the Pendred's syndrome gene (Pds) suggests a key role for pendrin in the inner ear. Proc Natl Acad Sci USA 96:9727–9732

Fauser C, Schimanski S, Wangemann P 2004 Localization of beta1-adrenergic receptors in the cochlea and the vestibular labyrinth. J Membr Biol 201:25–32

Johnsen T, Larsen C, Friis J, Hougaard-Jensen F 1987 Pendred's syndrome. Acoustic, vestibular and radiological findings in 17 unrelated patients. J Laryngol Otol 101:1187–1192

Kido T, Sekitani T, Yamashita H, Endo S, Masumitsu Y, Shimogori H 1991 Effects of carbonic anhydrase inhibitor on the otolithic organs of developing chick embryos. Am J Otolaryngol 12:191–195

Kimura RS, Schuknecht H 1965 Membranous hydops in the inner ear of the guinea pig after the obliteration of the endolymphatic sac. Pract Otorhinolaryngol 27:343–345

Konishi T, Hamrick PE, Mori H 1984 Water permeability of the endolymph-perilymph barrier in the guinea pig cochlea. Hear Res 15:51–58

Kumagami H, Loewenheim H, Beitz E et al 1998 The effect of anti-diuretic hormone on the endolymphatic sac of the inner ear. Pflugers Arch 436:970–975

Lee MP, Ravenel JD, Hu RJ et al 2000 Targeted disruption of the Kvlqt1 gene causes deafness and gastric hyperplasia in mice. J Clin Invest 106:1447–1455

Marcus DC, Chiba T 1999 K^+ and Na^+ absorption by outer sulcus epithelial cells. Hear Res 134:48–56

Marcus DC, Wu T, Wangemann P, Kofuji P 2002 KCNJ10 (Kir4.1) potassium channel knockout abolishes endocochlear potential. Am J Physiol Cell Physiol 282:C403–407

Mondini C 1791 Anatomia surdi nati sectio: De Boroniensi scientarium et artium instituto atque acedemia commnetarii. Bononiae 7:419–431

Morgans ME, Trotter WR 1958 Association of congenital deafness with goitre; the nature of the thyroid defect. Lancet 1:607–609

Mori N, Yura K, Uozumi N, Sakai S 1991 Effect of aldosterone antagonist on the DC potential in the endolymphatic sac. Ann Otol Rhinol Laryngol 100:72–75

Pendred V 1896 Deaf mutism and goitre. Lancet ii:532

Pryor SP, Madeo AC, Reynolds JC et al 2005 SLC26A4/PDS genotype-phenotype correlation in hearing loss with enlargement of the vestibular aqueduct (EVA): evidence that Pendred syndrome and non-syndromic EVA are distinct clinical and genetic entities. J Med Genet 42:159–165

Royaux IE, Suzuki K, Mori A et al 2000 Pendrin, the protein encoded by the Pendred syndrome gene (PDS), is an apical porter of iodide in the thyroid and is regulated by thyroglobulin in FRTL-5 cells. Endocrinology 141:839–845

Royaux IE, Wall SM, Karniski LP et al 2001 Pendrin, encoded by the Pendred syndrome gene, resides in the apical region of renal intercalated cells and mediates bicarbonate secretion. Proc Natl Acad Sci USA 98:4221–4226

Royaux IE, Belyantseva IA, Wu T et al 2003 Localization and functional studies of pendrin in the mouse inner ear provide insight about the etiology of deafness in pendred syndrome. J Assoc Res Otolaryngol 4:394–404

Salt AN, Thalmann R 1988 Interpretation of endolymph flow results: a comment on 'Longitudinal flow of endolymph measured by distribution of tetraethylammonium and choline in scala media'. Hear Res 33:279–284

Salt AN, Thalmann R, Marcus DC, Bohne BA 1986 Direct measurement of longitudinal endolymph flow rate in the guinea pig cochlea. Hear Res 23:141–151

Scott DA, Karniski LP 2000 Human pendrin expressed in Xenopus laevis oocytes mediates chloride/formate exchange. Am J Physiol Cell Physiol 278:C207–211

Scott DA, Wang R, Kreman TM, Sheffield VC, Karniski LP 1999 The Pendred syndrome gene encodes a chloride-iodide transport protein. Nat Genet 21:440–443

Stankovic KM, Brown D, Alper SL, Adams JC 1997 Localization of pH regulating proteins H+ATPase and Cl-/HCO3- exchanger in the guinea pig inner ear. Hear Res 114:21–34

Takeda T, Takeda S, Kitano H, Okada T, Kakigi A 2000 Endolymphatic hydrops induced by chronic administration of vasopressin. Hear Res 140:1–6

Takeda T, Sawada S, Takeda S et al 2003 The effects of V2 antagonist (OPC-31260) on endolymphatic hydrops. Hear Res 182:9–18

Tsujikawa S, Yamashita T, Tomoda K et al 1993 Effects of acetazolamide on acid-base balance in the endolymphatic sac of the guinea pig. Acta Otolaryngol Suppl 500:50–53

Uziel A, Gabrion J, Ohresser M, Legrand C 1981 Effects of hypothyroidism on the structural development of the organ of Corti in the rat. Acta Otolaryngol 92:469–480

Vetter DE, Mann JR, Wangemann P et al 1996 Inner ear defects induced by null mutation of the isk gene. Neuron 17:1251–1264

Wall SM 2006 The renal physiology of pendrin (Slc26a4) and its role in hypertension. In: Epithelial anion transport in health and disease: the role of the SLC26 transporters family. Wiley, Chichester (Novartis Found Symp 273) p 231–243

Wall SM, Hassell KA, Royaux IE et al 2003 Localization of pendrin in mouse kidney. Am J Physiol Renal Physiol 284:F229–241

Wangemann P, Itza EM, Albrecht B et al 2004 Loss of KCNJ10 protein expression abolishes endocochlear potential and causes deafness in Pendred syndrome mouse model. BMC Med 2:30

Yoshida A, Taniguchi S, Hisatome I et al 2002 Pendrin is an iodide-specific apical porter responsible for iodide efflux from thyroid cells. J Clin Endocrinol Metab 87:3356–3361

Yoshino T, Sato E, Nakashima T et al 2004 The immunohistochemical analysis of pendrin in the mouse inner ear. Hear Res 195:9–16

DISCUSSION

Markovich: Is there a thyroid phenotype in your mice?

Everett: Not at the moment. Thyroid function tests are completely normal. If we look at the size of the thyroids, these are normal too. It is interesting that other thyroid parameters differ quite considerably among strains. We are interested in making a congenic onto C57BL/6J to see whether they have a thyroid phenotype.

Muallem: How can you get high K^+ within the endolymph and still lose membrane potential? What channels are missing?

Everett: In the pendrin knockouts, KCNJ10 has been shown to be missing (Wangemann et al 2004). The stria vascularis is made up of three cell types: the marginal, intermediate and basal cells. There is an intrastrial space in between which does contain a positive endocochlear potential but it does not have the high K^+. NKCC1, KCNE1 and KCNJ10 are involved. There are a whole range of transporters and channels which cause the endocochlear potential to become positive in the intrastrial compartment, but then the K^+ is pumped out to create a high level in the endolymph itself through the marginal cells.

Muallem: Since you are getting K^+ secretion you must be losing some kind of channel activity that is associated with pendrin to explain the loss of membrane potential.

Everett: The problem is, in the pendrin knockout most of the changes in endolymphatic composition are going to be secondary effects relating to the degeneration of the stria vascularis, so it is difficult to discern the primary effect of a lack of pendrin. In these mice there is clearly an endolymphatic hydrops. However, the fact that KCNJ10 is missing due to strial degeneration may actually be lessening the severity of the hydrops as a lack of KCNJ10 causes a partial collapse of the scala media. Without this secondary effect the hydrops could be worse. Perhaps a high K^+ level is retained because of the endolymphatic hydrops and the ion composition, owing to a lack of endolymph anion resorption by pendrin. If just KCNJ10 is missing there is collapse of the endocochlear compartment, loss of endocochlear potential and a reduced K^+ concentration.

Welsh: What generates the voltage? Is it related to the K^+ diffusion potential?

Everett: It is actively pumped into the endolymph.

Ashmore: There is meant to be an electrogenic pump between the marginal and interstitial cells. This is considered to be the main cause.

Welsh: Is the pump a K^+ pump?

Ashmore: It is an Na^+/K^+-ATPase.

Welsh: If that is still intact, based on the fact that you have high K^+, why is the voltage not there?

Muallem: There must be some kind of a channel present that gets inactivated. It would have to be a K^+ channel of some sort.

Everett: There are a lot of channels. There is KCNQ1 and NKCC1 on the marginal cell, and when these are knocked out the result is collapse of the endolymphatic compartment and loss of the endocochlear potential.

Muallem: You are seeing a completely different phenotype. NKCC1 knockout is deaf, but this is because it results in complete loss of endolymphatic K^+. Every time K$^+$ is lost there is deafness. It seems that you are maintaining all the ion transporters and losing the potential.

Ashmore: There are two steps. You need to have K$^+$ to load the scala media facing the hair cell stereocilia and on top of that you have to crank up the potential of the layer of cells that is facing the internal compartment of the stria vascularis. These two events seem to occur independently during development.

Alper: Is it a proton pump that does this?

Ashmore: There are several different models, but I don't think it is. It is a Na$^+$/K$^+$ exchanger.

Alper: Where is this located?

Knipper: At the basal cells of the stria vascularis. In the marginal cells I would guess that KCNQ1 and KCNE1 may be affected. This may be the reason the current is lost. Once you have this kind of defect, the high turnover of these K$^+$ ion channels, which passively drive K$^+$ into the endolymph, appears to be affected within hours.

Muallem: She is getting the K$^+$ into the endolymph. The K$^+$ looks normal. So the channel has to be there.

Oliver: If I remember correctly, the endocochlear potential is essentially a diffusion potential across the membrane of the intermediate cells. The intrastrial fluid space is very low in K$^+$ so there is a highly positive potential generated at the membrane of the intermediate cells by the potassium channel KCNJ10 (Kir4.1).

Knipper: I presume the problem is that as soon as you get rid of the inward rectifying Kir4.1 you lose the driving force. This is because the inward rectifying channel is the one responsible for keeping the K$^+$ low in the intermedial space. If the potassium in the intermedial space is high because the Kir4.1 ion channel is missing, there is no driving force for passive transport of K$^+$ from one side to the other.

Welsh: What role does pendrin play in this? Is it involved in absorption, for example?

Everett: There is no pendrin in the stria vascularis. It seems that most of the endolymphatic composition changes are resulting from the loss of stria vascularis and this presumably occurs as a secondary effect to the loss of pendrin. I assume that pendrin loss causes an early problem with endolymphatic homeostasis (and this precise defect will prove difficult to identify) and this then poisons the stria vascularis itself. It may be because of loss of the K$^+$ recycling mechanism.

Knipper: I don't understand. If you consider what the channel might do, you suggest that it transports iodine versus chloride in the thyroid. If I consider its expression pattern in the cochlear, it is striking how close it maps to deiodinase, the enzyme that catabolizes T4 to T3. Only T3 is active. If SLC26A4 is only there to get rid of the iodine, this would make sense. If this is the case, can you exclude that the deafness has perhaps nothing or only indirectly to do with SLC26A4, but instead, may be the result of a reduced level of thyroid hormone during pregnancy, which in turn is the result of reduced levels of local T3. The reduced levels of local T3, however, may be the result of a loss of driving force for the deiodination required for T4 to T3 conversion due to PDS gene deletion. This would fit nicely with the progressive hearing loss you see. During a critical time period the reduced level of thyroid hormone caused by the presumptive low function of deiodinases would lead to a failure of proper final differentiation of hair cells or other compartments and deafness.

Everett: I agree that sensory hair cell defects are seen in hypothyroidism.

Knipper: The phenotype you described would fit with a loss of function of deiodinases.

Everett: So you are postulating that the swelling of the endolymphatic hydrops is due to iodide.

Quinton: What is the concentration of chloride?

Everett: The only ion measurement study published so far is that of Dan Marcus's K^+ measurements (Royaux et al 2003).

Kere: Has pH been measured yet?

Everett: This is going to be vital. I am quite sure that pendrin could be transporting bicarbonate.

Alper: Wangemann has found that pendrin-expressing cochlear epithelial cells mediate Cl^-/OH^- and Cl^-/formate exchange, and that cochlear endolymph in the deaf pendrin knockout mouse is 0.4 units more acidic than wild-type endolymph.

Ashmore: So pendrin is actually a chloride–bicarbonate exchanger?

Alper: It can function in this way, yes.

Quinton: If it were functioning as a chloride channel, wouldn't this give you the positive potential?

Everett: I think it is involved in reabsorption of the endolymph. It is a secondary effect.

Quinton: How do you get that positive potential? Is this simply due to a proton ATPase?

Everett: In wild-type it is definitely created by the stria vascularis, and pendrin is not located here

Alper: I'd like to ask the nephrologists in the audience, could this be obstruction nephropathy? Is this a hydrostatic issue? Or is it going to be a pH and potential issue?

Aronson: I won't answer that directly, but I was struck that proton ATPase and pendrin are in the same cells with carbonic anhydrase. The proton pump mutations can cause deafness also. If one is acting as a chloride–bicarbonate exchanger and the other is pumping protons, they have opposite effects on pH. This suggests that they have to do something together; knocking out either one disrupts this function. In a sense, if you have a proton pump in parallel with a chloride–bicarbonate exchanger, the net effect is to have an inward chloride pump. This would in a sense be an electrogenic chloride uptake mechanism. This would also favour the transepithelial potential difference that is generated. Is it possible that you need a proton pump and a chloride–bicarbonate exchanger working together, and losing either one causes deafness?

Muallem: Does mutation of the proton pump also lead to a collapse of the membrane potential?

Aronson: I don't know.

Welsh: Then you might expect too little fluid. They are saying they have hydrops.

Soleimani: If there is a basolateral NKCC1 in these cells, then once you inhibit pendrin on the apical membrane, the intracellular chloride should drop which should activate the basolateral NKCC1, if indeed there is one. It is possible that if this happens, NKCC1 is going to bring the chloride in. Can we get a similar scenario if there is an inactivating mutation in pendrin?

Ashmore: The issue is that pendrin is in a different location from all the other transporters we are talking about. This is the complexity.

Aronson: This model for the proton pump working with the chloride–bicarbonate exchanger would have an obligate role for carbonic anhydrase. As the proton pump generates hydroxyls you need to form bicarbonate to work the pendrin. There is a metabolon here that has the net effect of mediating electrogenic uptake of chloride.

Kere: In the chloride diarrhoea, one of the things that has not been mentioned here is that not only is the output from the large intestine very high in chloride, but it is also acidic. The pH is entirely screwed up when A3 is not working.

Soleimani: CA2 mutants are not deaf.

Alper: There is a lot of redundancy. I have an off-the-wall question. The sodium iodide symporter NIS is also located in the mammary duct, and it is greatly upregulated during lactation. How does iodide exit the apical side of the mammary duct for secretion?

Everett: Pendrin is expressed in mammary gland.

Alper: Marlies Knipper was pursuing this issue earlier: one can certainly imagine a neonatal iodide deficiency in that context.

Everett: It is difficult to explain the deafness features, but I agree that pendrin could be acting as iodide transporter here.

References

Royaux IE, Belyantseva IA, Wu T et al 2003 Localization and functional studies of pendrin in the mouse inner ear provide insight about the etiology of deafness in pendred syndrome. J Assoc Res Otolaryngol 4:394–404

Wangemann P, Itza EM, Albrecht B et al 2004 Loss of KCNJ10 protein expression abolishes endocochlear potential and causes deafness in Pendred syndrome mouse model. BMC Med 2:30

The renal physiology of pendrin (SLC26A4) and its role in hypertension

Susan M. Wall

Renal Division, Emory University School of Medicine, 1639 Pierce Drive, NE Atlanta, GA 30322, USA

Abstract. SLC26A4 (pendrin, PDS) is a Na^+-independent, $Cl^-/HCO_3^-/OH^-$ exchanger that is expressed in the apical regions of type B and non-A, non-B intercalated cells within the cortical collecting duct (CCD), the connecting tubule and the distal convoluted tubule where it mediates HCO_3^- secretion and Cl^- absorption. *SLC26A4* is upregulated with aldosterone analogues and with Cl^- restriction. While under basal conditions no renal abnormalities are observed in mice and humans with genetic disruption of *SLC26A4* (Pendred syndrome), differences become apparent under conditions wherein the transporter is stimulated. Following treatment with aldosterone analogues, e.g. deoxycorticosterone pivalate (DOCP), weight gain and hypertension are observed in *Slc26a4*$^{+/+}$ but not in *Slc26a4*$^{-/-}$ mice. During dietary NaCl restriction, a model in which serum aldosterone is appropriately increased, urinary volume and urinary excretion of Cl^- are greater in *Slc26a4*$^{-/-}$ than in wild-type mice which results in apparent vascular volume contraction in *Slc26a4*$^{-/-}$ mice. Moreover, during NaCl restriction or following DOCP treatment, *Slc26a4*$^{-/-}$ mice have a higher serum HCO_3^- than wild type mice from an impaired ability to excrete OH^- equivalents. In conclusion, SLC26A4 regulates blood pressure and arterial pH, likely by participating in the renal regulation of net acid and Cl^- excretion.

2006 Epithelial anion transport in health and disease: the role of the SLC26 transporters family. Wiley, Chichester (Novartis Foundation Symposium 273) p 231–243

Aldosterone increases blood pressure by acting on the collecting duct to stimulate the absorption of NaCl and acts on vascular tissue to increase resistance. Aldosterone administration *in vivo* leads to increased absorption of Na^+ by the kidney primarily through activation of the Na^+ channel ENaC (Eaton et al 2001), the thiazide-sensitive NaCl cotransporter (TSC/NCC)(Kim et al 1998), and the Na^+/K^+-ATPase (Ewart & Klip 1995). The result is aldosterone-induced Na^+ retention and increased blood pressure. In rodents, administration of aldosterone analogues and a high NaCl diet has been used as a model of salt-sensitive hypertension (Kurtz & Morris 1983). In salt-sensitive hypertension, it has generally been assumed that intake of Na^+, rather than Cl^-, leads to increased blood pressure. However, in humans with salt-sensitive hypertension blood pressure varies more with intake of

Cl⁻ than with intake of Na⁺ (Kurtz et al 1987). Nevertheless, identification of aldosterone-sensitive Cl⁻ transporters and the contribution of these Cl⁻ transporters to salt-sensitive hypertension, has not been described fully (Ornellas et al 2002).

The collecting duct is the final site of regulation for renal Cl⁻ excretion. With administration of aldosterone *in vivo* transepithelial transport of both Cl⁻ and HCO_3^- increases dramatically along the collecting duct, particularly within the cortical collecting duct (CCD) (Knepper et al 1984, Star et al 1985, Garcia-Austt et al 1985, Hanley & Kokko 1978). In models of metabolic alkalosis, the CCD absorbs Cl⁻ and secretes HCO_3^- through transepithelial transport across intercalated cells, rather than across principal cells (Schlatter et al 1990). This process is mediated by an apical Cl⁻/HCO_3^- exchanger acting in series with a basolateral Cl⁻ conductance (Schuster 1993) and an H⁺-ATPase (Fig. 1) (Emmons 1993).

Intercalated cells represent the minority cell type within the CCD, connecting tubule (CNT) and distal convoluted tubule (DCT) and are classified as A, B, or

Intercalated Cell Subtypes

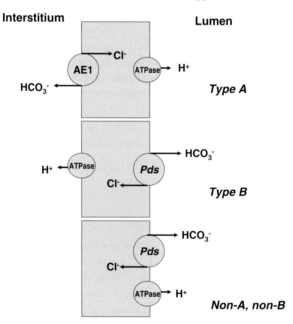

FIG. 1. Intercalated cell subtypes: classification of intercalated cell subtypes is based on the presence or absence of AE1 and the polarity of the H⁺-ATPase. Pendrin expression is high in the apical regions of non-A, non-B intercalated cells and type B intercalated cells (Wall et al 2003).

non-A, non-B based on the presence or absence of the anion exchanger, AE1, and the distribution of the H^+-ATPase within the cell (Alper et al 1989, Kim et al 1999, Schuster 1993, Teng-ummnuay et al 1996) (Fig. 1). Intercalated cells secrete or absorb net H^+ equivalents depending on whether the H^+-ATPase is expressed on the apical or the basolateral plasma membrane (Brown et al 1988, Emmons 1993). Type A intercalated cells secrete H^+ equivalents through the H^+-ATPase, which localizes to the apical plasma membrane and functions in series with the Cl^-/HCO_3^- exchanger, AE1 expressed on the basolateral plasma membrane (Teng-ummnuay et al 1996). In the type B intercalated cell, Cl^-/HCO_3^- exchange across the apical plasma membrane functions in series with the H^+-ATPase expressed on the basolateral plasma membrane (Alper et al 1989, Kim et al 1999, Teng-ummnuay et al 1996) to mediate secretion of OH^- equivalents, particularly during metabolic alkalosis (Schuster 1993). As will be discussed below, *SLC26A4* (*PDS*, pendrin) likely represents the gene responsible for apical anion exchange-mediated HCO_3^- secretion (Royaux et al 2001, Soleimani et al 2001, Frische et al 2003, Verlander et al 2003) since both localize to the apical plasma membrane of the type B intercalated cell (Emmons & Kurtz 1994, Royaux et al 2001, Wall et al 2003, Kim et al 2002), both transport Cl^- and HCO_3^- (Scott et al 1999, Royaux et al 2001, Soleimani et al 2001, Wall et al 2004) and both are regulated in parallel (Frische et al 2003, Verlander et al 2003). As will be discussed below, by altering excretion of Cl^- and OH^- equivalents, SLC26A4 is an important regulator of both arterial pH and blood pressure.

Pendred syndrome

In 1896 Vaughan Pendred described an Irish family in which 2 of 10 children suffered from congenital deafness and goitre (Pendred 1896). Pendred syndrome accounts for up to 10% of all hereditary hearing loss and thus is recognized as one of the leading causes of congenital deafness (Everett et al 1997). Patients with Pendred syndrome have multinodular goitre, although they are generally euthyroid (Everett et al 1997, Pendred 1896).

While pendrin is highly expressed in kidney, the physiological significance of pendrin in human kidney is not obvious since affected individuals and $Slc26a4^{-/-}$ mice (pendrin knockout mice) have no known acid-base or fluid and electrolyte abnormalities and appear to have normal renal function at least under basal, unstimulated conditions (Royaux et al 2001). However, under conditions that stimulate pendrin in normal rodents, such as with administration of aldosterone analogues, differences between $Slc26a4^{-/-}$ and $Slc26a4^{+/+}$ mice in acid–base and fluid balance become apparent (Verlander et al 2003), due to reduced capacity of $Slc26a4$ null mice to secrete HCO_3^- and absorb Cl^- (Royaux et al 2001, Wall et al 2004).

Cloning and the tissue distribution of SLC26A4

The molecular structure of SLC26A4 was first reported by Everett and colleagues (Everett et al 1997). In human, mouse and rat kidney, pendrin is expressed in the CCD, the CNT, the initial collecting tubule (iCT) and the DCT (Royaux et al 2001, Wall et al 2003, Kim et al 2002), where it localizes to the apical plasma membrane and apical cytoplasmic vesicles of both type B and non-A, non-B intercalated cells (Royaux et al 2001, Wall et al 2003, Kim et al 2002) (Fig. 1). Apical localization of pendrin to a subset of intercalated cells suggests that pendrin participates in the regulation of acid-base balance (Royaux et al 2001). However, intercalated cells also participate in the process of Cl^- absorption and secretion (Wall et al 2001, Verlander et al 2003).

Aldosterone analogues up-regulate pendrin expression

Slc26a4 is up-regulated in a variety of models of metabolic alkalosis, such as with aldosterone analogues (Verlander et al 2003) and following ingestion of $NaHCO_3$ (Frische et al 2003, Quentin et al 2004). Our laboratory has observed a modest increase in *Slc26a4* mRNA and total pendrin protein expression following the administration of the aldosterone analogue, deoxycorticosterone pivalate (DOCP) (Verlander et al 2003). However, pendrin is regulated in this treatment model primarily through subcellular redistribution rather than changes in total protein expression (Verlander et al 2003).

The subcellular distribution of pendrin protein is highly regulated by aldosterone analogues (Verlander et al 2003). In kidneys from control mice, pendrin immunoreactivity in type B intercalated cells was prevalent over apical cytoplasmic vesicles, with little immunolabels present along the apical plasma membrane. Following DOCP treatment, however, the apical plasma membrane label density of *Slc26a4* increased twofold in both type B intercalated cells and in non-A, non-B intercalated cells. Boundary length increased twofold in type B but did not change in non-A, non-B intercalated cells, whereas total protein expression per cell increased twofold in non-A, non-B, but does not change in type B intercalated cells. Thus, following the administration of aldosterone analogues, apical plasma membrane *Slc26a4* expression rose twofold in non-A, non-B intercalated cells, and rose sixfold in type B intercalated cells (Verlander et al 2003).

Aldosterone analogues administered *in vivo* might up-regulate *Slc26a4* indirectly such as through hypokalemia, volume expansion and/or metabolic alkalosis which are observed in this treatment model. However, since pendrin is up-regulated in many treatment models in which serum aldosterone levels are increased independent of changes in serum K^+, arterial pH and vascular volume status, aldosterone likely has a direct effect on *Slc26a4* expression. However, pendrin is also regulated through aldosterone-independent mechanisms such as with changes in Cl⁻ balance (Quentin et al 2004).

Slc26a4 mediates HCO$_3^-$ secretion and Cl⁻ absorption, which modulates acid–base balance, fluid balance and blood pressure

Because pendrin is a Na⁺-independent, Cl⁻/HCO$_3^-$ exchanger in heterologous expression systems, we explored the role of *Slc26a4* in the process of Cl⁻ absorption and HCO$_3^-$ secretion in native kidney tissue. Thus, *Slc26a4*$^{-/-}$ and *Slc26a4*$^{+/+}$ mice were treated with DOCP and given 50 mM NaHCO$_3$ in their drinking water for 7 days to up-regulate pendrin (Royaux et al 2001, Frische et al 2003, Verlander et al 2003). CCD tubules from these mice were perfused *in vitro* in the presence of symmetric, physiological, HCO$_3^-$/CO$_2$-buffered solutions. Cl⁻ absorption and HCO$_3^-$ secretion were observed in CCDs from wild-type mice, but were not observed in tubules from *Slc26a4* null mice (Royaux et al 2001, Wall et al 2004). Thus *Slc26a4* is a critical component of the Cl⁻ absorption and HCO$_3^-$ secretion observed in the CCD following administration of aldosterone analogues and ingestion of NaHCO$_3$. In mouse, *Slc26a4*-dependent Cl⁻ absorption and HCO$_3^-$ secretion are probably greater in the CNT than in the CCD since pendrin expression is much greater in the apical plasma membrane of the CNT than the CCD under both basal conditions and following the administration of DOCP (Wall et al 2003, Verlander et al 2003). Moreover, Na⁺ absorption is fourfold higher in CNT than CCD (Almeida & Burg 1982).

Although pendrin is highly expressed in kidney, under basal, unstimulated conditions, both *Slc26a4* null mice and individuals with genetic disruption of the *Slc26a4* gene (Pendred syndrome) have normal acid–base balance (Royaux et al 2001). However, following DOCP administration, arterial pH and serum HCO$_3^-$ are higher in *Slc26a4*$^{-/-}$ than *Slc26a4*$^{+/+}$ mice, consistent with the reduced HCO$_3^-$ secretion observed in the CCD of *Slc26a4*$^{-/-}$ mice, which should limit correction of a metabolic alkalosis (Royaux et al 2001, Verlander et al 2003).

Since pendrin is a Cl⁻ transporter, we explored the role of pendrin in salt and water balance. After consuming a NaCl-replete diet for 7 days, systolic blood pressure and weight change were similar in *Slc26a4*$^{-/-}$ and *Slc26a4*$^{+/+}$ mice (Fig. 2) (Verlander et al 2003). Seven days after administration of DOCP and a NaCl-replete diet, wild type mice gained weight and their systolic blood pressure rose from 111 ± 1 ($n = 7$) to 132 ± 3 mmHg ($n = 10$, $P < 0.05$) (Verlander et al 2003). However, *Slc26a4*$^{-/-}$ mice treated with DOCP and pair-fed the same NaCl-replete diet had no change in blood pressure and did not gain weight (Fig. 2). Thus the cumulative excretion of NaCl over the treatment period was probably greater in *Slc26a4*$^{-/-}$ than *Slc26a4*$^{+/+}$ mice.

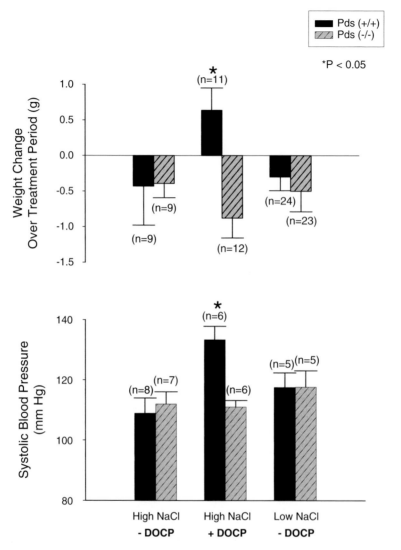

FIG. 2. Effect of DOCP and NaCl-restriction on weight and systolic blood pressure in *Slc26a4*[-/-] and *Slc26a4*[+/+] mice. Mice were pair-fed a balanced NaCl-replete (0.78 mEq/day NaCl) or a NaCl-restricted (0.13 mEq/day) diet. After 7 days of diet alone or diet plus DOCP, systolic blood pressure and weight change over the treatment period were measured (Verlander et al 2003; Wall et al 2004).

A critical component of aldosterone-induced hypertension (Jiang et al 2003) occurs through increased arterial sensitivity to vasoconstrictors (Tanoue et al 2002, Jiang et al 2003) and through structural changes in the vessel wall (Tanoue et al 2002). However, *Slc26a4* expression in vascular tissue has not been detected at the protein level. Thus, there is no evidence for a role of *Slc26a4* in the regulation of vascular tone. *Slc26a4* most likely modulates blood pressure, therefore, by mediating Cl⁻ uptake by the kidney which expands vascular volume.

NaCl restriction up-regulates pendrin expression

We asked whether pendrin is up-regulated in treatment models that involve neither metabolic alkalosis nor increased intake of OH⁻ equivalents. We evaluated pendrin expression in *Slc26a4*$^{+/+}$ mice ingesting either a NaCl-restricted (0.13 mEq/day NaCl) or a NaCl-replete (0.78 mEq/day NaCl) diet, which fall within the range of NaCl intake found in normal rodent diets. We observed that arterial pH was the same in NaCl-replete and NaCl-restricted *Slc26a4*$^{+/+}$ (wild-type) mice (Wall et al 2004), although *Slc26a4* expression varied in the two groups (Wall et al 2004). In type B intercalated cells, dietary NaCl restriction caused a redistribution of pendrin from cytoplasmic membrane vesicles to the apical plasma membrane, which led to a twofold increase in pendrin expression on the apical plasma membrane. To further study the physiological role of pendrin during NaCl restriction, we asked whether NaCl restriction unmasks a difference in acid–base or fluid and electrolyte balance between *Slc26a4*$^{+/+}$ and *Slc26a4*$^{-/-}$ mice. We observed that arterial pH was the same in NaCl-replete *Slc26a4*$^{+/+}$ and *Slc26a4*$^{-/-}$ mice. However, during NaCl restriction, metabolic alkalosis and apparent vascular volume contraction were observed in *Slc26a4*$^{-/-}$ mice but not in *Slc26a4*$^{+/+}$ mice (Wall et al 2004). Since *Slc26a4* null and wild-type mice had similar blood pressure under either basal conditions or following moderate NaCl restriction, we asked whether differences in blood pressure become apparent when mice are further NaCl restricted. When pair-fed a NaCl-deficient gelled diet (<0.1 mEq NaCl/day NaCl), systolic blood pressure in *Slc26a4*$^{+/+}$ mice was 102.6 ± 3.4 mmHg ($n = 4$) versus 90.0 ± 3.9 mmHg ($n = 4$) in *Slc26a4*$^{-/-}$ mice ($P < 0.05$) (Wall et al 2004). One interpretation of these data is that pendrin-mediated Cl⁻ absorption and HCO₃⁻ secretion are required to maintain normal arterial pH and also to maintain vascular volume, and hence blood pressure, during NaCl restriction. During NaCl restriction, the reduced ability of *Slc26a4*$^{-/-}$ to absorb Cl⁻ and secrete HCO₃⁻ in the CCD causes metabolic alkalosis and apparent vascular volume contraction.

Conclusions

Slc26a4 is expressed in the apical regions of type B and non-A, non-B intercalated cells. It is regulated by aldosterone analogues and through changes in Cl⁻ balance, independent of aldosterone. *Slc26a4* is regulated primarily through changes in subcellular distribution. In kidney, *Slc26a4* attenuates metabolic alkalosis and maintains vascular volume, at least in part, by mediating secretion of OH^- equivalents and uptake of Cl^-.

Acknowledgements

This work was supported by NIH grant, DK 52935 (to S. M. Wall).

References

Almeida AJ, Burg MB 1982 Sodium transport in the rabbit connecting tubule. Am J Physiol 243:F330–334

Alper SL, Natale J, Gluck S, Lodish HF, Brown D 1989 Subtypes of intercalated cells in rat kidney collecting duct defined by antibodies against erythroid band 3 and renal vacuolar H^+-ATPase. Proc Natl Acad Sci USA 86:5429–5433

Brown D, Hirsch S, Gluck S 1988 Localization of a proton-pumping ATPase in rat kidney. J Clin Investig 82:2114–2126

Eaton DC, Malik B, Saxena NC, Al-Khalili OK, Yue G 2001 Mechanisms of aldosterone's action on epithelial Na^+ transport. J Membrane Biol 184:313–319

Emmons C 1993 New subtypes of rabbit CCD intercalated cells as functionally defined by anion exchange and H+-ATPase activity. J Am Soc Nephrol 4:838 (abstr)

Emmons C, Kurtz I 1994 Functional characterization of three intercalated cell subtypes in the rabbit outer cortical collecting duct. J Clin Investig 93:417–423

Everett LA, Glaser B, Beck JC et al 1997 Pendred syndrome is caused by mutations in a putative sulphate transporter gene (PDS). Nat Genet 17:411–422

Ewart HS, Klip A 1995 Hormonal regulation of the Na^+-K^+-ATPase: mechanisms underlying rapid and sustained changes in pump activity. Am J Physiol 269:C295–311

Frische S, Kwon T-H, Frokiaer J, Madsen KM, Nielsen S 2003 Regulated expression of pendrin in rat kidney in response to chronic NH_4Cl or $NaHCO_3$ loading. Am J Physiol 284: F584–593

Garcia-Austt J, Good DW, Burg MB, Knepper MA 1985 Deoxycorticosterone-stimulated bicarbonate secretion in rabbit cortical collecting duct: effects of luminal chloride removal and *in vivo* acid loading. Am J Physiol 249:F205–212

Hanley MJ, Kokko JP 1978 Study of chloride transport across the rabbit cortical collecting tubule. J Clin Investig 62:39–44

Jiang G, Cobbs S, Klein JD, O'Neill WC 2003 Aldosterone regulates the Na-K-2Cl cotransporter in vascular smooth muscle. Hypertension 41:1131–1135

Kim G-H, Masilamani S, Turner R, Mitchell C, Wade JB, Knepper MA 1998 The thiazide-sensitive Na-Cl cotransporter is an aldosterone-induced protein. Proc Natl Acad Sci USA 95:14552–14557

Kim J, Kim Y-H, Cha J-H, Tisher CC, Madsen KM 1999 Intercalated cell subtypes in connecting tubule and cortical collecting duct of rat and mouse. J Am Soc Nephrol 10:1–12

Kim Y-H, Kwon T-H, Frische S et al 2002 Immunocytochemical localization of pendrin in intercalated cell subtypes in rat and mouse kidney. Am J Physiol 283:F744–754

Knepper MA, Good DW, Burg MB 1984 Mechanism of ammonia secretion by cortical collecting ducts of rabbits. Am J Physiol 247:F729–738

Kurtz TW, Morris RC 1983 Dietary chloride as a determinant of 'sodium-dependent' hypertension. Science 222:1139–1141

Kurtz TW, Hamoudi AAB, Morris RC 1987 'Salt-sensitive' essential hypertension in men. N Engl J Med 317:1043–1048

Ornellas DS, Nascimento DS, Christoph DH, Guggino WM, Morales MM 2002 Aldosterone and high-NaCl diet modulate ClC-2 chloride channel gene expression in rat kidney. Pflugers Arch 444:193–201

Pendred V 1896 Deaf-mutism and goitre. Lancet 2:532

Quentin F, Chambrey R, Trinh-Trang-Tan MM et al 2004 The Cl-/HCO3- exchanger pendrin in the rat kidney is regulated in response to chronic alterations in chloride balance. Am J Physiol 287:F1179–1188

Royaux IE, Wall SM, Karniski LP et al 2001 Pendrin, encoded by the pendred syndrome gene, resides in the apical region of renal intercalated cells and mediates bicarbonate secretion. Proc Natl Acad Sci USA 98:4221–4226

Schlatter E, Greger R, Schafer JA 1990 Principal cells of cortical collecting ducts of the rat are not a route of transepithelial Cl⁻ transport. Pflugers Arch 417:317–323

Schuster VL 1993 Function and regulation of collecting duct intercalated cells. Annu Rev Physiol 55:267–288

Scott DA, Wang R, Kreman TM, Sheffield VC, Karniski LP 1999 The Pendred syndrome gene encodes a chloride-iodide transport protein. Nat Genet 21:440–443

Soleimani M, Greeley T, Petrovic S et al 2001 Pendrin: an apical Cl⁻/OH⁻/HCO₃⁻ exchanger in the kidney cortex. Am J Physiol 280:F356–364

Star RA, Burg MB, Knepper MA 1985 Bicarbonate secretion and chloride absorption by rabbit cortical collecting ducts: role of chloride/bicarbonate exchange. J Clin Investig 76:1123–1130

Tanoue A, Koba M, Miyawaki S et al 2002 Role of the α_{1D}-adrenergic receptor in the development of salt-induced hypertension. Hypertension 40:101–106

Teng-ummnuay P, Verlander JW, Yuan W, Tisher CC, Madsen KM 1996 Identification of distinct subpopulations of intercalated cells in the mouse collecting duct. J Am Soc Nephrol 7:260–274

Verlander JW, Hassell KA, Royaux IE et al 2003 Deoxycorticosterone upregulates Pds (Slc26a4) in mouse kidney: role of pendrin in mineralocorticoid-induced hypertension. Hypertension 42:356–362

Wall SM, Fischer MP, Mehta P, Hassell KA, Park SJ 2001 Contribution of the Na⁺-K⁺-2Cl⁻ cotransporter (NKCC1) to Cl⁻ secretion in rat outer medullary collecting duct. Am J Physiol 280:F913–921

Wall SM, Hassell KA, Royaux IE et al 2003 Localization of pendrin in mouse kidney. Am J Physiol 284:F229–241

Wall SM, Kim Y-H, Stanley L et al 2004 NaCl restriction upregulates renal Slc26a4 through subcellular redistribution: role in Cl⁻ conservation. Hypertension 44:1–6

DISCUSSION

Markovich: Would Pendred syndrome patients suffer from hypertension? Assuming that the mouse model is representative of the human, you wouldn't think so.

Wall: This hasn't been studied.

Welsh: They might be relatively protected. Have you looked at heterozygote animals?

Wall: Not for blood pressure.

Welsh: There are some large genetic studies underway that are attempting to find genes that predispose to hypertension and also genes that protect. Has anyone studied the people who are heterozygous for pendrin?

Wall: Not for blood pressure.

Kere: What do we know about the expression of other SLC26 family members in the intercalated cells in humans?

Soleimani: We haven't looked at humans. We have seen A7 in mouse, mostly in the outer medullary collecting duct but a little bit in the cortical collecting duct as well. A6 is absent. A11 is expressed in the collecting duct (personal observation) but we don't know about its membrane localization. It doesn't seem to co-localize with pendrin.

Wall: The distribution of pendrin is similar in human and rodent kidney (Royaux et al 2004).

Kere: We have a recent paper showing A6 and A7 both being present in the intercalated cells of the distal segment (Kujala et al 2005). In human kidney A6 is expressed in distal segments of proximal tubules, the thick and thin ascending loops of Henle, macula densa, distal convoluted tubules and a subpopulation of intercalated cells of collecting ducts. A7 is expressed in extraglomerular mesangial cells and a subpopulation of intercalating cells of the collecting duct.

Soleimani: Susan, what happens to H-ATPase in response to DOCA in pendrin knockout mice? These animals get alkalosis, presumably because of the imbalance of H-ATPase and pendrin.

Wall: Under basal conditions there is no renal phenotype in *Slc26a4* null mice or in persons with Pendred syndrome (Royaux et al 2001, Wall et al 2004). Therefore, we asked whether the absence of a renal phenotype in these mice is because pendrin is not functionally important in kidney under basal conditions or because pendrin is important, but powerful compensatory mechanisms prevent changes in acid–base and/or fluid and electrolyte balance. Therefore we measured H^+-ATPase expression in kidneys from wild-type and *Slc26a4* null mice under basal conditions. We observed a marked decrease in H^+-ATPase expression in type B and non-A non-B intercalated cells from *Slc26a4* null mice (Kim et al 2004). This adaptive response would be predicted to limit the cellular alkalinization expected in the absence of pendrin-mediated OH^- secretion. We have not determined if renal H^+-ATPase expression differs in kidneys of wild-type and *Slc26a4* null mice following DOCP treatment.

Soleimani: You are using the model of DOCA treatment in mouse for 6 or 7 days. Fred Luft uses the same protocol and he monitors the cardiac output, an index of vascular volume. He doesn't see any increase in cardiac output in response to DOCA, yet he sees that mice develop hypertension. The idea that the volume should go up and eventually the resistance is contradicted by his studies showing that DOCA-induced hypertension is not associated with increased volume. There is an additional aspect to these studies. The aldosterone receptor is heavily expressed in the arteries which you can readily detect by RT-PCR. The vascular smooth muscle (VSM) cells express the sodium–hydrogen exchanger NHE1. On the basis of what we know about aldosterone, is it possible that there is a chloride-bicarbonate exchanger in VSM cells, such as pendrin, that co-localizes and functions in parallel with NHE1 and is regulated by aldosterone? In such a scenario aldosterone would be increasing Na^+ and Cl^- influx into the vascular wall, therefore initiating stretch or proliferation, leading to increased resistance and hypertension. This is our hypothesis at the moment and it is based, in part, on the extrarenal distribution of aldosterone (mineralocorticoid) receptor.

Wall: Aldosterone-induced hypertension has large vascular and neural components (Passmore et al 1985, Reid et al 1975, Takeda & Bunag 1980). However, pendrin protein expression has not been detected in either vascular or neural tissue.

Alper: Manoocher Soleimani, what mediates the chloride–formate exchange in the vascular smooth muscle that you once described?

Soleimani: After Peter Aronson showed the chloride–formate movement in kidney, we looked at heart and VSM. In both tissues we detected chloride–formate exchange. Presumably this component (chloride–formate exchange) in the cardiac cells is mediated via A6, but in arteries we find abundant A7, which potentially could be the transporter mediating the chloride–formate exchange in VSM cells. We have looked for pendrin but we haven't been successful in detecting it in the arteries.

Alper: Have you compared the Dahl salt-sensitive and salt-resistant rat strains? Have you sequenced pendrin or loaded them and looked at variable pendrin response to salt?

Wall: No, but this would be an interesting salt-sensitive hypertension model.

Alper: This is one model for which one candidate gene after another has been proposed and then dropped by the wayside. I have a question for both of you. Fisher rat thyroid cells are a favourite cell line for Na^+ physiologists. Is pendrin there, and would Fisher rat thyroid cells without any manipulation serve as a good transepithelial iodide transport system?

Welsh: We and others use them because they behave almost like pavement: they don't seem to perform much transepithelial electrolyte transport. This could be a nice model.

Everett: Pendrin is in FRTL5 cells.

Soleimani: I thought pendrin expression was very low in the rat thyroid.

Alper: These are cultured cells.

Soleimani: Various studies have shown that in the rat thyroid the expression of pendrin is almost non existent compared with human, whereas we know that in the kidney it is expressed in both human and rat. Rat thyroid expresses very little if any pendrin.

Everett: Len Kohn's group have performed extensive thyroid studies in rat. They showed low expression. It is up-regulated by thyroglobulin (Royaux et al 2000).

Knipper: Is it known how the receptor manages to transfer the PDS to the membrane? You said that an agonist manages to get membrane expression of PDS?

Wall: The signalling cascade that regulates pendrin-mediated transport has not been studied.

Turner: Have you looked at the time scale of the insertion?

Wall: No, but we plan to.

Turner: It is probably fast.

Wall: The time course of activation of pendrin by aldosterone is not known. In rabbit cortical collecting tubules perfused *in vitro*, addition of aldosterone to the bath increases Na^+ absorption within 3 hours (Wingo et al 1985).

Kere: We have A6 present apically in aquaporin 2-negative and H-ATPase positive cells, whereas A7 is basolateral in the ducts.

Wall: That is what you found: that A7 is basolateral.

Soleimani: Yes. I'd like to return to the issue of hypertension and DOCA. I am not sure we know why aldosterone causes hypertension, even though we talk about increased volume reabsorption in the kidney. Patients that have primary aldosteronism have normal vascular volume. Some people have suggested that they might have had increased vascular volume initially. When Fred Luft monitors cardiac output in mouse for seven days, he shows that hypertension develops, but vascular volume doesn't change. On the basis of the presence of the aldosterone receptors in the resistant arteries, we have done a clinical study on patients on dialysis. The majority of these patients have high plasma aldosterone levels and high blood pressure. We put them on aldosterone antagonist spironolactone, which should not have any effect on the vascular volume since these patients have no functioning kidneys. What we observe is that a good number of these patients show reduced blood pressure in response to the aldosterone antagonist. It has also been published that aldosterone antagonists reduce heart fibrosis in patients with congestive heart failure. I suspect there are extra-renal components to blood pressure lowering effects of aldosterone antagonists.

Wall: Aldosterone-induced hypertension has large vascular and neural components (Passmore et al 1985, Reid et al 1975, Takeda & Bunag 1980). So, clearly there are extra-renal components of aldosterone-induced hypertension.

Soleimani: In the model where they get hypertension, contrary to mouse the rat needs uninephrectomy to see the effect of DOCA on blood pressure.

Wall: The DOCP-treated mice we studied were not uninephrectomized.

Aronson: I should like to comment further on this interesting observation that pendrin has a role in chloride retention. Eladari's group has done an additional model in which dietary chloride is varied but sodium is kept the same, doing this in a way where aldosterone is moving in the other direction, by varying KCl intake. He had an almost perfect correlation between chloride excretion and pendrin expression (Quentin et al 2004). Giving high KCl, which would suppress pendrin, would cause aldosterone to go in the other direction. There was some way that the chloride was being detected and turned into a change in pendrin expression.

References

Kim Y-H, Verlander JW, Matthews SW et al 2004 H^+-ATPase expression is markedly reduced in type B and non-A, non-B intercalated cells of Pds$^{-/-}$ mice. J Am Soc Nephrol 15:70A (Abstr)

Kujala M, Tienari J, Lohi H et al 2005 SLC26A6 and SLC26A7 anion exchangers have a distinct distribution in human kidney. Nephron Exp Nephrol 101:e50–58

Passmore JC, Whitescarver SA, Ott CE, Kotchen TA 1985 Importance of chloride for deoxy-corticosterone acetate-salt hypertension in the rat. Hypertension 7:I115–120

Quentin F, Chambrey R, Trinh-Trang-Tan MM et al 2004 The Cl^-/HCO_3^- exchanger pendrin in the rat kidney is regulated in response to chronic alterations in chloride balance. Am J Physiol Renal Physiol 287:F1179–1188

Reid JL, Zivin JA, Kopin IJ 1975 Central and peripheral adrenergic mechanisms in the development of deoxycorticosterone-saline hypertension in rats. Circ Res 37:569–579

Royaux IE, Suzuki K, Mori A et al 2000 Pendrin, the protein encoded by the Pendred syndrome gene (PDS), is an apical porter of iodide in the thyroid and is regulated by thyroglobulin in FRTL-5 cells. Endocrinology 141:839–845

Royaux IE, Wall SM, Karniski LP et al 2001 Pendrin, encoded by the pendred syndrome gene, resides in the apical region of renal intercalated cells and mediates bicarbonate secretion. Proc Natl Acad Sci USA 98:4221–4226

Takeda K, Bunag RD 1980 Augmented sympathetic nerve activity and pressor responsiveness in DOCA hypertensive rats. Hypertension 2:97–101

Wall SM, Kim Y-H, Stanley L et al 2004 NaCl restriction upregulates renal Slc26a4 through subcellular redistribution: role in Cl^- conservation. Hypertension 44:1–6

Wingo CS, Kokko JP, Jacobson HR 1985 Effects of *in vitro* aldosterone on the rabbit cortical collecting tubule. Kidney Int 28:51–57

Interaction of prestin (SLC26A5) with monovalent intracellular anions

Dominik Oliver, Thorsten Schächinger and Bernd Fakler

Physiologisches Institut, Universität Freiburg, Hermann-Herder-Str.7, 79104 Freiburg, Germany

Abstract. Outer hair cells (OHCs) of the mammalian cochlea are equipped with a specific form of cellular motility that is driven by changes of the membrane potential. This electromotility is a membrane-based process generated by the membrane protein prestin (SLC26A5). Current models suggest that prestin undergoes a force-generating conformational transition upon changes of the membrane potential. The voltage dependence of prestin needs to be mediated by a charged particle within the protein, a 'voltage sensor', that can move through the membrane electrical field to trigger these conformational rearrangements. Indeed, voltage sensor translocation can be measured as electrical charge transfer. Here, we review and extend data indicating that charge movement by prestin and consequently electromotility depend on the presence of small monovalent anions such as chloride and bicarbonate at the cytoplasmic side of the membrane. The voltage dependence of prestin varies with concentration and species of the anion present, consistent with a partial translocation of the anion through the membrane. Thus anions may act as extrinsic voltage sensors. These conclusions suggest that charge movement and subsequent conformational rearrangements may relate to anion transport by other SLC26 members. Insights into molecular properties of prestin may provide clues to common mechanisms of anion transport by SLC26 proteins.

2006 Epithelial anion transport in health and disease: the role of the SLC26 transporters family. Wiley, Chichester (Novartis Foundation Symposium 273) p 244–260

Outer hair cell (OHC) electromotility (Brownell et al 1985) comprises cycles of contraction and elongation of the cylindrical cell body in response to depolarization and hyperpolarization, respectively (Ashmore 1987, Santos-Sacchi 1989). The length changes are extremely rapid and can follow sinusoidal electrical stimulation up to frequencies of at least 70 kHz in a cycle-by-cycle manner (Frank et al 1999). It is believed that this high-frequency motion can boost mechanical vibration within the cochlea, thereby providing a major component of the so-called 'cochlear amplifier', an active mechanical process underlying the high sensitivity and frequency selectivity of the mammalian inner ear (Dallos 1992, Dallos & Evans 1995, Liberman et al 2002).

In contrast to conventional types of cellular motion that are based on motor proteins acting on cytoskeletal elements, OHC electromotility is an intrinsic membrane process. Motility persists even if the cytoskeleton has been digested with proteases (Kalinec et al 1992, Huang & Santos-Sacchi 1994) and can be evoked in membrane patches isolated from the OHC (Kalinec et al 1992, Gale & Ashmore 1997b). As a structural correlate of electromotility, the lateral membrane of OHCs contains a high density of particles (Forge 1991, Kalinec et al 1992), which are interpreted as individual elementary motors. Central to the understanding of how distributed elementary motors in the membrane may generate movement is the observation that electromotility has an 'electrical signature': it is associated with the transfer of electrical charge across the membrane that can be measured as a 'gating current' at the onset of a voltage jump, equivalent to gating currents of ion channels (Ashmore 1989, Santos-Sacchi 1991, Dallos & Fakler 2002).

Charge transfer and motility share the same voltage dependence, suggesting that the former results from the translocation of a charged particle (voltage sensor) within a membrane motor protein, which in turn directly triggers a force-producing conformational change. Voltage dependence is well described by a two-state Boltzmann function, again consistent with the voltage-dependent redistribution of an elementary motor between two conformational states. The charge movement is equivalent to a bell-shaped voltage dependent capacitance (e.g. Fig. 1A), usually termed non-linear capacitance (NLC), which can be measured conveniently using patch-clamp methods (for details see Dallos & Fakler 2002). Since motility and NLC share the same voltage dependence and kinetics, NLC is often used to characterize the properties of electromotility.

Using a subtractive OHC cDNA library Zheng, Dallos and colleagues (Zheng et al 2000) isolated a cDNA that was enriched in OHCs but absent from the non-motile inner hair cells. The protein product, which they termed prestin, is highly and exclusively expressed in the lateral membrane of the OHC (Belyantseva et al 2000, Zheng et al 2001). Prestin turned out to be the fifth member of the SLC26 family of anion transporters (SLC26A5). Analysis of the amino acid sequence shows overall similarity to all other members: hydrophilic N- and C-termini and a largely hydrophobic core region that may contain 10 to 12 transmembrane spans. Both N- and C-termini are located intracellularly (Ludwig et al 2001, Zheng et al 2001). When expressed heterologously prestin generates all hallmarks of electromotility: cells transfected with prestin are motile when stimulated electrically and exhibit the corresponding electrical signature, i.e. gating charge movement and NLC (Zheng et al 2000). Vice versa, OHCs from prestin knockout mice completely lack electromotility. Thus prestin is the motor protein of the OHC membrane.

Little is known about the molecular structure of prestin and the mechanisms that underlie force generation. Consistent with the primary sequences of all other SLC26 members, two sequence motifs are identified by homology. The N-terminal half contains a highly conserved sequence stretch, the SLC26 transporter's signature (Prosite entry PS01130), and the cytoplasmic C-terminus contains a STAS (sulfate transporter and anti-sigma antagonist) domain (Prosite entry PS50801) (Zheng et al 2000, Mount & Romero 2004). Their role is as yet unknown. The work shown here represents first steps towards the elucidation of the molecular mechanisms that underlie the generation of electromotility, focusing on the identity of the voltage-sensing mechanism of prestin.

Methods

Electrophysiological and optical methods used for the analysis of charge movement, non-linear capacitance and cell motility have been detailed elsewhere (Oliver & Fakler 1999, Ludwig et al 2001, Oliver et al 2001, Dallos & Fakler 2002). Briefly, NLC was measured using lock-in patch-clamp methods in native OHCs from rat, in patches excised from OHCs and from recombinant rat prestin expressed heterologously in CHO cells. To quantify the voltage dependence of NLC, the experimental bell-shaped capacitance-voltage curves were fitted with the derivative of a first-order Boltzmann function,

$$C(V) = C_{lin} + \frac{Q_{max}}{\alpha e^{\frac{v - v_{1/2}}{\alpha}} \left(1 + e^{-\frac{v - v_{1/2}}{\alpha}}\right)^2}$$

where C_{lin} is residual linear membrane capacitance, V is membrane potential, Q_{max} is maximum voltage-sensor charge moved through the membrane electrical field, $V_{1/2}$ is voltage at half-maximal charge transfer and α is the slope factor of the voltage dependence. C_{lin} is subtracted from the traces shown. $V_{1/2}$ is the voltage at maximum NLC and in the two-state model half of the molecules reside in either state. The slope factor α signifies the steepness of the voltage dependence; the larger α, the more shallow is the voltage dependence.

Results and discussion

Structure–function studies

In voltage-gated ion channels, the positively charged fourth transmembrane helix (S4 segment) serves as the voltage sensor (Stuhmer et al 1989, Papazian et al 1991). Movement of this sensor across the membrane is thought to generate gating charge and to trigger opening and closing of the channels. A similar mechanism might operate in prestin, thus we initially sought charged residues that could contribute to the voltage sensor. Because the closest homologue of prestin, SLC26A6, did not exhibit a comparable charge movement (Oliver et al 2001), the voltage sensor residues of prestin should be unique, i.e. not conserved between SLC26A5 and A6. We thus neutralized all non-conserved positive and negative charges located in the putative transmembrane region of prestin either individually or in clusters by site-directed mutagenesis and tested the mutant prestin forms for NLC (Oliver et al 2001). All mutations neither abolished charge-movement nor reduced the apparent sensor charge, as indicated by an unchanged steepness of NLC. However, a subset of charge-neutralizing mutations shifted the voltage dependence to either hyperpolarized or depolarized voltages (Oliver et al 2001). This shift disagrees with a direct involvement in voltage sensing but may be explained by an electrostatic interaction of these residues with the voltage sensor or a local deformation of the electrical field. The additional mutations shown in Fig. 1 support this interpretation. Combination of the mutations D342Q and KKR233/235/236QQQ that individually shift the voltage dependence to depolarized and hyperpolarized voltages, respectively, yields the wild-type phenotype, indicating mutual compensation of both neutralizations. Vice versa, increasing the charge by adding an additional charged residue at a neighbouring position instead of neutralization at a position that affected voltage dependence, led to a shift of $V_{1/2}$ opposite to neutralization. For example, V341D shifted $V_{1/2}$ negatively to $-166\,mV$, while the corresponding charge neutralization D342Q shifted $V_{1/2}$ positively to $0\,mV$ (Fig. 1B). These findings would be consistent with a purely electrostatic interaction of the charged residues with the voltage sensor.

Although these findings suggest that the voltage sensor is not made up from charged residues within the core region of prestin, it may still comprise conserved amino acids not tested. Further, the apparent sensor charge (≈ 0.8 elementary charge) is much smaller than in ion channels and may well be supplied by several partial charges present in the membrane-embedded parts of the protein.

Anion dependence

Instead of being an intrinsic property of the protein, voltage dependence may as well be conferred by an external charged particle. In particular, electrogenic interaction of an anion with prestin appeared feasible, given that prestin is a member of the SLC26 anion transporter family. We thus tested for the requirement of anions for charge movement/NLC. Both in recombinant prestin expressed in culture cells and in patches excised from native OHCs, NLC was abolished completely and reversibly when monovalent intracellular anions (usually Cl⁻ under control conditions) were replaced by divalents (sulfate) or large organic anions (Oliver et al 2001). In contrast, replacement of extracellular anions did not affect NLC.

Besides chloride at the intracellular side, other small inorganic anions, such as halides or bicarbonate, could support NLC (Oliver et al 2001). The amount of

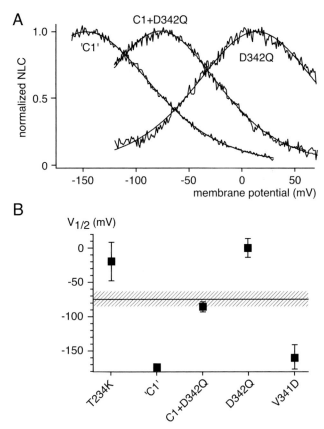

FIG. 1.   Shifts of voltage dependence induced by mutational neutralization of charged residues are additive. (A) Charge neutralization at position D342 results in a shift of NLC to positive potentials, whereas neutralization of a cluster of positively charged residues ('C1', i.e. K233Q/K235Q/R236Q) results in a shift towards negative potentials. Combination of both neutralizations yields an intermediate (wild-type) phenotype. Note the differences in $V_{1/2}$ (voltage at peak NLC) without changes of the steepness of the curve. NLC was measured in CHO cells transfected with the mutant version of rat prestin indicated. Data are normalized to peak NLC and smooth lines are fits of the derivative of a Boltzmann function to the measured data (see Methods). (B) $V_{1/2}$ values obtained from experiments as shown in panel A (error bars indicate SD). Hatched area indicates $V_{1/2}$ of wild-type prestin (mean ± SD). Combination of two oppositely shifting mutants yields wild-type phenotype ('C1 + D342Q'); addition of a charge at a neighbouring sequence position (e.g. V341D instead of D342Q) yields a shift opposite to the corresponding neutralization.

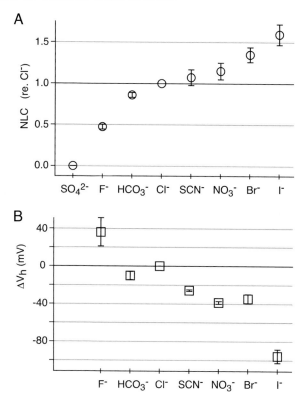

FIG. 2. Dependence of NLC on anion species present in the intracellular solution. NLC generated by native prestin was measured in inside-out patches excised from rat OHCs in the presence of various inorganic anions as the only anion present at the cytoplasmic face of the patch. (A) Peak NLC (mean ± SD) recorded with the different intracellular anions (150 mM) is shown normalized to the peak NLC obtained with 150 mM Cl^- in each patch. (B) Shift in $V_{1/2}$ obtained with the different anions is plotted relative to the $V_{1/2}$ with Cl^- in each patch. Data were not corrected for changes in liquid junction potentials (<20 mV).

charge moved and, equivalently, the amplitude of the NLC varied with the anion species present, with iodide yielding the largest and fluoride the smallest NLC of the anions tested (Fig. 3A). While the steepness of NLC was constant for all anions, $V_{1/2}$ strongly varied with the ion species. The shift of $V_{1/2}$ along the voltage axis was well correlated with the amplitude of NLC supported by each ion, with the most negative voltage dependence occurring at the largest NLC (iodide, Fig. 3B). NLC decreased with reduction of the anion concentration, yielding dose–response curves that could be well fitted will a Hill equation. The affinity derived from such fits was 6 mM for Cl^- and 44 mM for bicarbonate (Oliver et al 2001).

In contrast to the constancy of slope observed with small anions, replacement with organic anions of increasing size, such as carboxylic acids, yielded decreasing steepness with increased molecule size (Oliver et al 2001). A different example obtained with alkyl-sulfonate ions of different alkyl chain lengths is presented in Fig. 4 (cf. Rybalchenko & Santos-Sacchi 2003). Similarly, the bulky organic anion

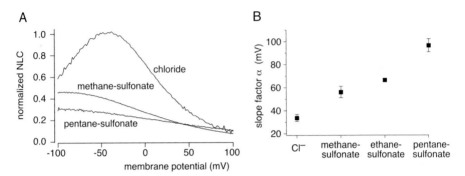

FIG. 3. The steepness of voltage dependence of NLC decreases with increasing size of the intracellular organic anions. (A) NLC obtained from excised patches as in Fig. 2 with Cl⁻, methane-sulfonate or pentane-sulfonate (150 mM) at the intracellular side of the patch. Curves are shown normalized to peak NLC for each patch. (B) Slope factor of NLC obtained from Boltzmann fits (see Methods) in the presence of the various alkyl-sulfonates indicated.

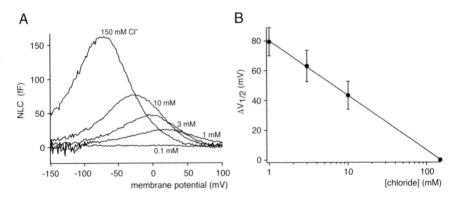

FIG. 4. Dependence of $V_{1/2}$ on the Cl⁻ concentration. (A) NLC measured in an excised patch as in Fig. 2 with the [Cl⁻] as indicated present at the intracellular side of the patch. Note that both peak NLC and $V_{1/2}$ vary with [Cl⁻]. Sulfate was substituted for Cl⁻ such that a constant ionic strength was maintained. (B) Summary of $V_{1/2}$ data from 4 patches measured as in (A). Straight line indicates a fit of a logarithmic function to the data. $\Delta V_{1/2}$ is the difference in $V_{1/2}$ with respect to the value obtained with 150 mM Cl⁻ in each patch.

salicylate can support NLC, although with a very shallow voltage-dependence (Oliver et al 2001). Competition with chloride for the interaction with prestin can therefore explain the inhibition of normal electromotility by salicylate. In the simplest model of a voltage sensor, the slope of the voltage dependence is proportional to the total charge of the sensor and to the fraction of the membrane electric field traversed by this charge (Woodhull 1973). The decrease of slope with increased anion size (at constant charge) is most easily explained by a reduced swing of the voltage sensor. This conclusion is entirely consistent with the anion itself being translocated through the electric field, acting as an external voltage sensor. Thus bulky anions may be sterically hindered from reaching as far through the field as small anions or may require larger voltage changes to be pushed through.

We conclude that intracellular anions may act as the voltage sensor, being translocated through most of the membrane electrical field by voltage changes (Oliver et al 2001). It is thus tempting to hypothesize that this anion translocation is equivalent to a partial reaction in the anion transport/exchange cycle of other SLC26 proteins. Notably, the anion selectivity derived from the amplitude of NLC with the different anions is similar to the transport selectivity determined for SLC26A4 (pendrin) (Scott & Karniski 2000). However, because extracellular anion concentration and species have no influence on NLC, it appears that the ion cannot unbind from prestin at the extracellular side.

One caveat to this simple model is that reducing the chloride concentration shifted the voltage-dependence to more positive potentials as demonstrated in Fig. 4 (Oliver & Fakler 2002, Rybalchenko & Santos-Sacchi 2003, Song et al 2005). In other words, at a lower anion concentration less negative intracellular potential is necessary to push an anion towards the extracellular side, contrary to what is expected for a simple anion translocation. Thus more complex reaction schemes may underlie the charge transfer and charged protein domains of prestin may contribute to it as well. An alternative model has been suggested in which anions bind to one or more distinct binding sites to regulate charge translocation allosterically and thus regulate motility (Rybalchenko & Santos-Sacchi 2003, Song et al 2005). In this model, anion binding shifts the position of the NLC curve along the voltage axis. However, such a model predicts two populations of molecules either without or with Cl^- bound which differ in the set point of voltage dependence. At any Cl^- concentration NLC should then be the sum of two bell-shaped curves each with a fixed $V_{1/2}$ at the extreme voltages. This is not observed experimentally (Oliver & Fakler 2002, Rybalchenko & Santos-Sacchi 2003, Song et al 2005) (Fig. 4).

In any case, the essential role of anions for prestin's function directly provokes the question of the molecular identity of the anion binding site. Future research may focus on the molecular mechanisms of prestin's function. Does the NLC indeed result from physical translocation of anions through the electrical field?

Alternatively, how much of the sensor charge is supplied by movable residues of the protein? And how does the sensor translocation subsequently trigger a gross conformational rearrangement that can even generate macroscopic cellular forces? Which are the domains of prestin that actually move upon changes in membrane voltage? An intriguing question is whether prestin may actually transport anions. Although extracellular anions do not appear to affect NLC (Oliver et al 2001) and preliminary transport studies did not reveal measurable ion fluxes (Dallos & Fakler 2002), complete anion translocation may still occur at rates much lower than the extremely fast forward and reverse rates of charge transfer, under certain ionic or electrical conditions that are permissive for transport, or in the presence of an as yet unknown substrate.

Finally, answers to these questions may prove helpful for understanding the currently unresolved mechanism of anion transport by other SLC26 transporters.

References

Ashmore JF 1987 A fast motile response in guinea pig outer hair cells: the cellular basis of the cochlear amplifier. J Physiol 388:323–347
Ashmore JF 1989 Transducer motor coupling in cochlear outer hair cells. In: Kemp D (ed) Mechanics of hearing. Plenum, New York, p 107–113
Belyantseva IA, Adler HJ, Curi R, Frolenkov GI, Kachar B 2000 Expression and localization of prestin and the sugar transporter GLUT-5 during development of electromotility in cochlear outer hair cells. J Neurosci 20:RC116
Brownell WE, Bader CR, Bertrand D, de Ribaupierre Y 1985 Evoked mechanical responses in isolated cochlear outer hair cells. Science 227:194–196
Dallos P 1992 The active cochlea. J Neurosci 12:4575–4585
Dallos P, Evans BN 1995 High frequency motility of outer hair cells and the cochlear amplifier. Science 267:2006–2009
Dallos P, Fakler B 2002 Prestin, a new type of motor protein. Nat Rev Mol Cell Biol 3:104–111
Forge A 1991 Structural features of the lateral walls in mammalian cochlear outer hair cells. Cell Tissue Res 265:473–483
Frank G, Hemmert W, Gummer AW 1999 Limiting dynamics of high-frequency electrome-chanical transduction of outer hair cells. Proc Natl Acad Sci USA 96:4420–4425
Gale JE, Ashmore JF 1997b An intrinsic frequency limit to the cochlear amplifier. Nature 389:63–66
Huang G, Santos-Sacchi J 1994 Motility voltage sensor of the outer hair cell resides within the lateral plasma membrane. Proc Natl Acad Sci USA 91:12268–12272
Kalinec F, Holley MC, Iwasa KH, Lim DJ, Kachar B 1992 A membrane based force generation mechanism in auditory sensory cells. Proc Natl Acad Sci USA 89:8671–8675
Liberman MC, Gao J, He DZ, Wu X, Jia S, Zuo J 2002 Prestin is required for electromotility of the outer hair cell and for the cochlear amplifier. Nature 419:300–304
Ludwig J, Oliver D, Frank G, Klocker N, Gummer AW, Fakler B 2001 Reciprocal electrome-chanical properties of rat prestin: the motor molecule from rat outer hair cells. Proc Natl Acad Sci USA 98:4178–4183
Mount DB, Romero MF 2004 The SLC26 gene family of multifunctional anion exchangers. Pflugers Arch 447:710–721

Oliver D, Fakler B 1999 Expression density and functional characteristics of the outer hair cell motor protein are regulated during postnatal development in rat. J Physiol (Lond) 519:791–800

Oliver D, Fakler B 2002 Functional properties of prestin—how the motormolecule works work. In: Gummer AW (ed) Biophysics of the cochlea. From molecules to models. World Scientific, Singapore, p 110–115

Oliver D, He DZ, Klocker N et al 2001 Intracellular anions as the voltage sensor of prestin, the outer hair cell motor protein. Science 292:2340–2343

Papazian DM, Timpe LC, Jan YN, Jan LY 1991 Alteration of voltage-dependence of Shaker potassium channel by mutations in the S4 sequence. Nature 349:305–310

Rybalchenko V, Santos-Sacchi J 2003 Cl⁻ flux through a non-selective, stretch-sensitive conductance influences the outer hair cell motor of the guinea-pig. J Physiol 547:873–891

Santos-Sacchi J 1989 Asymmetry in voltage dependent movements of isolated outer hair cells from the organ of Corti. J Neurosci 9:2954–2962

Santos-Sacchi J 1991 Reversible inhibition of voltage-dependent outer hair cell motility and capacitance. J Neurosci 11:3096–3311

Scott DA, Karniski LP 2000 Human pendrin expressed in Xenopus laevis oocytes mediates chloride/formate exchange. Am J Physiol Cell Physiol 278:C207–211

Song L, Seeger A, Santos-Sacchi J 2005 On membrane motor activity and chloride flux in the outer hair cell: lessons learned from the environmental toxin tributyltin. Biophys J 88:2350–2362

Stuhmer W, Conti F, Suzuki H et al 1989 Structural parts involved in activation and inactivation of the sodium channel. Nature 339:597–603

Woodhull AM 1973 Ionic blockage of sodium channels in nerve. J Gen Physiol 61:687–708

Zheng J, Shen W, He DZ, Long KB, Madison LD, Dallos P 2000 Prestin is the motor protein of cochlear outer hair cells. Nature 405:149–155

Zheng J, Long KB, Shen W, Madison LD, Dallos P 2001 Prestin topology: localization of protein epitopes in relation to the plasma membrane. Neuroreport 12:1929–1935

DISCUSSION

Muallem: In the old days, when people were working on the pumps' turnover cycle, it was possible to know whether there was just charge movement, and looking at actual transport you could determine if the ions are occluded during the pulse of exposure to the liganded ions. Is it possible to apply two fast pulses to see whether the time-course of the second pulse is the release from capacitance? The only one you showed going up and coming down was the 2 s time course. If the ion gets occluded, you should be able to apply one pulse, see the change in capacitance, and the second pulse should have a delay.

Oliver: If you assume occlusion of the ion, then the jump back to the initial voltage should produce a smaller charge. We don't see anything like this; it is very symmetric. Jumping back, we always get exactly the same amount of charge returning.

Muallem: It must leave very quickly.

Oliver: This makes sense for the operation of prestin, of course.

Ashmore: It might make more sense for the 'slower' prestins—the zebrafish ones. Is that possible?

Oliver: We also don't see that.

Muallem: What happens with fructose?

Oliver: We have never tried it.

Alper: How much of an individual's or species' frequency range is determined by the arrangement of the BK_{Ca} channels (or some other determinant of tonotopic hearing) versus the range over which the amplification occurs as determined by zebrafish prestin's narrower and lower frequency spectrum and rat prestin's broader and higher frequency?

Oliver: Tuning in the mammalian cochlea is all mechanical. In lower vertebrates such as amphibia there is an electrical tuning mechanism. A stimulus of the right frequency induces an electrical ringing in the cell. There is nothing like this in the mammalian cochlea.

Turner: It is an interesting idea that a member of a transport family can have a transport-related but different function. People who have worked with facilitated transport systems like this have for many years thought about what they called a 'rocker model': in one conformation the protein sits with its substrate binding site exposed on one side of the membrane, rather like an open clam. Then domains of the protein move or 'rock' relative to one another so that the substrate binding site is exposed on the other side of the membrane—so the clam simultaneously closes on one side and opens to the other. Recently, some of these transporters have been crystallized and they seem to look this way. Of course, we may see them that way because it is what we are expecting. But there have been crystals made of proteins that are thought to be locked in a certain conformation, and in the crystal structure the transmembrane helices are arranged in such a way that you see what looks like a substrate binding vestibule on one side of the membrane. But this kind of rocking motion doesn't result in an increased volume of the protein; the protein just gets larger on one side of the membrane or the other. Your model implies that the protein gets bigger or at least stretches. Can you reconcile these ideas?

Oliver: Apparently, the prestin proteins produce a conformational change that makes a difference for the area occupied in the membrane. They would then have to work differently from a simple symmetrical rocker model. One important experiment in this direction was done in Jonathan Ashmore's lab: they showed that it is possible to do the reverse thing, pulling on the membrane and thereby producing an electrical charge movement (Gale & Ashmore 1994). This is nice evidence that both components are present: the charge movement can induce this conformational change in the area of the membrane and, vice versa, changing the conformation can induce the charge movement. This reciprocal relation between the mechanical event and the electrical charge movement is very good evidence for the area model. If we think in terms of a rocking motion, this then needs to produce different areas in both conformations.

Turner: It looked like the diameter of the hair cell didn't change; it was just the length.

Oliver: It does. The idea is that the volume of the cell is constrained. There are no volume changes. As the cell shortens, its diameter has to enlarge.

Welsh: As I understood your initial question, if there is a mechanism like that, then one might expect bending of the membrane. You don't see this, do you? If the molecule were changing shape, then a bend should be generated in the membrane?

Turner: If you think of the prestin molecules as forming a belt around the cell, you could imagine that this sort of motion would constrict or relax the belt by expanding the volume of the outer or inner lipid leaflet.

Ashmore: I agree. The reason why we hearing types are hanging out with you transporter guys is because we thought that the rocker motion of a transporter is exactly what we need for a motor to drive outer hair cell motility. I think a rocker mechanism works as an area motor if the rocker is actually asymmetrical. It opens more in one direction than it does in the other. Then an asymmetrical rocker essentially becomes an area motor.

Welsh: Won't this cause bending?

Ashmore: It could cause bending, but it depends what the buffering around the protein actually is. If the proteins are really crowded up close together then there may be no problems at all.

Welsh: It would be interesting to take some of amphipathic molecules that insert themselves into membranes from either one leaflet or the other and use them in your cells. Chin Kung has some interesting data with large mechanosensitive channels in bacteria and testing how bending influences their opening and closing.

Ashmore: Bill Brownell's lab has been looking at specific molecules that distort the shape of the membrane. There is an interaction between the voltage dependence of prestin and the dosage of those compounds into the membrane.

Welsh: Do some of the molecules used for treating schizophrenia cause tinnitus or changes in hearing?

Ashmore: I am not sure. I think that tinnitus arises from a central rather than a peripheral mechanism.

Alper: Chlorpromazine is used as a red cell crenating agent, but it has many actions.

Welsh: Does chlorpromazine do anything to hearing?

Oliver: In fact, chlorpromazine has some impact on prestin's voltage dependence, whether direct or indirect, but currently I don't see how this will help understand how prestin works.

Romero: I have an offbeat question. It is well appreciated, even in the work you show with zebrafish prestin, that there are differences between dynamic hearing ranges between organisms. Have you thought about looking at species homologues of animals that have frequency domains that are radically skewed from what we have seen in rodents or zebrafish?

Oliver: I need to clarify: I am not saying that the charge movement I see with the zebrafish clone has anything to do with active electromechanical processes occurring with hearing in the zebrafish. As far as we know, there is no electromotility in any other auditory hair cell except for the mammalian outer hair cells (OHC). My idea was rather that in these other vertebrates or insects these SLC26s might have function as true transporters, and that what we are seeing here is a mirror effect of any aspect of the transport activity.

Muallem: Even in zebrafish you are finding these SLC26 transporters in mechanosensing cells, so they are probably going to have some mechanosensing function.

Oliver: On the other hand, prestin's function has to have evolved from somewhere. Before prestin and active electromotility, there may have been an SLC26 member in this cell serving another function.

Romero: The fact that you see readily identifiable homologues in zebrafish and *Drosophila* implies that this protein was present pretty early on, for there to be something as readily identifiable as the prestin homologue. I know from looking in *Drosophila* myself, all of the other members of the larger SLC26 family are not readily identifiable as anything that looks like the mammalian gene products. I think you are right: there is conservation of a function that has been amplified on through the course of evolution. It looks like a mechanical or frequency property that has been preserved.

Kere: You mentioned that the volume remains constant for those cells. This should mean that the stretching of prestin is unidimensional, rather than expansion in two dimensions. If it were to expand in two dimensions the circumference of the cell would also increase as well as length, resulting in an increase in volume.

Oliver: Are you talking about the volume of the cell or the volume of the molecule?

Kere: Both. If they are arranged like a belt, if you expand the belt then the diameter of the cell expands as well as the length of the belt.

Oliver: One basic idea is that the area change of prestin is isometric. The anisotropy that you see with the whole cell is then probably due to specific cytoskeleton and structural constraints that the cell imposes onto the membrane.

Ashmore: It is a skin. Imagine a football full of water: you can squeeze it and turn it into a rugby ball without much effort but the volume remains the same.

Turner: When the belt expands the cell has to shorten to keep the volume constant.

Ashmore: It does.

Kere: That means it has to know what orientation it is in. If the belt (circumference) becomes longer and also on the perpendicular length dimension the cell becomes bigger, then the volume expands.

Oliver: The simple answer is that the volume can't expand that fast. We are talking about tens of kHz. You can't have water fluxes this fast. Then you can make many different kinds of shapes with the same area of cell surface. The change of the shape depends on the structural constraints of the cell.

Alper: What is your estimate of the single molecule gating charge, and the proportion of the electric field that traverses? Is any of this pH dependent?

Oliver: We have no independent measures of the fraction of the field that is traversed and the charge that is moved. We get the electrical field times the charge that is moved, which gives the steepness of the Boltzmann slope. The experimental result is usually about 0.8–0.9 elementary charges moved, which may mean that one electrical charge is transferred through 80–90% percent of the whole electrical field. With regard to pH, we can shift the voltage dependence along the voltage axis by changing either intracellular or extracellular pH.

Alper: Has anyone used group-specific reagents on whole OHC or transfected cells?

Oliver: Cysteine reactive reagents have been used in a recent study by Joe Santos-Sacchi (Santos-Sacchi & Wu 2004) and yielded moderate inhibitory effects on prestin-mediated non-linear capacitance.

Welsh: How does the balance sheet add up? You have an estimate of the number of molecules in the cell membrane. You say that charge movement is 1. How does that look compared with the number of molecules you estimate to be present in the cell surface?

Oliver: It fits within a factor of two. I am not aware of the actual numbers, but there is no real disagreement between these estimates.

Ashmore: A factor of two is pretty good is it not?!

Welsh: These kinds of studies are so fraught with assumptions, that if you come away with two, that's the same as saying equal.

Oliver: Perhaps this is far-fetched, but I would say that at least a part of the reactions in this transport cycle should be voltage dependent or electrogenic.

Welsh: I was surprised by your mutagenesis studies. Do they suggest that it is just electrostatic; that it is just affecting the approach or avoidance of anions to the membrane? Are you sure it is not a vestibule effect, or a specific effect on the binding site?

Oliver: It might just deform the electrical field that is seen by the anion. But we don't know much about this.

Ashmore: When you do the mutagenesis studies, do you get the shift in direction that is compatible with it just being a surface charge effect on either the cytoplasm or externally?

Oliver: We don't know on which side of the membrane these mutations are. There is no spatial pattern that we could extract from the presumed position of these charges. This is why it would be helpful to have a good topological model of the transporters.

Turner: Isn't it necessary that those charges have to be able to move? If you mutated a static site on the molecule, one that didn't move across the membrane, then you wouldn't expect an effect.

Oliver: We would. If we removed a charged residue that undergoes voltage-dependent reorientation through the electric field and therefore contributes to the observed charge movement then we should abolish the charge movement, but this is not what we see: we rather see that the voltage dependence of the charge that is actually moving is shifted to some degree.

Welsh: What if you go to higher divalent cation concentrations? Let's say you raise magnesium or calcium concentration. How does this affect the voltage dependence?

Oliver: I guess that it would shift the voltage dependence. There are a huge number of different manipulations that can shift the voltage dependence of prestin along the voltage axis. If you change the tension in the membrane, for example, this will have an effect. We don't understand yet why the $V_{1/2}$ should shift with the different anions that we use.

Muallem: It occurs to me that these mutations that shift the voltage dependence by almost 50 mV are telling you that this is not likely to be a transport mode, but rather a charge movement. This incredible shift seems to me much more compatible with something that relates to a capacitance change.

Oliver: Do you mean that the size of the effect would be too large to have to do with transport?

Muallem: How do you get a transporter to cause such an incredible difference in voltage dependence of transport, by 50 mV on either side, and then you neutralize these charges and all of a sudden you get the same transport with the wild-type voltage dependence? It is not clear to me how you can do this with a transporter.

Turner: Isn't the voltage driving the conformation of the transporter in one direction or the other?

Thomas: Maybe we are showing our bias here. We seem locked into talking of this as a partial reaction of a transporter. Perhaps it is something different and more dramatic; perhaps the protein is in one conformation, and when it binds an anion it is locked into that conformation. Without the anion it is relaxed, it simply can not stably populate the ordered conformation. Thus it could be more like an order–disorder reaction.

Quinton: You could take it further than that. This sounds a lot like a piezoelectric effect in which there are no ions involved; you may be just shoving electrons around. Perhaps these effects have something to do with the ease by which the electrons can be displaced, and this is purely a 'piezoelectric' effect at the protein level.

Oliver: It might be. At least we know that we need these anions here.

Kere: Can you measure the membrane tension in the OHCs?

Oliver: I am not sure we could measure its absolute values, but we can change it.

Ashmore: In principle your experiment is possible. There are techniques using atomic force microscopy. The problem is that we don't know what holds these proteins together in the membrane. Can I come back to a point discussed earlier? The simplest form of this prestin model is that it is just a sort of vestibule which is then undergoing a deformation as the anions move in and out. Is there any evidence that any of the other SLC26 family members have a vestibule on the cytoplasmic surface? This is the implication of this model, unless the prestin topology is completely different from all the other members of the family.

Romero: You are asking whether there is a gating current. I showed results in my paper that would argue for that for A9 at least. To do the experiment correctly we would want to do something that really blocked the channel-like activity of the transporter, so you could get a much better look at the gating currents. But the current fall off after shutting down the clamp is the only way these can be described.

Ashmore: Do the transients you showed in your data have the characteristics of a gating charge?

Romero: They were a lot slower.

Oliver: That was the whole oocyte, so you may not clamp fast enough to see the kinetics.

Alper: Is there a voltage dependence of the off-rate for the prestin inhibitor, salicylate?

Oliver: We didn't look at the gating charge with salicylate but only at the non-linear capacitance, which gives us the steady-state distribution of the conformations but not the kinetics.

Ashmore: We tried to look at this a long time ago. The salicylate binding curve looked as though it was voltage independent. In the light of new data we want to do those experiments again.

Welsh: What are the lipids like in the OHC membranes?

Knipper: They have an unusually high amount of cholesterol. I don't have the numbers.

Alper: What is the temperature dependence?

Ashmore: It is virtually nothing. It has a Q10 of about 1.2. It looks like a diffusion-limited step.

References

Gale JE, Ashmore JF 1994 Charge displacement induced by rapid stretch in the basolateral membrane of the guinea-pig outer hair cell. Proc Biol Sci 255:243–249
Santos-Sacchi J, Wu M 2004 Protein- and lipid-reactive agents alter outer hair cell lateral membrane motor charge movement. J Membr Biol 200:83–92

Final discussion

Chan: Many of the experiments we have discussed during the meeting have been done in oocytes. There have been noticeable differences between various labs. No one has mentioned that oocytes have a whole array of endogenous channels. Have people compared the expression of endogenous channels after injection or transfection? It would be interesting to compare expression profiles of candidate genes that could be activated by the introduction of exogenous genes. Shmuel Muallem demonstrated that SLC26A3 and A6 could activate CFTR. They could also activate endogenous channels. The other possibility would be that depending on the time the oocyte is collected, different genes could be expressed at different stages of oocyte maturation. This could account for some of the differences seen.

Muallem: The fundamental properties of the transporters must be the same independent of expression system. My view is that these transporters do the same before and after transfection. I don't think they switch mode of transport just because we are expressing the genes in this cell or another. I think expressing the transporters only in oocytes will always be problematic. Therefore it is important to verify what we see in oocytes in mammalian cells like HEK293 cells. If we can see current with similar characteristics in two expression systems, we tend to believe that this is not an oocyte artefact. One of the ways to solve the problem is to switch clones between labs and repeat the same experiments in different labs but using the same rigorous techniques. Unfortunately, in the case of A7 this didn't work: Manoocher Soleimani sent us his clones and using his clones we saw the same thing we saw with our clones.

Soleimani: Your oocytes also have a very low membrane potential at baseline.

Alper: We have heard from people who have baseline potentials of −60 mV. We get around −40.

Soleimani: We always get −50 to −55 mV at baseline.

Muallem: This is hand waving! Transporters are transporters. We have to see the same thing; it's an important point. We need to express them in another cell type and see whether they show the same transport characteristics.

Alper: Dr Chan, you have raised an important point about subunits. Shmuel Muallem's work has contributed the knowledge that the SLC26s can act as subunits of multitransporter complexes. This could be thought of as the old MinK, or it could be thought of as SUR1 or SUR2—it is missing something; like the G-protein regulated inward rectifier K channels (GIRKs) where subunits missing a required binding partner encounter that endogenous subunit in the *Xenopus* oocyte.

Romero: This missing partner is probably the best argument. The only thing that would argue against this is to propose clones doing similar things, and which have voltage changes in opposite directions. Shmuel's data would imply that such clones interact with at least one other partner but in the same way. And his data would indicate that the partner interacts with them in a similar way to increase their function.

Knipper: I have a problem considering similarities or differences in the whole family. One aspect we haven't discussed directly is whether in development all the members of the family are up-regulated or targeted in different cells prior to the time when the cell or organ starts to function. Or do they start working long before organs such as the kidney and pancreas are functioning? In human diseases this may be of functional relevance for describing the phenotype and the pathology. They probably only work *in vivo* when they are properly targeted. So all mechanisms involved in targeting may have functional implications and may lead to similar phenotypes. This kind of developmental aspect is something that I would be curious about.

Kere: We know that we don't know a lot! What diseases should we be looking for with the different transporters? What would be our guesses for diseases for those SLC26's that are still without a phenotype.

Case: Chronic pancreatitis.

Soleimani: It seems that eventually one has to apply the knowledge that is gained *in vitro* to the *in vivo* situation. Perhaps once we get into *in vivo* some of these controversies may be addressed. For example, if you look at A6, just by looking at chloride–oxalate exchange you will not guess that it can transport oxalate-stimulated NaCl unless this is studied *in vivo*. My point is that, similar to this, we can take A7, and once we delete it we can go back to the animal model and see what functions are missing. Depending on the tissue, we have seen that the same transporter may function predominantly in one mode in one tissue and in another mode in a different tissue. One should also take into consideration the effects of overexpressing these genes in tissues.

Knipper: If you look to the kind of difference between the human and rodent phenotype, the mutations may either affect the complete function of the organ, or perhaps a simple lowering in the number of correct targeted proteins in the membrane at the right time. Thus mutations within the gene, but also mutations in genes involved in targeting and scaffolding, may exhibit different redundancy in humans and rodent, and therefore be responsible for differences in the phenotype.

Kere: We would not be sitting here without genetics—without knowing first that in human A2 mutations lead to diastrophic dysplasia, A3 leads to congenital chloride diarrhoea and A4 leads to Pendred syndrome. I appreciate that it is important to understand what these things are doing in cells, but we could use another kind of science as a shortcut, getting help from genetics to try to understand what these might be doing in different parts of the body.

Muallem: There are many cases of idiopathic pancreatitis that turned out to be linked to mutations in CFTR or trypsin, for example, and I would not be surprised if we were to find a significant linkage to A6 or even A3. On the basis of Min Goo's data, A6 is the one where I would expect to see mutations affecting the function of the pancreas leading to chronic pancreatitis. One would think that there are going to be mutations in A9 linked to respiratory diseases. A9 is a potent Cl^- channel that is expressed in the surface epithelium that is exactly where cystic fibrosis-like symptoms develop.

Soleimani: In the mouse if you knockout CFTR it does not develop any lung phenotype. This could pose a major problem for A9 knockout. In other words, one cannot be sure if A9 deletion is going to have any discernible phenotype in mouse lung if collaboration with CFTR is essential for the generation of lung phenotype, as it is in human.

Knipper: Juha Kere, could you not speculate about how to best proceed to get more successful genetically diagnosed cases of SLC26?

Kere: It is straightforward: we could, with reasonable cost, take all these SLC26 family members. For three genes we already know what the diseases are. We could tag all these genes with single nucleotide polymorphisms (SNPs), look for common variations in populations and then do gene association studies with patient and control cohorts. They would have to be large enough to have statistical power. The genetic experiment is not too complicated, it is just a little expensive.

Alper: I would like to return to the difference between electrogenic ion exchange and whatever it is that you are calling A7. In the recent paper on A7 (Kim et al 2005) the activity has properties of electrogenic anion exchange, yet Shmuel went to some length not to call it this. What is it instead?

Muallem: It seems to me that it is a Cl^- channel that has finite anion or HCO_3^- permeabilities. The reason we define it as an ion channel is that when we expressed it in oocytes we saw a strong current and absolutely no movement of HCO_3^- of any sort. We saw movement of other anions. We were still not completely comfortable with that so we put it into HEK293 cells to look at currents. We got currents in excess of 800 pA. I am not aware of any transporter that functions as a coupled transporter that can generate close to 1 Na^+ current in a small cell like HEK293 cells! Then we looked at the effect of pH and noticed regulation by intracellular pH, with the channel activated by acidification. We can reproduce all of these findings in both oocytes and HEK293 cells. All the HCO_3^- movement looks pretty much miniscule when we go into the oocyte system, which has huge buffering capacity. We came to the conclusion that if you generate so much current you cannot generate it by coupling to something else. If it looks like a channel, it is a channel.

Soleimani: This is important. We have examined A7. I also looked at Dr Muallem's recent paper (Kim et al 2005). One important point is that in HEK cells when lumenal chloride is removed there is an intracellular alkalinization, and when chloride is returned the alkalinization disappears. Independent of whether this exchanger is electrogenic, this observation is consistent with the coupling between chloride and bicarbonate.

Ashmore: I have one word: pharmacology. This hasn't entered our discussions much. Have people started to think about targeted drugs for this particular family of transporters?

Case: All of our lives have been made more difficult by the fact that we don't have animals that produce anion blockers.

Welsh: It is time to bring our discussions to a close. I think this has been a wonderful meeting. This is an absolutely fascinating topic. What makes it particularly interesting is that this field is so young. We are right at the beginning of understanding, and this makes things fun. Also, there has been tremendous and lively participation by the people in this room, and I thank you for your contributions.

Reference

Kim KH, Shcheynikov N, Wang Y, Muallem S 2005 SLC26A7 is a Cl- channel regulated by intracellular pH. J Biol Chem 280:6463–6470

Contributor index

Non-participating co-authors are indicated by asterisks. Entries in bold indicate papers; other entries refer to discussion contributions.

Subject index